AIR AND SATELLITE COMMUNICATIONS

Edited by Lee M. Paschall
Foreword by John L. McLucas

AFCEA/*SIGNAL* MAGAZINE C^3I SERIES

AFCEA International Press
Washington, D.C. • Brussels • Manila

Library of Congress Cataloging in Publication Data
Main entry under title:

Air and satellite communications.

(AFCEA/SIGNAL magazine C^3I series; v. 2)
Compilation of articles most of which first appeared in Signal.
Includes bibliographies.
1. Command and control systems — United States — Addresses, essays, lectures. 2. United States — Armed Forces — Communication systems — Addresses, essays, lectures. 3. Artificial satellites in communication — Addresses, essays, lectures. 4. Electronics in military engineering — Addresses, essays, lectures. I. Paschall, Lee M. 1922- . II. Signal (Falls Church, Va.) III. Series.
UB212.A37 1985 358.4'133041'0973 84-72110
ISBN 0-916159-03-5

Contents

PART III: OTHER SATELLITE COMMUNICATIONS APPLICATIONS

PART IV: COMMUNICATIONS FOR AIR OPERATIONS

PART V: ELECTRONIC WARFARE

Preface

This volume of the AFCEA C^3I series deals with air and satellite communications together with a section on electronic warfare (EW). It contains a compilation of articles appearing (for the most part) in *SIGNAL* over the past 10-15 years. Anyone seeking either a fundamental understanding or a historical perspective of the evolution of C^3I will find much of interest in the pages of that valuable journal. Given the amount of material published in *SIGNAL* on satellites, air communications and EW, it would seem a simple task for an editor. The opposite is true. There are so many valuable articles that the selection process was a real challenge. The selection criteria sought out those articles which contain fundamental ideas and particularly those which first enunciated new ideas which later came into general acceptance. Space limitations precluded the inclusion of many interesting articles which failed the editor's somewhat arbitrary criteria.

Part I deals with the early years of satellite communications and the foresight of those who made space based communications platforms a reality. Those ideas which were frequently considered visionary have become commonplace today. This section of the book concludes with what is considered visionary today in an article on the concept of space based defense. Twenty or so years hence, that too may be a reality. One ought never underestimate our accelerating ability to exploit technology.

Part II emphasizes military communications satellite architecture and systems while Part III contains a sampler of other satellite based communications applications. Common threads running throughout are the importance of requirements definition, total system planning, continuity or replenishment planning, evolutionary change and the system control elements. Neglecting any of those fundamentals poses great risk.

Part IV focuses on three major forms of air operations — strategic, tactical and airlift — and their communications requirements. The long haul, tactical and base communications required to satisfy those and other missions must continually evolve in response to the needs of the operating forces for communications to support their command, control and management systems.

Part V continues articles which principally emphasize the need to plan for and to conduct electronic warfare activities as a normal part of military operations. That represents the major thrust of the EW thinking in the United States as it has emerged during the last decade.

Certainly not all of the fundamental ideas we have tried to capture herein are immutable. All have been challenged at one time or another — many will be again. However, any such challenges should be based upon an understanding of how they came into being and why their distinguished authors believed them to be important.

The editor gratefully acknowledges the creative effort of those whose articles are contained herein and of Dr. McLucas who prepared the Foreword. This book would not exist were it not for them. The AFCEA staff responded quickly and graciously to all of the editor's whims. Three ladies at the American Satellite Company—Margaret Gould, Rose Yelle and Sharon Rawls—contributed greatly to this book and, in particular, to the editor's peace-of-mind.

Lee M. Paschall

Foreword

Those of us who have lived through World War II and the technological explosion following that cataclysm may be forgiven for trying to categorize the major milestones and turning points in our evolving but sometimes revolutionary technology based history.

I find it intriguing that within a few weeks of each other we had the explosion of the first air-delivered nuclear weapon at Hiroshima and the publication of Arthur Clarke's description of a worldwide communication system based on three geostationary satellites. Equally intriguing is the rapid obsolescence of "truths known to be self-evident." The communications satellite, which was known in 1945 to be impractical, was rendered practical just three years later by the invention of the transistor.

This book is not about nuclear war, nor even about peace enforced by nuclear deterrence. But it does embrace most of the other technologies forming the basis for our modern age — whether it be termed the information society, the postindustrial society or the era of modern communications.

Asked to name the most important events in recent history — or the most important elements of our technology based society, one's biases quickly show through. As a practitioner in the fields of aerospace technology, I have no trouble compiling my list.

1. The growth of a worldwide transportation network based on long-range fan-jet aircraft, a phenomenon made possible by universally accepted communications and navigation devices and systems, together with the right of innocent passage through vast reaches of air space.

2. The opening of the space age, followed by the putting in place of the INTELSAT worldwide communications satellite system, a cooperative of 109 countries bringing modern communications to the people of more than 140 countries and which last year (1984) beamed the Los Angeles Olympic Games to more than two billion people (viewers).

The space age has brought equally dramatic improvements to military communications for forces on land, at sea and in the air.

3. Following the development of the transistor, the now-accepted computer revolution which has brought information processing within the reach of all businesses large and small and is moving rapidly toward doing the same in the American home.

4. The evolving merger of communications systems and computers to permit the rapid expansion of, and almost universal access to, instantaneous worldwide telephone service and data distribution, and the rapidly growing fields of electronic funds transfer, teletext and all forms of information processing.

5. The domestic cable TV network which, with fiber optic cable and domestic communications satellites, soon will cover the United States with highly interconnected, countrywide local area networks. While its role in entertainament is well known, its potential for education and data services is just beginninng.

This book traces the development of many of the key elements in this new world of rapid communications and the role of those elements in both civil and military life. Just as the history of commercial aviation owes an invaluable debt to military aviation, commercial communications is similarly indebted to the military. The role of NASA in communications development was essential. But I was distressed recently to hear a NASA official render paeans of praise to NASA for its role in Syncom and Advanced Technology Satellites (ATS) without once mentioning key efforts of the military such as USAF's TACSAT, the LES series and the development of Atlas, Titan and other space boosters. As much as some of us would like to separate these developments into neat civil and military categories, their histories are highly interwined.

This book embraces both civil and military communications satellite development and stresses the important role played by *SIGNAL* Magazine in chronicling and encouraging the emerging technology of the last 20 years. For many of those years, the important thing was to design and emplace systems that would work reliably. As I sat in the Pentagon trying to foster the creation of reliable satellite communications, I sometimes fretted that our friends in civil communications seemed to be having more success than we in exploiting satellite technology. Reliability seemed to come more readily to the civil side than to the military. Then in 1977, I moved to COMSAT and played host to a group of officials from the Air Force Space Division (then known as SAMSO) who sought the secret of our success in achieving a reliable INTELSAT system. Fortunately many of the factors cited then are of historical interest only, but it may be a fact of life that military systems will always achieve somewhat lesser reliability than their civil counterparts because of penalties paid for resistance to jamming, improved command security and the like.

Five years ago I directed a Pentagon study of "enduring communications," one of a series designed to improve satellite systems' reliability and survivability. Two factors which I believe may be important legacies of that study are the need for the military to:

- Abandon the search for the one true ultimate survivor concept in favor of achieving survivability through the interconnection of redundant systems, at least some of which are likely to survive, and
- Develop plans to use the highly proliferated commercial communications systems of the United States for military purposes in the event of crisis or conflict.

The extent and variety of commercial systems are phenomenal. There is no way the military could afford to develop anything of comparable capability.

Over the last five years a large number of military planners have come to appreciate that C^3 or C^3I is just as much a key ingredient of military capability as are the major weapons and forces. This belated realization is very welcome. Whether it is dignified by the term "force multiplier" or not is perhaps immaterial if we begin to accept the fact that something so important cannot be considered merely a corollary to adequate force planning but is, in fact, an essential component of military capability.

Part III—"Other Satellite Communications Applications" contains some of the more intriguing chapters in this book. The number two military and space power in the world is, of course, the Soviet Union. While we placed no impediment in the way of the Soviets joining INTELSAT, they chose not to do so and instead created INTERSPUTNIK, a system which not surprisingly operates much less autonomously than INTELSAT and which — for that and other reasons — has not played a major role in worldwide communications. The Soviets have participated in SARSAT/KOSPAS, a search and rescue system based on satellites operated by East and West, and appear to be proud (justifiably) of their public service role in this system. They are also proud of their participation in INMARSAT (a system regrettably not included in this book) which caters to maritime interests.

Also covered here are weather satellites, educational applications and optical fiber communications. We will soon see keen competition between satellites and optical fiber systems — a competition which I expect to extend through many ramifications and transformations to the ultimate benefit of all who expext to see a growing need for expanded data, entertainment, education and information processing systems.

Not covered under "other applications" are surveillance satellites and LANDSAT, all of which are special purpose communications satellites. The former, sometimes called "national technical means" of arms control verification, are among the most important products of the space age. LANDSAT and its predecessors and successors could well be of enormous significance in the future, especially as picture quality improves.

The inclusion of a section on electronic warfare is inevitable. If information processing is the key to modern warfare, its interruption becomes terribly important. The never ending battle between electronic countermeasures (ECM) and counter countermeasures (ECCM) is still with us and I believe we can count on witnessing this recurring seesawing contest out into the indefinite future.

The introduction to Part I refers to the concept of space as sanctuary. Those of us in civil space operations recognize the need for a benign space environment if civil space commerce is to flourish. We thus tend to applaud efforts to keep weapons out of space principally through continuing discussions with the number two space power, the Soviet Union. We all hope these discussions will continue indefinitely, thus delaying the stationing of weapons in space. However, we all recognize the need to do R&D on space weapons such as the President's initiative of March 1983. The ideal combination in my view matches R&D on the Strategic Defense Initiative's (SDI's) antimissile defense with arms control negotiation so that space weapons never are fired in anger. This would be a wonderful legacy of the opening of the space age.

Dr. John L. McLucas

Part I: Communications Satellites — The Early Years

Editor's Introduction

Communications by means of satellite repeaters on a regular, non-experimental basis is only 20 years old, beginning with the first live television coverage of the 1964 Tokyo Olympic Games via the Syncom III satellite. The subsequent impact of this new technology on societies worldwide has been incalculable. The geosynchronous satellite first suggested by Arthur C. Clarke in a 1945 article in *Wireless World* dominates much of our thinking about satellite communications, and indeed, it is the most commonplace orbit used today.

There are, of course, several polar orbiting systems used for earth sensing and special purpose telecommunications, particularly in far northern latitudes, but geosynchronous orbits dominate. It was not always so, as there was an active debate early in the satellite communications era about the relative merits of medium altitude and geosynchronous satellites.

The Military Operations Subcommittee of the House Committee on Government Operations of the 88th Congress, after reviewing government operations in the field of satellite communications during the spring and early summer of 1964, issued its report in October of that year. The introductory section of that report summarizes the debate as seen principally by the military community. An excerpt from the committee report is included here to indicate the early (and continuing to this day) concern on the part of the military about survivability against jamming and spoofing as well as the need for reliability in military satellite communications.

Cynics have said that in stressing the case for medium altitude satellites, the military was simply trying to defend having its own system rather than being required to use the forthcoming commercial systems. Although certainly a consideration, the principal reasons were the unique military requirements which had to be satisfied. Another excerpt from the same committee report describes those distinctive military requirements as seen in 1964. There is remarkably little difference with those seen in 1984 by other authors of articles contained in this book.

Many of the early experiments with both active and passive satellite repeaters in space are almost forgotten today. Some seem somewhat quaint by today's standards. Many appeared to be failures, but lessons were learned and theories confirmed by these early experiments, most of which were sponsored by the military. "Communications Satellite Experiments" (Annex I of the 1964 House Government Operations Committee report) describes these early experiments. It is included here principally for its historical value. Few such concise compendia exist.

The early experimental days involved both the National Aeronautics and Space Administration (NASA) and the military. NASA's involvement was largely through the Goddard Space Flight Center whose contributions towards the advancement of communications satellite technology, particularly for commercial usage, are legion. Thus, a 1976 article by Drs. John F. Clarke and William N. Redisch describing Goddard's role and, in particular, the Applications Technology Satellite (ATS) series of satellites, is of genuine interest. The authors, in their final paragraph, refer to a 1973 decision by NASA to phase out of communications satellite research. That decision was based on a premise that the commercial sector would develop the necessary advanced technology required to advance the state-of-the-art for commercial use of space for telecommunications. Given the very large investment required to construct and deploy a communications satellite, the commercial operators seek rather conservative evolutionary improvements — not the revolutionary change characterized by the ATS experiments. Fortunately, the 1973 decision was reversed, and in recent years, NASA has again embarked on some important new development activities in the Advanced Communications Technology Satellite (ACTS) program.

A brief account of the early years with emphasis on the communications satellite would not be complete without taking into account the development of the means used to get into orbit. A 1974 article "USAF's Conquest of Space" by Kenneth W. Shultz, then Commander, Space and Missile System Organization (SAMSO), which appeared in *AIR FORCE Magazine,* vividly describes the impetus for development of long-range ballistic missiles, the tough early years and the emergence of space boosters. Those many early failures were followed by increasingly successful launch vehicles which have made possible the routine use of space for many purposes in a short two decades.

That the world has been drawn closer together by communication among diverse people is a tribute to the pioneering efforts of those who participated in the early years. One can expect that the future will bring even greater cultural understanding and tolerance, particularly if the world can be made safer by the continued deterrence of nuclear war.

There are some who consider any use of space by the military as bad. Others accept military use of space for warning, verification and communications as desirable but oppose further military activity in that environment. Still others, particularly those who believe that a strong U.S. national security posture is vital to the deterrence of nuclear war, realize the importance of space as a medium. The debate over future military use of space will certainly continue for years to come, as will the development of the technology associated with potential military uses of space.

In a speech to the National Space Club in 1972, Grant L. Hansen, then the Assistant Secretary of the Air Force for Research and Development (R&D), describes how space contributes to the accomplishment of various military tasks. It is instructive to compare that speech with the one following, entitled "The New Space Command," by Gen. James Hartinger, USAF (Ret.). One can see the rapid evolution of Air Force thinking about space in a short decade. The Space Command was established by the U.S. Air Force in September 1982 and marks a crucial milestone in the evolution of military space operations. Its first commander and an articulate advocate of its creation describes the command and what it will do.

Finally, looking further into the future is the address entitled "Space Based Defense" by George A. Keyworth, II, Science Advisor to President Reagan. This is a particularly useful discussion of the President's March 1983 proposal for a new space based defense initiative, particularly in light of the strong reactions it provoked, which resulted in much press treatment of the concept as a "Star Wars" fantasy.

Satellite Communications: Military-Civil Roles and Relationships

New Means of Global Communications

Space satellites offer promising means to extend and improve global communications. Civil and military agencies of government and sectors of industry are actively engaged in studies and experiments to exploit space potentials for this purpose. The issues have been dramatized for the public by the Telstar and Relay demonstrations that live television programs can be beamed to and from other continents. Within this decade, it is clear, satellites will play an increasingly important role in man's ability to communicate quickly with his fellow men in faraway places.

Undersea cables now provide communications links to Europe and elsewhere, and high frequency radio transmissions also are used for transocean communications. Cable circuits are limited, however, in capacity and coverage; and radio channels in the higher frequencies are crowded, while reception is subject to noise, blackouts and other interferences. By placing satellites in space orbits to serve either as reflecting or relay stations, we can overcome limitations imposed by the curvature of the earth and other of nature's obstacles to transmission of radio waves. The satellite station promises new gains in the ease, economy, dependability and geographical coverage of all forms of long-distance communications — telephone, teletype, television, photo facsimile and data exchange.

**This article originally appeared in a report of the Military Operations Subcommittee of the House Committee on Government Operations in 1964.*

High Versus Medium Altitudes

Generally, communications coverage increases with satellite altitude. To achieve true synchronous orbit, a satellite must be boosted into circular orbit at a precise altitude of 19,300 nautical (22,300 statute) miles in the earth's equatorial plane. Then it will revolve in unison with the earth and appear to be a fixed point in the sky.

Three synchronous satellites equally spaced around the Equator could provide continuous worldwide coverage except for the polar caps. Mutual visibility at such a great height makes it possible for one satellite to cover South America, the east coast of the United States and major parts of Europe and Africa. Outside the polar regions, any station on earth could communicate with any other station by the use of no more than two satellites.

A synchronous satellite system, if successfully deployed, has this advantage, then, that only three satellite positions are needed for almost complete worldwide coverage, in contrast with lower altitude satellites, which would be numbered in the dozens for almost equivalent coverage. The synchronous system also would simplify greatly the ground station requirements, not only because the satellites would be fewer, but in their ''stationary'' positions there would be no complicated problems of tracking and ''hand over'' from one satellite to the next.

To realize these and other advantages of the synchronous system in practice, some military authorities state that the technical demands are many and exacting. To achieve their great distance from the earth and unique orbital position, synchronous satellites require large boosting power; and in orbit, they must be stabilized with their directional antenna pointed toward the earth to get and transmit enough signal strength. To prevent drift, their orbital speed must be precisely controlled. These controlled systems require the use of special devices such as spinning wheels, small gas jets or rocket motors.

A synchronous stationary satellite requires many more electronic parts than a medium altitude satellite, and the problems of reliability grow in proportion. Consequently, the theoretical advantages of worldwide coverage with just three satellites might be lessened or lost by the need to have backup satellites in orbit or always ready for launch — in either case a difficult and expensive process. Military authorities have favored for several years a medium altitude system involving a score of satellites in random orbit to provide continuous worldwide coverage and avoid the possibility of complete blackout for large geographic areas which would result from the failure of a single synchronous satellite.

In the military view, also, the synchronous satellite is more susceptible to enemy interference. The control devices to keep it in fixed orbital position are commanded from the ground, while the medium altitude satellite in random orbit needs no station keeping. Furthermore, an enemy can pinpoint a stationary synchronous satellite more easily for jamming, while multiple medium altitude satellites roving at random compound the enemy's problem. Those satellites that are over the horizon are not in his line of sight, and those that come into view cannot be simultaneously jammed without great difficulty and cost.

Another advantage of the medium altitude system, perhaps of lesser moment to the military but of considerable interest to commercial carriers, is the fact that it poses fewer problems of time delay and echo in voice communications. The synchronous satellite is so far from the earth that voice transmissions take about 0.6 second in a round trip through the satellite. Long-distance transmission also generates echoes, which, coupled with time delays, could be a source of annoyance to users. Techniques have been developed to suppress echoes in conventional long-distance telephone service. However, echo suppressors may cause chopped up conversations when the time delay factor is substantial, as in the case of the synchronous satellite. Experience with operational satellites and results of research into user reactions will determine whether time delay is a serious or significant handicap in voice communications.

Although time delay, because of the sheer physical distance of the synchronous satellite, cannot be overcome, only adjusted to, in another sense the advocates of the synchronous concept believe that time is on their side. They anticipate that advances in technology may overcome limitations or disadvantages now associated with the synchronous concept. For example, larger, more powerful boosters, such as the Titan III, could inject many satellites simultaneously into synchronous stationary orbits, providing sufficient redundancy in coverage to compensate for outages, hostile jamming and other contingencies. Improvements in components and careful quality control promise more simplicity and longer life of satellites. Even the factor of ground command for station keeping, which the military sees as a disadvantage, conceivably could be eliminated by onboard control devices still to be developed.

Active Versus Passive Satellites

The synchronous and medium altitude concepts each have their advocates and defenders, but both have in common a preference for active rather than passive satellites. In the passive system, the satellite receives radio signals from one ground station and merely reflects them back toward other stations, serving as a "radio mirror in the sky." An active satellite receives, amplifies and sends radio signals; and for these functions it must carry its own receiver, transmitter and power supply.

Passive satellites have some advantages. They do not need onboard electronic components and conceivably could last much longer in space. Unlike the active satellites presently designed, passive satellites could be used simultaneously by many communicators, and methods of modulation or bandwidths could be changed after the passive satellite is placed in orbit.

Yet the passive satellite, as a reflector rather than an amplifier of electromagnetic energy, has an enormous efficiency loss. Transmitting and receiving antenna must be very powerful, which means large ground station size and costs. In the Echo balloon concept, the sphere itself must be huge, rigid enough to endure in the space environment and inflated after injection into orbit, which makes launching awkward and difficult. These are some of the offsetting factors against the comparative simplicity and flexibility of passive satellites.

Since passive satellites offer limited communications channel capacity, and active satellites have considerable growth potential in communications capacity, the far greater promise lies with the active satellite concept, which is commanding major government and industry attention and effort. As technology in these fields progresses, passive satellites may be shown to be useful for specialized and supplementary communications purposes rather than as competitive alternatives to active satellite systems.

Satellite Communications: Distinctive Military Requirements

For several years, through its Advent and other projects, the Defense Department had been studying the potentials and conducting experiments looking toward future operational systems of satellite communications. Prior to 1961, there had been stated military requirements to Advanced Research Projects Agency (ARPA), but no Joint Chiefs of Staff (JCS) approved ones. In late 1961, the JCS validated a requirement for a military communications satellite capability to be integrated with the Defense Communications System. The approved plans stemmed from a document known as Minimum Essential Survivable Communications (MESU). Although criticized for its lack of precision, this document was the key to prescribing a workable military system of satellite communications. For the first time requirements were stated that could be satisfied without inordinately high power within the satellite or complicated stabilization and command control schemes.

The time scale was initial capability by April 1963 and full system operation by August 1964. This requirement took on added urgency, since the "Year of the Quiet Sun" was predicted for 1964, introducing a period of reduced solar activity which could disrupt long-distance radio transmissions. Microwave satellite relays would not be affected by these phenomena.

For an operational system, the military has posed these distinctive requirements:

**This article originally appeared on a report of the Military Operations Subcommittee of the House Committee on Government Operations in 1964.*

1. *Positive operational control:* The system must be under military command and control at all times without dependence on foreign companies or governments. A commercial system requires negotiations, agreements and cooperative working arrangements with many nations and their carriers. Although military base rights would have to be negotiated with sovereign governments to permit installations when and where needed for operation of a U.S. military system, command and control would not be affected.

2. *Mobility and remote area access:* Military communications cannot be limited to fixed stations and points of high density traffic. A commercial system can employ large, elaborate ground stations to handle the traffic burden of populous areas, but a military system must be able to penetrate remote and sparsely populated areas as emergencies dictate. Relatively simple, transportable ground stations are required to quickly establish communications among many terminal locations and to move out on short notice. Also, communications must be maintained with flying aircraft and ships at sea, and communications satellites hold forth the promise of capabilities in this direction.

3. *Protection against physical attack:* Whereas a commercial system is designed for normal peacetime operations, a military system demands special protective measures both for the ground stations and the satellite repeaters. Hardening, dispersal and mobility are protective factors for ground stations; the space repeater system should be designed to operate even if some of the repeaters are destroyed by hostile action.

4. *Protection against electronic countermeasures:* Even in peacetime, a military system must be able to overcome jamming action. This requires the ability to switch from one frequency to another and entails a much larger ratio of radio frequency bandwidth to informtion bandwidth than would be normally used in a commercial system. Different modulation techniques and higher transmitter powers also may be required to protect against jamming.

5. *Low capacity and secret message transmission:* Unlike commercial systems, which are designed for capacities of hundreds of standard voice channels or one or two television channels operating primarily on a trunk basis, a military system requires relatively low capacity for a few voice, teletype or digital data channels. However, redundant or multiple channels are important in case of countermeasures. Also cryptographic security requires bandwidth and transmission features not normally required of commercial systems.

6. *Separate frequencies for military use:* The choice of frequencies for use in satellite communications poses difficult problems. The military now have bands of frequencies assigned for its own use in the United States. It also intends to take full advantage of the special bands set aside in the Geneva radio revisions for satellite communication services. International agreement on the use of civilian bands now shared among diverse types of users for military satellite communications purposes might be difficult or impossible under a commercial satellite system encompassing military requirements.

The concept of a military communications satellite system, as formulated in general terms by the Defense Communications Agency and approved by the Joint Chiefs of Staff, was to seek a minimum esssential capability in the earliest possible time within cost limitations. In the short range, the satellite communications capability was viewed as a "redundancy" or additional resource to back up or supplement existing long-haul military communications, which are accomplished by various means and combinations, including high frequency circuits, ionospheric and tropospheric scatter circuits, cables and microwave systems. As time went on, and technology advanced, it was expected that the satellite communications system would grow in reliability, coverage and versatility and would acquire the full range of military characteristics advanced in justification for a separate military system.

The rationale for a separate or unique military system was developed by the Defense Department both preceding and following the enactment of the Communications Satellite Act of 1962, since there were recurrent questions by members of Congress as to the need for separate systems which might be duplicative and, in any case, would be very costly.

In making their case the military authorities did not deny that commercial systems could be designed to accommodate, in lesser or greater degree, their special requirements, providing the extra costs, the technical complications and the inconveniences were accepted. Considering, however, the widely divergent needs of a commercial system committed to economic operations in a friendly peacetime environment and a military system which must be prepared for a whole range of contingencies in both friendly and hostile environments, there seemed to be no likelihood that the two would merge. Indeed, a RAND study concluded that a separate military system not only would best meet our security needs but actually would help to promote the success of the commercial system and our foreign policy objectives. Other countries would be more inclined to participate in a commercial system not intermixed with military operations.

Communications Satellite Experiments

I. Passive Satellite Experiments

Moon Relay

The passive satellite concept is exemplified in the use of the moon itself as a relay in a telecommunications link. As early as January 1946, Army Signal Corps scientists bounced radio signals off the moon. Not a communications experiment as such, it proved that radio transmission across space was feasible with moderate power.

Navy experiments with the Moon Relay were authorized by the Chief of Naval Operations in 1950. By 1954, the Naval Research Laboratory had transmitted the first voice message over an earth-moon-earth path, and the following year it effected the first cross-continental transmission, Washington to San Diego, via the moon.

Using an 84 foot antenna and the moon as reflector, the Navy established a two-way operational link at 400 and 450 megacycles during 1959 between Washington, D.C. and Hawaii. The two stations were manned during several hours each day when the moon was visible from both stations.

The Communications Moon Relay, as the experiment is called, was really the first "operational" space communications system, transmitting radio, teletype, microwave, voice and facsimile traffic. Though presently in standby rather than active use, at times when severe ionospheric disturbances cause the shutdown of normal medium and high frequency traffic, Communications Moon Relay is the only operational link between the two points. One feature of the experimental program is the demonstration of ship-shore communications. In 1962, the first demonstration was made of two-way, ship-shore transmissions at 2300 megacycles.

**This article originally appeared in Annex I of the Military Operations Subcommittee of the House Committee on Government Operations 1964 report.*

MISER

In 1961 the microwave space relay (MISER) program was initiated. Its purpose is to develop a ship-ship and ship-shore terminal operating at microwave frequencies (8000 megacycles) and using both natural and artificial passive reflectors or repeater type satellites for long-range relay. The experiments involve studies of signal characteristics and devising optimum means for transmitting information.

Project Echo

A man-made passive satellite for communications — a balloon in orbit to reflect signals — is Project Echo, sponsored by the National Aeronautics and Space Administration (NASA). After a launching failure in May 1960, a second attempt injected Echo I into a low orbit about 1,000 miles from the earth on August 12 of that year. The launching vehicle was a three stage Delta. The balloon was a 100 foot mylar aluminum sphere weighing 157 pounds, which inflated after orbit. Experiments included voice and crude television transmissions between the East and West coasts[1] and voice signals between the United States and Great Britain.

Echo I, now almost four-years-old, misshapened and wrinkled due to loss of inflating gases after puncture by micrometeorites, continues in a fluctuating orbit. Its communications capacity is much reduced.

To overcome the deficiencies and structural weaknesses in Echo I, NASA planned to launch Echo II, a larger and more rigid sphere of mylar aluminum laminate, sometime in 1963. Problems were encountered, delaying the project, and two suborbital ballistic flights in 1962 were necessary to test satellite injection and inflation techniques. An associated project, Rebound, to launch Echo II spheres into low earth orbit during 1963-64 from spacecraft with multilaunching capabilities, was canceled as premature in view of the demonstrated need for more developmental work in space technology and fabrication of passive satellites.

Echo II was successfully launched into orbit with a Thor-Agena vehicle on January 25, 1964. The satellite, 135 feet in diameter and weighing about 550 pounds, was placed in a circular high angle (near polar) orbit ranging between 600 and 800 nautical miles. Television pictures were made of the satellite as it was erected and inflated after orbit.

Experiments are being conducted to determine the shape of Echo II and its usefulness as a reliable reflector of radio waves. The Department of Defense is participating in communications tests of the rigidized sphere by providing some of the ground stations and other equipments.

Echo II also is the subject of communications experiments jointly conducted by the United States and the Soviet Union pursuant to a bilateral space

---

[1] In April 1962, Echo I made possible the first transmission of a television picture via a communications satellite. This occurred between ground stations in Massachusetts and California, a distance of 2,700 miles, in an experiment of the Lincoln Laboratory, Massachusetts Institute of Technology.

agreement reached at Geneva on June 8, 1962, between NASA and the Soviet Academy of Sciences. This bilateral space program followed an exchange of views between President Kennedy and Chairman Khrushchev on cooperation in space activities after John Glenn's successful orbital flight in February 1962.

NASA reported to the subcommittee that scientific agencies in the Soviet Union were kept informed of all phases of the Echo II launch operation, including launch schedule data and types of optical observations that would be helpful in evaluating satellite inflation and pressurization. The Radio Astronomical Observatory of Manchester University at Jodrell Bank in the United Kingdom participated in the experiments by sending transmissions at 162 megacycle per second to the Zimenki Observatory at the Gorki State University, the U.S.S.R. cooperating ground station. The quality of the transmissions was measured under certain conditions, and speech, telegraphy and facsimile experiments were made. Recent test results, including photographic data, are being analyzed in the joint program.

There are no plans presently at NASA for flight programs of new passive communications satellites. The subcommittee was advised that NASA will continue "a modest laboratory program" relating to materials development and studies of structures to insure that technical potentials of passive satellites are considered if breakthroughs occur.

Project West Ford

Still another concept in passive satellite communications is Project West Ford conducted for the Air Force by Lincoln Laboratory of the Massachusetts Institute of Technology. In this project, begun in 1959 as Project Needles, and renamed after a Massachusetts town, 400 million tiny copper fibers, called dipoles, were dispersed in orbit from a special dispenser package. The dipoles were packaged in four decks of 100 million each, each deck being one dipole length in thickness. The dipoles were held together by a volatile binder made of naphthalene. Solar heating and the space vacuum caused the binder to gradually evaporate. The cylinder was ejected from the parent satellite and caused to spin about its axis. The dipoles were released over a period of many hours. They spread along a circular orbit path about 2,000 miles in altitude and 40,000 miles in circumference, forming a thin, narrow ring or belt around the earth.

Each dipole fiber, 0.7 inch long and 0.0007 inch in diameter, one-third the thickness of a human hair, acts as a tiny passive communications satellite, tuned to reflect radio signals at about 8,000 megacycles per second. The orbital scatter of thin fibers provides a large reflecting area with very little mass.

A test dispenser was launched on October 21, 1961, riding piggyback on a Midas, part of an Air Force early warning research project. An Atlas-Agena B was the launching vehicle. Though the Midas was orbited, the canister packed with 50 pounds of copper dipoles failed to spin properly upon ejection, and instead of dispersing, the dipoles apparently remained in several clumps.

An interim experiment was carried out early in 1962, when six small wires in a canister, again riding piggyback with a parent military satellite, were orbited to test for "charge drag." Then another launch in May 1963 successfully ejected the whole dipole test package, weighing 40 pounds. The fibers first formed a compact cloud around the canister, circling the earth every 166 minutes in a near-polar orbit. This cloud spread out slowly in both directions along the orbital path, making a complete ring around the earth after several months. About 2 weeks after dispersion commenced, it was possible to send radio signals from coast to coast, transmitting voice, teletype and high speed digital data.

The Lincoln Laboratory scientists see in this concept "a number of unique and interesting advantages for reliable military communications." These include continuous worldwide communication with two 8,000 kilometer altitude belts (one polar and one equatorial) which can be placed in orbit by two launches; high reliability because all orbiting elements are passive; inherent invulnerability to enemy physical attack because any holes blown in the belt would quickly fill with fibers traveling in separate orbits at different velocities; excellent antijamming protection; simultaneous use of hundreds of communication circuits; and simplified tracking.

However, as far as Project West Ford is concerned, the major effort has been concluded. Continuing measurements will be made of changes in the orbital parameters of the experimental belt. The expected lifetime of the belt, which is based upon the orbit selected for the experiment, is a maximum of five years. Radiation pressure from sunlight will force the copper fibers down to lower altitudes, and they will burn up in denser atmosphere.

Project West Ford has evoked considerable opposition and protest in scientific circles here and abroad. Radio and optical astronomers particularly have been concerned that the seeding of space with metallic fibers might cause interferences with their scientific work. For this reason, and in accord with Presidential policy, full disclosure of the experiment has been made to the scientific community. The National Academy of Sciences, through its Space Science Board, has been informed of the project from its inception. The Board's final report in February 1964 concluded that "the present West Ford experiment has not been harmful to either optical or radio astronomy."

II. Active Satellite Experiments

Project Score

The active communications satellite concept was demonstrated in an early phase by Project Score (signal communication orbiting relay experiment), a 150 pound satellite — heaviest to that date — launched on December 18, 1958. It was boosted by an Atlas 10—B (experimental booster) into an orbit ranging between 110 and 920 miles. The project was sponsored by the Advanced Research Projects Agency, with the Army Signal Corps responsible for the communications equipment and the Air Force Ballistic Missile Division for the launch vehicle.

This relatively primitive experiment was a communications first. Its purpose was to demonstrate the feasibility of voice and teletype transmission and reception at high frequencies between the satellite and its ground station. It had two complete and duplicating communications systems, and it operated at a frequency of about 150 megacycles per second with a power output of eight watts.

Score was a delayed repeater satellite, although it was capable of transmitting real time data; that is, information transmitted from one ground station through the satellite to another ground station which is in sight of the satellite at the same time. In the delayed repeater mode, when Score passed over one of four ground stations, a coded signal directed the satellite to receive and record transmissions from the ground. When the satellite passed over a second ground station, the stored information was transmitted to the ground.

It was by means of this two-way delayed repeater experiment that a prerecorded Christmas message by President Eisenhower was broadcast to the world on December 19, 1958, the first time a voice message was received from outer space. In ensuing demonstrations, voice and teletype messages were transmitted to the satellite, recorded, stored and relayed individually to ground stations in Georgia, Texas, Arizona and California. The equivalent of some 140,000 words was transmitted in the delayed repeater mode; also 11 real time relays between California and Georgia.

Voice signals were terminated December 31, 1958, and on January 21, 1959, the satellite entered the earth's atmosphere and burned up. Score is credited with demonstrating the practical operation of a satellite radio relay system capable of spanning intercontinental distances.

Project Courier

An advanced version of Score was Courier, designed for both delayed repeater and real time operation on command. A 500-pound satellite, called Courier 1A, was launched by a Thor-Able-Star from Cape Kennedy on August 18, 1960, but the vehicle exploded several minutes after liftoff when the first stage prematurely shut down. Courier 1B (marking the 100th launch of the Douglas Thor) was placed in orbit ranging from 600 to 760 miles on October 4 of the same year. After completing one orbit, it received and recorded a transcribed message to the United Nations by President Eisenhower sent from Fort Monmouth, New Jersey, and retransmitted to Puerto Rico.

The Courier satellite, operating in the microwave range at about 2,000 megacycles per second, had a power output of four watts. It was designed to handle up to 360,000 words per station at each pass for a total of four ground stations. A great volume of simulated traffic was handled, and though not operating at full capacity, the satellite made possible the transmission of more than 50 million words of teletype data between ground stations at Fort Monmouth, New Jersey, and Salinas, Puerto Rico. Delayed and real time voice, teletype and facsimile tests were conducted between these two stations during the first 17 days in orbit. On October 21, 1960, the satellite communications link failed and was not reestablished. The satellite remains in orbit.

Project LOFTI

To demonstrate the feasibility and determine the configuration of a satellite for communications between surface ships or shore stations and submerged submarines using radio signals in the very low frequency (VLF) range, the Navy organized Project LOFTI (low frequency transionospheric investigation). LOFTI I rode piggyback on Transit III-B in a Thor-Able-Star launched on February 21, 1961, to a 103/597 mile orbit. Planned experiments included measurments of the following types: Penetration of 18 kilocycles VLF signals into the ionosphere from ground based transmitters; signal intensity (magnetic and electric components); time and phase delay; earth and solar aspect data; and Doppler shift.

Some of these measurements could not be made since the electric dipole antenna failed to extend. The satellite decayed into the earth's atmosphere 36 days after launch. However, it was the first demonstration of VLF reception in the ionosphere.

LOFTI II was launched January 24, 1962, in a five satellite spacecraft but failed to orbit because of insufficient second state thrust. LOFTI II-A was launched successfully on June 15, 1963, as part of another five satellite spacecraft. It operated in a 69.9 degree inclination orbit with 109 mile perigee and 550 mile apogee for 33 days and reentered on July 18. It made measurements on two VLF frequencies, 18 and 10.2 kilocycles. The results are being analyzed and are expected to provide valuable data for use in further studies. The program is under the direction of the Naval Research Laboratory.

Project Telstar

Telstar I, the first communications satellite developed by a commercial carrier, was successfully launched into a 593/3,503 mile orbit by a Delta vehicle on July 10, 1962. This was a cooperative venture between Bell Telephone Laboratories of the American Telephone & Telegraph Company (AT&T) and the National Aeronautics and Space Administration, with the government agency setting the environmental test specifications for the spacecraft, procuring the launch vehicle, launching the company owned satellite, providing tracking data and assisting in arrangements for participation by foreign communications agencies. NASA was reimbursed by the company for identifiable project costs.

The 170 pound satellite, though operable only 10 percent of the time because of limited power supply, successfully conducted various communications tests — black and white and color television, telephone, photo facsimile and high and low speed data — between Bell Telephone Laboratories stations in Maine and New Jersey, and British and French government stations. Significant data also were gathered on radiation decay relating to one U.S. and two U.S.S.R. nuclear test detonations.

Telstar I was silent for more than five weeks following November 23, 1962, when radiation damaged a transistor in the command decoder. The trouble was temporarily overcome by laboratory analysis and ground commands which led to the removal of all voltage from circuitry during eclipses, permitting recovery from radiation effects. Telstar I commenced operating again on January 3, 1963, but fell silent the next month for the second time. It is still in orbit.

Telstar II was placed in a high elliptical orbit on May 7, 1963, with a Thor-Delta three stage rocket. An improved, more powerful Delta rocket made it possible to place this satellite in an orbit with a greater apogee than Telstar I — about 3,000 miles farther out in space. Consequently, Telstar II enlarges greatly the area of mutual visibility between the United States and Europe, occasionally including Japan. Its first communications assignment was to transmit a television program from the AT&T station at Andover, Maine, to receiving stations in Great Britain, France and Italy. Tests also have included communications to a station in Japan.

Weighing 175 pounds, Telstar II is but slightly heavier than its predecessor. However, it carries electronic equipment more resistant to radiation and operates with its higher apogee (6,700 miles) in areas of less intense radiation. Telstar II can send its telemetry reports by microwave as well as on the very high frequency (VHF) beacon used by Telstar I.

For four weeks in July-August 1963, Telstar II also failed to respond to radio command signals. This failure apparently was caused by spurious radio signals interfering with normal command control. The condition cleared on August 12, 1963, and Telstar II has operated satisfactorily since that time.

Project Relay

NASA's Relay I satellite, weighing 172 pounds, was launched on December 13, 1962, in a 819/4,612 mile orbit. Just as improvements in the Delta vehicle gave Telstar II a higher boost than Telstar I, so they also accounted for Relay's higher orbit. Its transmitter was four times as powerful as that of Telstar I, so that the communications capacity was considerably expanded.

Relay was designed to be the first repeater satellite linking three continents — North America, South America and Europe. Telephone, television, radio, teletype and photofacsimile tests were conducted between the United States and European stations, and radio tests between the United States and Brazil. Ground stations services in the United States for these tests are provided to NASA by contract with AT&T and International Telephone & Telegraph Co.

Hours after the Relay launching, communications experiments were halted because on-board battery power leaked away and was too low to operate the transponders. By use of a backup transponder, Relay came to life on January 3, 1963, about the same time as Telstar I, which had been silenced since November 23. Each of Relay's transponders is capable of relaying a standard television channel or equivalent high speed data. The spacecraft also has the availability for multichannel, two-way telephone service between selected ground stations. Equipment is installed for measuring energy levels of space radiation in its orbital path and for determining the extent of radiation damage to solar cells and electronic components.

Relay I is still in orbit, and communications tests will continue, though all planned demonstrations and experiments, numbering over 2,000, were achieved.

Relay II, also weighing 172 pounds, was launched successfully by a Delta vehicle on January 21, 1964. Incorporated in the vehicle were design modifications based on operating experience with Relay I in order to improve lifetime and performance. This launch substantially completed the Relay program. A third Relay spacecraft, earlier scheduled for launch, remains on the shelf in mothballs, with no plans to launch it.

Project Syncom

NASA's Syncom I was the first try for synchronous orbit. This 86 pound satellite (160 pounds including the apogee "kick" motor) was launched from Cape Canaveral on February 14, 1963, by a three stage Thor-Delta. The satellite was boosted into a highly elliptical orbit, carrying an internal rocket motor to be fired at the apogee, providing the extra spurt of power which would make the orbit circular. During the five hours it took to reach apogee, preliminary transmissions showed the communications equipment was in working order. But after the on-board rocket was fired, telemetry signals stopped. Syncom I has been silent since that time. Telescopic observations established that Syncom is in a near synchronous orbit, in accord with the flight mission, but the communications experiments could not be carried out.

The next try, with Syncom II was successful. It was launched July 26, 1963, in the same inclined (33°) orbit. It was the first successfully operating synchronous satellite and it is still working successfully. According to NASA officials, high quality, two-way voice transmissions through the satellite have been demonstrated. Changes in drift and orientation have been commanded from ground stations, which are operated by the U.S. Army.

Syncom III (known as Syncom C until successful orbiting) was launched on August 19, 1964, and placed in synchronous orbit on August 20, the following day. It achieved a nearly perfect equatorial orbit, with an inclination of less than one-half a degree at present. It was "walked" into a stationary position over the Pacific near the international date line and is to be used for live television transmissions of the Olympic games to be held in Tokyo during the fall of this year. Preparatory arrangements are being made for this experiment under the technical direction of the Communications Satellite Corporation, responding to a request of the State Department.

In addition to the achievement of a nearly true synchronous orbit, the Syncom III experiment will test the performance of new critical components at synchronous altitude and the performance of new ground station equipment and crews. The thrust augmented Delta (TAD) booster was used by NASA for the first time in the Syncom III launch.

Satellite Communications at the Goddard Space Flight Center

John F. Clark and William N. Redisch

The beginning of the satellite era was triggered by the dramatic launch of Sputnik I in 1957. The use of space as a medium of communications started with the 1958 launch of the U.S. Army Score, an experimental, short-lived, store and forward repeater satellite of simple, off the shelf hardware that successfully tested the feasibility of relaying narrowband voice and teletype from low orbit.

In those early days of space flight, the National Aeronautics and Space Administration's (NASA) Goddard Space Flight Center (GSFC) became one of the pioneers in satellite communications, starting with the 1960 Delta launch of the first Echo satellite, a 30 meter (100 foot) diameter, balloon shaped passive reflector.

The Relay Program, begun at GSFC in 1960, demonstrated the feasibility of relaying real time wideband communication signals such as television between ground stations and low orbiting satellites. Relay I was launched in December 1962 and Relay II in January 1964 by the Thor-Delta. The experiments conducted by these two satellites were very successful, but perhaps the most

---

**This article originally appeared in the March 1976 issue of SIGNAL.*

significant contribution to space communications from the Relay Program was the knowledge gained of how to design, test and operate communications spacecraft.

Project Syncom

It was recognized early in the space age that a satellite in a geostationary orbit would have great advantages for communications compared to a low orbiting satellite. A spin stabilized communications satellite was first proposed by Hughes Aircraft Company in the autumn of 1959, and Project Syncom was initiated as a joint NASA/Department of Defense (DOD) development effort at Goddard. The major objectives of the program were to develop the capability of launching satellites with the Delta launch vehicle into Earth synchronous orbit and to demonstrate the utility of this orbit for satellite communications.

Three satellites were launched: Syncom I, II and III. Syncom I was launched in February 1963 and achieved orbit, but all communications were lost when the apogee motor fired. Syncom II successfully achieved an inclined synchronous orbit in July 1963, and Syncom III was placed into a geostationary orbit in 1964 by a thrust augmented Delta. All of the program objectives were satisfied by the numerous experiments and demonstrations that were conducted. Once these objectives were achieved, the satellites were given to DOD in early 1965 to provide an operational communications capability serving the Far East, the Pacific Ocean and the Western United States.

During the 1960s, all aspects of communications satellite systems were studied extensively, with emphasis on the use of active satellites in geostationary orbit with frequencies in the 1 to 10 GHz band.

Thereafter, the Goddard effort turned to developing high gain satellite antennas, spacecraft stabilization approaches, multiple access techniques applicable at these frequencies and high reliability, long life components. At the same time, the necessary development of compatible ground terminal technology spawned low noise receive systems; threshold extension demodulators allowing detection down to low values of signal to noise ratio; accurate satellite ranging and orbit predicting techniques, so that aimed high gain antennas could be used; and highly reliable operational performance. By late 1963, launch vehicle technology, especially the Goddard managed Delta program, had progressed so that satellites of a reasonable size could be injected into synchronous orbit.

By mid-1966 high satellite Effective Isotropic Radiated Power (EIRP) (due to greater available prime power, advanced TWTs and higher gain antennas) made it possible to provide communications and position location at very high frequency (VHF) for aircraft, ships, remote data platforms and mobile land terminals.

The ATS Program

The Advanced Syncom program was established to further advance communications satellite technology. This program quickly evolved into the Applications Technology Satellite (ATS) program that incorporated Advanced Syncom techniques and meteorology, science and navigation experiments. A multitude of experiments and demonstrations were designed to advance communications satellite technology.

The design philosophy of the ATS-1 through -5 satellites was aimed at the development of multiple mission satellites having large and adaptable volumes for experiment payloads while employing basic satellite configurations appropriate for either spin or gravity gradient stabilization.

ATS-1, a spin stabilized, active repeater satellite with VHF and C-band frequencies, was launched December 1966 into geostationary orbit and positioned over the Pacific Ocean where it has remained. Since completion of its initial program of GSFC communications and propagation experiments (including the feasibility demonstration of an electronically despun antenna at C-band and VHF), individuals and organizations, both government and private, have been encouraged to use the satellite for approved communications experiments, as is the continuing practice today.

ATS-3 was successfully positioned over the Atlantic Ocean in a geostationary orbit on November 7, 1967. The original experiments conducted with the satellite were highly successful, and the satellite has been used since for many additional experiments similar to those of ATS-1. A major contribution by ATS-3 to satellite communications was the demonstration of the feasibility of mechanically despun antennas.

Intelsat began to establish a third generation satellite system in late 1968. By early 1970 five successful Delta launches had provided a worldwide system providing service over the Atlantic, Pacific and Indian oceans. These satellites were a direct outgrowth of the technology and experience developed in the Syncom and early ATS programs.

Studies at Goddard were expanded during the mid 1960s to include consideration of higher frequencies for providing various communications services. Interest developed in employing L-band frequencies for communications and position location in avionic and maritime applications.

After launch vehicle problems resulted in the unsuccessful orbiting of ATS-2 and -4 in 1967 and 1968, ATS-5, designed for gravity gradient stabilization, was launched in 1969. The Atlas Centaur launch vehicle performed properly but control of the satellite was lost during the initial spin stabilizing period. As a result, the spacecraft was left spinning about a longitudinal axis so that it could not be despun and the gravity gradient boom deployed.

In spite of improper stabilization, the L-band and millimeter wave experiments included on this satellite were performed, and valuable data was and is being obtained that contributes to the utilization of these frequency bands.

The ATS-6 Project

As the initial program for the ATS satellites proceeded and Goddard began to look to the future, the program was directed to advancing techniques for bringing satellite communications to small users having multiple access to the satellite system; for broadcast satellite applications, both radio and television; for more effective utilization of the frequency spectrum; and for developing communications spacecraft components and techniques. The ATS-6 project was the next (and last) step of the ATS program and resulted in the Titan IIIC launching of a very versatile spacecraft into a geostationary orbit on May 30, 1974.

The principal communications objective of the ATS-6 program was to demonstrate the operation of a high powered communications spacecraft in orbit that could relay color television and other high quality communication signals at multiple frequencies to relatively simple and inexpensive ground receivers located at various points on the Earth's surface. The ATS-6 spacecraft deployed a 9.1 meter (30 foot) diameter, parabolic reflector in space, performed maneuvers to point the reflector to selected points on the Earth's surface upon command from the ground and provided the necessary communications links and power to receive and rebroadcast communications signals transmitted from the ground.

ATS-6 has accurate three axis stabilization using inertia wheels and hydrazine thrusters with error signals derived from a variety of sensors including a C-band interferometer and three frequency monopulse system. It has a multiple frequency integrated transponder that transmits at UHF and receives and transmits in very high frequency (VHF), L-band, S-band and C-band.

Experiments have investigated the feasibility of employing a synchronous communications satellite as a relay for tracking other satellites and transferring digital signals between them and a ground station; aircraft position location and air to ground communications using a synchronous satellite; frequency modulation (FM) television reception to small S-band and UHF ground terminals; uplink radio frequency interference (RFI) at six GHz; and wave propagation at 13, 18, 20 and 30 GHz.

Upon completion of the second year of operation at 35° East longitude, ATS-6 will be positioned over the Western Hemisphere for third year experiments. Communications and general societal experiments being considered for the third year of operations include:

1. Hospital emergency medical training for training physicians quickly in emergency room care and cardiovascular pulmonary resuscitation;

2. Video broadcasting from the Rosman Ground Station via ATS-6 to Ecuador in South America to demonstrate the use of satellites to upgrade the training of people in the field and to improve the standard of living in rural areas;

3. Substitution of communications for transportation. The University of Montana has field centers that have difficulty exchanging information. People drive 200 to 300 miles to exchange data or exchange it by mail. During winter, that 200 miles is a very long drive, so ATS-6 will be used to demonstrate the exchange of information between the field centers and negate the need for travel back and forth.

Current GSFC Efforts

The policy announced in January of 1973 that NASA would phase out of communications satellite programs involved the cancellation of the future ATS and other communications satellites. The communications effort at the Goddard Space Center now concentrates on:

1. Operating the ATS spacecraft in orbit and the planned experimenting in the 12 to 14 GHz frequency band with the Communications Technology Satellite — a joint NASA/Canadian 200-watt TWT, direct broadcast, three axis stabilized spacecraft launched by a Delta 2914 vehicle in January 1976;
2. Providing technical consulting services using the space communications expertise acquired over the last 15 years to support other government agencies such as the Federal Communications Commission and the Office of Telecommunications Policy;
3. Developing cost effective application services with Federal Government mission agencies by using existing technology for applications such as search and rescue and data collection.
4. Servicing the communications needs of NASA's satellites via the Spaceflight Tracking and Data Network, including development of the Tracking and Data Relay Satellite System.
5. Undertaking spectrum conservation research such as propagation characterization and electromagnetic field measurements; and
6. Undertaking advanced communications research to develop components and techniques for possible future communications satellite applications.

Some components under development are:

1. A 200 watt CW, 50 percent efficient S-band electron bombarded semiconductor as a low cost replacement for the conventional traveling wave tube amplifier (TWTA) for both spaceborne and airborne systems;
2. A 75° Kelvin noncryogenic Ku-band parametric amplifier for ground system applications.
3. A 300 megabit per second data relay receiver with ±700 MHz Doppler tracking and a 500,000 hour lifetime CO_2 laser transmitter and hemispherical optical antenna for potential future low altitude to synchronous or synchronous to synchronous satellite wideband data relay systems.
4. Schottky barrier mixers from 80 to 200 GHz and submillimeter maser amplifiers, oscillators operating from 150 to 1000 GHz and other millimeter and submillimeter equipment to be used both for obtaining propagation information for opening new spectral bands and for potential application for remote microwave sensing; and
5. Large adaptive multibeam phased array antennas for spacecraft and large thinned ground based conformal arrays for radio holography techniques, for mobile radiodetermination and position location applications.

This work and the effort to explore new frequency bands for future commercial telecommunications will be continued.

Applications

Applying the space communications technology already developed to new and potentially beneficial government and commercial applications is a major thrust of Goddard's ongoing applications program.

In situ measurement data collected directly from multiple locations and relayed by satellite into central processing facilities complement and enhance satellite remote sensing efforts for meteorology; earth, ocean and wildlife resources; environmental monitoring; and geology. Navigation can become more economical and as a result allow wider use for applications such as monitoring hazardous cargo shipments and keeping track of mobile vehicles by urban and local enforcement agencies: Satellites at a high vantage point can locate and relay distress alarms over broad expanses, no longer limiting search and rescue units to the use of electronic beacons now being heard only by ground or aircraft line-of-sight receivers. Hazard warning satellites can directly alert the public when impending disasters are near. Better health and educational services can be provided directly to the remote areas in this country. Satellite teleconferencing techniques can be developed to reduce expensive and energy consuming travel, and satellite information retrieval and transfer, already under way in the financial world, could be developed for libraries, universities, hospitals, law enforcement agencies and many others.

These types of applications are now being studied as NASA reexamines its 1973 communications satellite phase-out policy. NASA, and GSFC especially, can bring its many years of expertise to bear on today's communications applications problems and also can take advantage of the coming capabilities of the space shuttle in the 1980s to help keep this country in the forefront of satellite communications technology for peaceful applications.

USAF'S Conquest of Space

Kenneth W. Schultz

On the surface, 1954 seemed a year likely to generate little historic impact. The war in Korea was over and Americans were settling down to a life style in which the automatic transmission was the new "gimmick" to have in a car, and a first-class stamp cost three cents. An Englishman named Roger Bannister ran the first under four minute mile, and the Pulitzer Prize for biography went to Charles Lindbergh for his *Spirit of St. Louis.* President Eisenhower signed a bill providing for the establishment of an Air Force Academy. The first production model of the North American F-100 came off the line. The Air Force awarded a contract for study and development of a new supersonic bomber, the B-58.

Then, in the course of three tests in the Pacific, the hydrogen bomb advanced from experiment to practical reality.

In a sense, the Pacific tests were the "Open Sesame" to the space age. They proved the feasibility of an effective warhead small and light enough to be delivered by a ballistic missile of the power then within our grasp. That proof came at a time when the Soviets were known to be pressing development of long-range ballistic missiles that could give them an enormous strategic advantage over this country. And a concerned American people and Congress were solidly behind the government's determination to retain the strategic superiority that had so far deterred the eruption of the cold war into World War III.

**Reprinted by permission from AIR FORCE Magazine, August 1974. Copyright © 1974, the Air Force Association.*

The Beginning

On July 1, 1954, the Western Development Division of the Air Force Air Research and Development Command was established in an abandoned schoolhouse in Inglewood, California. Its top priority mission: to develop in the shortest time possible a United States long-range ballistic weapon system.

There have been a number of organizational and name changes since that time, but the basic mission has remained that of providing the United States with the best ballistic missile strategic capability that evolving technology would provide. As the missiles made possible the first actual moves into space, that mission expanded to include the exploration and development of the total potential of space for strengthening the country's defense.

Today, this initial epicenter of the U.S. space effort is named the Space and Missile Systems Organization (SAMSO) of the Air Force Systems Command. Its mission, in broadly simplified terms, has six major facets. SAMSO is responsible for planning, developing, and deploying the U.S. land based intercontinental ballistic missile force. It manages research and development of military space systems. It develops space launch systems and launches space payloads for military and other government agencies. It operates worldwide facilities for tracking and command and control of U.S. satellites and those of its allies. In its ABRES program, it conducts triservice development, test and evaluation of all Department of Defense advanced ballistic reentry systems. And it carries out continuing and extensive programs for identifying and developing technologies needed for missile and space programs of the future.

In the twenty years since the Western Development Division (WDD) was created, Americans have walked on the moon; probed the nature of Mars, Jupiter, Venus and Mercury; and lived and worked in space for a total of almost 22,000 man-hours, or two and a half years. We have orbited a growing galaxy of satellites to provide communications in war and peace; navigation through the treacherous ice of the Arctic; warnings of killer storms, nuclear detonations or enemy attack; surveys of natural resources, pollution, plant diseases and pests, forest fires, urban blight and scores of other terrestrial concerns and conditions that can be grasped in full only from the observation platform of a satellite orbiting thousands of miles in space.

As the primary missile and space projects have advanced, they have left a rich wake of peripheral benefits for the whole civil sector, benefits already beginning to have a major impact on the living and working patterns of our society. The space program has generated new products; new industrial and business techniques, tools, management methods; a host of health and safety innovations; new approaches to transportation, housing, the control of crime and development and management of massive municipal projects. The nation's energy administrators reviewing satellite findings on new sources of fuels, and the man determining his income tax on a minicomputer are both using tools given them by the space program.

Challenges, Blowups and Successes

The 20 years between the schoolhouse and such spinoffs, between the order establishing WDD and space successes like VELA, the Initial Defense Satellite Communication System, Mariner-10 or Skylab have not been all triumphs. Despite proved U.S. world leadership in aviation, the country did not have, in 1954, matured industrial resources for missile research, development and production. Some rare pioneers had stubbornly kept the breath of life in ballistic missile research through the lean post-World War II and Korean War years. But, in general, research and industrial resources for the mission had to be brought into being, nurtured and welded into a nationwide team with energies focused on the high priority objective.

The Air Force itself lacked sufficient in-house experience or management methodology to integrate and orchestrate so unprecedented, enormous and highly compressed a research and development effort. From the Guided Missile Research Division of Ramo Wooldridge, it created a kind of alter ego, a nonprofit pooling of civilian talent to act as intermediary and interpreter between the Air Force and contractors and to provide systems engineering and technical direction for the program. It was necessary, in short, to create both the research and industrial capabilities and the management methods for directing and controlling them.

The early years had their full share of educative hardware failures, too. Old hands still remember some of the standard jokes to fend off the mood of defeat: "What was that you guys launched yesterday — the missile or the gantry?" "How is a ballistic missile like having your brother-in-law in your business? It won't work, and you can't fire it." We looked at the captured German World War II films of V-2s blowing up on the pads at Peenemünde and were reminded that they, too, had their failures.

But the tide did begin to turn. In September 1957, after only two years' development time, the Thor intermediate-range ballistic missile passed its first completely successful test flight. By the end of that same year, the Atlas long-range missile also took that first hurdle successfully. In early 1959, the Titan I was first flight tested. In the summer and fall, the first Thor operational squadron was transferred to the Royal Air Force in the United Kingdom, and the first operational Atlas complex was turned over to the Strategic Air Command (SAC).

The Big Dig — Site Activation

Concurrent with development of the missiles and related system hardware, the enormous job of constructing launch sites was begun. In retrospect, the true dimensions of that project seem even more awesome than they did at the time the Air Force/Corps of Engineers/industry partnership tackled the job.

The earliest sites were "soft" ones above ground, but later models of the Atlas, Titan and, finally, Minuteman systems were hardened — buried under 25 to 35 feet of reinforced concrete and earth. The sites on which the individual squadrons and launch complexes were spaced out were enormous. Total area of the bases was about 125,000 square miles, almost as large as the combined areas of the states of Washington, Alaska and Hawaii. Work sites were remote requiring the importation of large members of construction workers, more than 3,000 at a site at peak construction. Climatic conditions ranged from the ice and snows of Northwestern winters to the deserts of the Southwest.

The work, both site construction and interface of the complex silo equipments, was without precedent, requiring the development of new approaches, techniques and skills. And the missiles themselves were evolving constantly under urgent force feeding of the state of the art, making necessary constant modifications of the ground environments.

One harried construction contractor, testifying about program difficulties before a congressional committee, characterized the job as building whole underground cities and filling them with advanced, integrated equipments that had to function with the precision of a Swiss watch.

The electrical output of diesel generators for a Titan I site was, indeed, sufficient to supply a city of 40,000 people. The nearly 4,000 miles of electrical circuitry in an Atlas control center was enough to provide a telephone exchange for a city the size of Cheyenne, Wyoming. The first Minuteman bases in Montana required about 120,000 cubic yards of concrete — a load that would fill some 5,300 railway freight cars. *Time* Magazine reported that "the missile base construction programs made the pyramids look like a Tinker Toy exercise."

Early in 1961, the Minuteman, first solid-fueled, "instant response" missile, was a phenomenal success in its initial flight test. In December, the first operational missiles, rushed to completion, were turned over to SAC at Malmstrom Air Force Base, Montana. The first complete wing was activated the following February.

The Minuteman system, progressing from one improved version to another, has since become the mainstay of our strategic missile force, which also still includes 54 updated Titan II missiles.

The Atlas and Titan I systems were retired from the active weapons inventory in the mid 1960s, the Atlas to continue its role as a space booster. We are currently modifying the bases for the 1,000 Minuteman force, shaping it to an integrated mix of Minuteman II and III systems. This Minuteman force, and our submarine-launched missile elements and long-range bomber units constitute the U.S. Triad of strategic defense.

The Space Program Evolves

While the missile side of the mission pushed forward rapidly with the advancing state of the art, missile adaptations used as launch vehicles were shaping a national space program. In January 1958, Explorer I, launched by a Jupiter missile, was the first U.S. space system to achieve orbit. The program will also be remembered for its discovery of the Van Allen radiation belts. In general, however, the last years of the 1950s were a time of discouraging failures in space. By 1959, the Thor, with an Agena or Able upper stage, had already become the workhorse of our space launch program, but results still gave little indication of the high reliability the missile was to achieve as a space booster in the 1960s.

By 1960, we had begun at last to break even. In that year, 16 launches out of a total of 29 were successful. Among them were Thor launches of the first successfully orbited meteorological satellite, TIROS-1; the first navigation satellite, TRANSIT-1B; and COURIER, the first active repeater communications satellite.

The Atlas also made the transition to space booster and contributed to the slowly but steadily climbing percentage of successes. Among its early noteworthy payloads were the first of the Mariner space probes, and the early VELA series development to monitor the Nuclear Test Ban Treaty.

In 1961, President John F. Kennedy announced the national space goal of putting a man on the moon and returning him safely to earth before the end of the decade. The Atlas was chosen as launch vehicle for the first step in the man in space endeavor — the Mercury program. In February of the next year, an Atlas-D launched the Mercury-6 in which John Glenn, the first American to circle the globe in space, orbited for nearly five hours.

The more powerful Titan II, specially "man rated," was selected to orbit the two-man Gemini series of spaceflights begun in 1965. It was still doing a flawless job in that program when development work began on the Titan III, the first Air Force missile to be developed specifically as a space launch system, and the largest and most powerful of present Air Force boosters. In one launch, the Titan III, with a 25,000 pound payload capability and an extremely sophisticated upper stage, can put as many as eight different payloads into separate and distinct orbits. We have used it to launch a large number of communications and nuclear detonation detection satellites, among other experimental payloads from a whole spectrum of government agencies.

For the final three-man thrust at the moon, NASA developed the giant Saturn, with a total thrust of more than nine million pounds. And, in July 1969, well before President Kennedy's deadline, the astronauts of Apollo-11 became the first men to set foot on the moon.

In this twentieth anniversary year, it is difficult to realize that there were years — much of the decade of the 1960s — in which we had to prove that space could be useful to us, prove it in the face of the skepticism of others and our own inexperience. Today, space is more or less taken for granted as a viable dimension of our world.

Justice Oliver Wendell Holmes once said, "Man's mind, stretched to a new idea, never goes back to its original dimensions." Surely none of us who saw the awe inspiring vision, relayed back for the first time by Apollo-8, of our world whirling half a million miles away, incredibly beautiful in the immensity of space, can ever go back to quite the prespace perspective.

And space itself will never be the same. Since 1957, man has orbited a total of some 1,570 spacecraft. Slightly fewer than half of those were U.S. efforts. Six hundred and fifty-seven were still in orbit at the beginning of 1974, 53 percent of them belonging to the United States, 39 percent to the Soviet Union and the remaining eight percent to other nations or international groups such as Australia, Canada, France, Germany, Japan, the North Atlantic Treaty Organization (NATO), INTELSAT, or the European Space Research Organization (ESRO).

The Military/Civilian Space Duo

The United States is unique among space conscious nations in its dual program — the careful distinction made between military and civilian space developments. NASA was created by Presidential act in 1958. It absorbed the National Advisory Committee on Aeronautics (NACA), originally established in 1915, and was charged with responsibility for the civilian space program, which could then be seen as an inevitable outgrowth of military space developments already well under way.

Sharp definition of separate roles for military and civilian space efforts has not always been easy. In actual fact, the two programs have worked in close and economical cooperation, sharing specially qualified manpower, advancements in technology and the ever broadening expertise that has come with experience. That cooperation was extremely close and functional in the Mercury and Gemini programs, in which NASA's manned space vehicles were launched by Air Force boosters and crews.

All but one of the 15 astronauts who participated were members of the armed forces. Head of the Apollo lunar landing program was a military man — MGen. Samuel C. Phillips on loan from the Air Force. Close cooperation has continued in present development work on the Space Shuttle, the reusable space launch system being designed to meet both military and civilian requirements.

Though the Air Force did at one time have responsibility for the Manned Orbiting Laboratory, a precursor of the NASA Skylab program, our principal emphasis today — except for the Shuttle — is on unmanned military space systems.

Secretary of the Air Force John L. McLucas recently summed up the military space mission in this way:

> Under the Space Treaty, the United States has agreed not to place weapons of mass destruction in outer space or in orbit around the earth. Rather, we are using the medium of space to increase our alertness to danger and to assist in maintaining an effective deterrent posture. Our space activities are the result of continuing reviews of our national defense needs and the medium where they can best be satisfied, be it land, sea, air or space . . . . Through the use of space systems we can give our defense forces improved communications, more accurate navigational assistance, better weather information and more reliable, earlier warning of attack.

One of our earliest and most prolific uses of space has been for military communications. Our first global system, composed of 26 satellites, was originally orbited as a research effort but proved so successful that it was converted to an operational system in 1967. It gave us good service during the war in Southeast Asia, and about half of the satellites are still working. In 1971, we launched two follow-on systems with which we encountered problems, but performance of the second pair of Phase II satellites, launched early this year, indicates that the difficulties have been corrected. A final launch of the remaining two satellites in this system is scheduled for fall of 1974. We are also into the second generation of similar military communication systems for NATO and the United Kingdom.

We have had less operational experience with tactical satellite communications designed for use with small, mobile ground terminals. But our first, TACSAT I, the largest communications satellite yet launched by the United States, did relay operational communications for all three military services for 17 months after it became operational. This was considerably beyond its expected lifetime.

We have developed and tested several satellites and a number of receivers for aircraft, ships, trucks, jeeps and a man pack to be carried by a three-man team. The first of the fully operational tactical systems scheduled for deployment is the Fleet Satellite Communications System that will provide improved control for Navy ships. The same satellites will provide communications capability for the worldwide Air Force communications system for command and control of Air Force strategic forces.

Weather, Warning and Navigation

Another military use being made of space is to gather data for analysis and forecasting vital information to defense operations. This has been the concern of our Defense Meteorological Satellite program. For more than a year, now, this information has been made available to the U.S. public and to other nations through the National Oceanic and Atmospheric Administration and its National Weather Service facilities in Maryland. The system consists of an integrated combination of satellite infrared and visual sensors, communications and ground processing facilities. The two types of sensors provide resolutions as fine as one-third of a nautical mile. The system is a valuable forecasting tool, especially useful because it can give us data on areas of the world where conventional weather observations are not available.

Our ground processing system has been designed to provide data to the user within minutes after it is collected in space. The system can convert pictorial images to a digital format that can be computer processed, and we have developed mobile, air transportable vans that permit data readout on the spot by military commanders anywhere in the world. Japan and the European community are planning geostationary meteorological satellites to continuously monitor weather conditions over their areas.

Early warning of missile attack and detection of nuclear explosions are other major missions for satellites that, from synchronous altitude of 20,000 statute miles, command an overview of almost an entire hemisphere. First elements of one of our earliest operational systems, the VELA, were orbited in 1963 to monitor the Nuclear Test Ban Treaty. Today, we have several systems devoted to the early warning mission, including different types of ground radars as well as space sensors.

We have developed satellites that can detect and track intercontinental ballistic missiles and submarine launched ballistic missiles almost from the moment they are fired. In addition, these warning satellites have the capability to detect nuclear explosions above the ground. In the near future, the current VELA systems will be replaced by the even more sophisticated capabilities now being developed. The creation of the satellite surveillance system has greatly improved the United States overall warning network.

Space based systems also offer important potential advantages in navigation. For defense applications, we need a system that can accurately and almost instantaneously provide data on position, course and speed any place in the world for a variety of users, and with the simplest possible user equipment.

For a number of years, SAMSO has been studyng such a system. The most recent proposal to develop the Global Positioning System has been approved, and Phase I contractual actions are now under way. This Global Positioning System will consist of 24 satellites in three different orbital planes. It will provide accuracy in the tens of feet in position and altitude to meet the needs of all users. And it can be available to any civilian users who purchase receiving equipment. The initial satellite launch will be in 1977 to facilitate the development of user equipment for all services. Additional elements will be in place by 1981, and the system will be fully operational by 1984.

In very broad terms, these are our principal areas of emphasis in military space activity today.

Civilian Space Programs

On the civilian side of our national space program, NASA is still analyzing and applying the wealth of data accumulated in the now completed manned space programs, Apollo and Skylab. It also has many other important programs providing a great variety of scientific data and practical benefits to the country, and indeed the world. Such interplanetary exploration programs as Pioneer and Mariner are continuing. From these space probes and the manned ventures, we now have newly revealing photographs and other data concerning five celestial bodies — the moon, Mars, Mercury, Venus and Jupiter — to compare with our own in attempting to understand the origins and nature of our own planet.

NASA also has many technology application and scientific programs involving satellites in earth orbit. In ten years, the civilian communications satellite applications have progressed so rapidly that satellites now handle more international telephone traffic than undersea cables, and the cost of a transatlantic call has been cut about 40 percent in the last three years.

The NASA Earth Resources Technology Satellite (ERTS-1), orbited in 1972, has become a kind of universal space tool for many nations in addition to our own. ERTS can map 100 million square miles of the earth's surface each week and passes over and photographs approximately the same 115 mile strip at the same time of day 20 times a year. Its photographs can be purchased by anyone, for prices ranging from $1.25 to $27, depending on the size and nature of the reproduction requested.

Among the almost infinite variety of ERTS projects are development of an agricultural map of the United States; studies of smog, earthquake faults and crop pests in California; a study of ecological changes taking place on the east coast of the United States; snow surveys to assess the risk of spring flooding in Norway; a study of land use and soil erosion in Guatemala; identification of oil, gas and mineral resources in Alaska; and detection of potential locust breeding sites in southwest Saudi Arabia.

A second Earth Resources Technology Satellite will be orbited in 1975.

Breakthrough in Space Economics — The Space Shuttle

As the second decade of America's space effort draws to a close, a new catalyst is in the making that could prove a major factor in shaping our national space effort for decades to come. The Space Transportation System, or Space Shuttle, a program for which NASA has primary responsibility, but in which the Air Force is actively participating, promises to be a most significant breakthrough in the economics of space. Essentially a reusable booster, the Space Shuttle can dramatically reduce the cost of putting space payloads into orbit.

The Shuttle will consist of two stages — a booster for launch from earth, and an airplane-like manned, reusable orbiter for flight into orbit, where it will conduct space missions. The Shuttle will be launched vertically. The orbiter will separate from its booster and go into orbit under its own power. When it finishes its mission, the pilots will fire its rockets to slow it down, fly it through the atmosphere and land it like an airplane on a jet sized airstrip. The delta winged manned orbiter will be about the size of a DC-9 airplane, with a 15 by 60 foot cargo compartment. The orbiter will be able to carry a combined passenger and cargo weight of 65,000 pounds.

Delivery and placement in earth orbit of payloads by means of propulsion stages, or "space tugs," will allow placement of satellites into very high earth orbits, such as geosynchronous orbit at an altitude of 20,000 nautical miles.

Space in the Military Process

Grant L. Hansen

I welcome this opportunity to discuss space activities in the Air Force and how these activities fit with the missions of the Air Force.

You will note that I did not use the words "Air Force Space Program." Many people do, but I consider them to be wrong. The National Aeronautics and Space Administration (NASA) has a space mission, the Department of Defense (DOD) does not. Our mission is defense, with such contributing submissions as strategic force management, tactical control, airlift, sealift, sea control, air defense, training, early warning, intelligence gathering, rescue, command and control and many others. In the Air Force, the fact that something can be done in space is a necessary but not sufficient condition. Space must be the best way to accomplish the function.

I have chosen the subject "Space in the Military Process" because it gives me the opportunity to put Air Force space activities into perspective. I would like for you to visualize the block diagram of a generalized, closed-loop military system. The principal elements are an array of technical and management systems for: data gathering; data processing, assessment and display; decision making; force employment; and performance assessment. These elements are tied together by command, control and communications systems which relay data and decisions and provide control signals for the entire operation.

Let us break into the loop at the data gathering element of the cycle, the sensors. The output of the sensors feeds into some data processing and then into data transmission. After reception of that data, there is more data processing and then more data transmission to a data display of some kind. Utilizing the displayed data, a situation analysis is made, from which a

---

**This article was originally presented at the National Space Club on November 30, 1972.*

command decision is made. This decision must be communicated to operating forces for command execution. The results of command execution must be assessed and fed back to the commander. The necessary assessment is done by the sensors, thus picking up the cycle at the point where we started. The process through this closed loop cycle is repeated until the commander learns from the sensors that the situation is what he wants it to be.

This highly oversimplified visualization of the military process applies to a local tactical command as well as to the National Command situation. As is true for any operating system, this one has noise with the signals. Also, each functional block and connecting line must be examined for both capability and survivability.

Each function must be performed using the most effective method. This is the way that space capabilities can be brought into the military system equation. Rube Goldberg machines are highly amusing, but our national defense interests demand that each function be done in the most direct, simple, reliable, survivable, cost effective way we know.

Sensor Function

Now let us follow through another cycle at that closed loop military system to examine where space fits in. The sensor function is one where space does offer many effective and unique opportunities. Many sensing functions are done better from airborne platforms, on the ground or under water. Some sensing is done better by man than by any hardware. Because of the tremendous area of coverage from a satellite, its capabilities are superior for sensing functions requiring broad coverage, such as cloud cover and meteorology, earth resource surveillance over land and water and detection of missile launches. Sensing may be in any part of the frequency spectrum, depending upon the characteristics of the information to be sensed. A classic example of the use of wide coverage space sensing is the early warning satellite which detects and characterizes missile launchings. Now in routine operational use, this spaceborne sensor is performing an important military information gathering function.

Satellites may also be used to sample and read out data from airborne or ground sensors, again taking advantage of the unique simultaneous line-of-sight coverage from space to a very large area of the earth. A reverse of this process is the Navigation Satellite System in which sensors on the earth, on the sea or in the atmosphere can use satellites as objects to be sensed in determining the accurate position of the sensor. Here, again, the wide area of earth coverage and the stability of satellite orbits are unique capabilities of space systems.

Another example of sensing for which space is uniquely capable is that of sensing nuclear explosions and solar radiation. The lack of atmosphere couples with tremendous area coverage to provide unique capability for nuclear burst detection, to monitor treaty compliance and to measure solar effects for prediction of atmospheric radio communication performance.

Thus, unique space capabilities are finding major application in sensing. Where other methods are better, however, such as in regional tactical surveillance, space capabilities are not applicable. There the job is done better with aircraft, drones, ground observers, picket ships, etc.

Normally, data processing of the sensor data would take place at ground facilities or in an airborne command post. To simplify the process, however, considerable attention is being given to selected processing of the collected data on board satellites. This preprocessing removes much of the background data that is not essential to the decision process which, in turn, greatly reduces the demands on the data links and the ground and airborne data processing systems. Our concern, of course, is with the reliability and cost of the spaceborne processing unit, and the amount of reduction in the data rate and bandwidth permitted in the data link.

Command, Control and Communications

Another major role of space systems in the military process is in the command, control and communications area.

The system to provide this command, control and communications and data relay is necessarily hybrid. Data relay systems must be capable of providing secure continuous high data rate communications directly from the data gathering sources to the data processing terminals. Use of data relay satellites permits the rapid and reliable transfer of data from these sources without dependence on overseas ground facilities. Backup or redundant systems are employed, as necessary, to give ample assurance that all essential data is available to the decision maker.

Communication systems are required for the transfer and distribution of information and status and for relaying commands and directives. The use of satellites for communications support is already widespread. In the military process, communication satellites perform many important functions. They enable the decision makers to relay directives to the force commanders and their elements. They can be used to provide positive control over the forces during all phases of their utilization and for reporting force status and, finally, they are used to support routine day-to-day operations. The Air Force Tactical Satellite Communications (TACSATCOM) program has demonstrated the feasibility of using spaceborne repeaters for communications between mobile force elements and their commanders. The Defense Satellite Communications Systems (DSCS) is used routinely for communications between fixed communication centers.

Other Factors and the Future

The military process applies equally to strategic and tactical situations. A major effort confronting us today is to convert from a loose knit confederation of individual mission oriented elements into a single, tightly organized system. For the most part, these mission elements currently are linked inadequately and not subordinated to a central system. To remedy this situation, the DOD has recently revitalized the Worldwide Military Command and Control Systems (WWMCCS) concept. The goal of WWMCCS is to provide improved flexibility and control and to increase the options available to our decision makers for dealing with crises and hostilities at different levels. It will provide a capability for an appropriate response with a feedback on the operational effectiveness of that response.

Many factors must be considered in determining the various elements involved in the military process. Each element must have its own unique capabilities with respect to performance and survivability. Speed, controllability, flexibility and precision are important if the process is to be exploited fully and be able to adjust rapidly to different or changing situations. It also must be responsive to moves intended to counter its effectiveness; for example, jamming of communications and sensors. Effective use and exploitation of the voluminous data from the various sensor systems may necessitate near real time and fully automated data processing. Software is a critical element. Information security is vital. Technology and cost will, as always, play an important role in determining which systems are developed and their specific capabilities.

In the future, the demands on the military process will be even greater than they are today. The desire for more options and even greater flexibility will impose more stringent demands on the individual elements and on the process itself. Maturing space technology will be applied in new ways to increase the capability of the military process. Sensors systems will increase in number, capability and application. Even greater efforts will be made to correlate the data from dissimilar sensors to improve the timeliness and credibility of the data provided the decision maker. Greater reliance will be placed on data relay satellites and communication satellites. Survivability of these systems will be essential. Serious consideration even may be given some day to deployment of a space command post where data processing, situation analysis, command decision and communication functions can be performed.

Survivability and Cost

Having discussed some actual and potential contributions of space based systems to performing military tasks, I now would like to briefly highlight two of the key factors in making this a reality. These factors are survivability and cost.

A frequent criticism of satellites as candidates to perform a military function is their alleged vulnerability. I want to counter that criticism, which I think is unwarranted.

When our national security policy was massive retaliation, satellites were depended upon only for peacetime operation. The presumption was that if deterrence failed, satellites would be lost but so would everything else.

With today's policy of flexible response, the survivability of the military system process must be assured. Major attention has been given to satellite survivability. An Air Force Scientific Advisory Board Summer Study at Woods Hole last August concluded that more work is needed to assure survivability of the ground segments of space systems than to assure survivability of the satellites themselves.

While I do not want to be too specific here, I can point out that there are many techniques available to make the destruction or negation of our satellite capability a most difficult and unprofitable task. Do not forget, it costs a lot of money just to get something into space. Speaking of money, that leads me to my second key factor.

As I stated earlier, not only is the performance of a system important, including such things as reliability, maintainability, survivability, etc., but also its life cycle cost is important. Military space systems to date have made excellent use of the technology that emerged from our early ballistic missile programs and from NASA space programs. Our current family of launch vehicles can trace their origin to missile systems both in concept and technology. These vehicles are launched from a vertical position at a fixed location and are expended for each launch but, in spite of this old concept, have developed a high degree of reliability and excellent performance. The payloads are also expended for each mission, though design improvements for reliability and long life have given us a mean time before failure of three to five years or more. This is our current generation of space systems for military applications.

As we look to the future, we must plan to do our present known applications better and at lower cost. Based on the current NASA Space Shuttle development, we can foresee a recoverable launch system that makes payloads as well as launch vehicles recoverable. The Space Shuttle and orbital tug should make military systems in space more economical and more competitive to non-space based systems, opening up a whole new era of military space concepts only dimly seen at the present time.

The volume, weight, power and soft ride capabilities of the Shuttle will result in significant reduction in the cost of space payloads only when these features enable a much greater degree of simplicity and standardization than we are accustomed to. We must reach the philosophy in which a new space payload is mostly a new assemblage of standardized proven components. Design to cost techniques are highly applicable to space hardware.

Space does have several unique and valuable characteristics for accomplishing necessary military functions. As the survivability of space hardware is more accepted and costs come down, I expect space to win more competitions with other ways of doing things, and the military uses of space will expand quite dramatically.

The New Space Command

James V. Hartinger

As the significance of communications has grown, so has the role of the Armed Forces Communications and Electronics Association (AFCEA). I know that the cooperation and interchange fostered by AFCEA will contribute to the success of Space Command and North American Air Defense Command (NORAD). We know that the dialogue between the military and industry will always be critically needed.

Establishment of the Space Command on September 1, 1982, marked a crucial milestone in the evolution of military space operations. The Space Command will give the Air Force an organization to both address the many challenges and to take advantage of the limitless opportunities that space affords the military.

I would like to give you my views on why a Space Command, why in Colorado Springs, what the Space Command will look like and what the Command will do.

Why a Space Command?

First, why a Space Command and why now? Our perception of space has changed. It is now seen that space is a place — like the land, and the sea and the air — another dimension. And, it was just a matter of time until we started treating it as such. Our view is that this move will substantially contribute to deterrence, and thus international stability, by reducing uncertainty.

**This article originally appeared in the March 1983 issue of SIGNAL.*

We have had an ongoing study in the Air Force looking at when we would need an operational space command and several factors converged in 1982 which led to that decision.

The Soviets have a major military space program. Over the past decade, they have launched four to five times as many satellites as we have. In 1981, they launched about 75 military related satellites and we launched eight.

They have the world's only space weapon system, an orbital antisatellite (ASAT) system which they have demonstrated for years and which is a threat to our low orbiters. We watch them test it. Also, they are working at a large booster which could provide the thrust for a weapon to threaten our geostationary satellites by the end of the decade.

They have a solid ground based electronic warfare system and their high energy laser program is three to five times the U.S. level of effort.

The Soviets also have a very active Manned Space Program. Since 1977, their 33 cosmonauts have amassed over five years in space, compared to our 60 man days. Obviously, they recognize the potential military value of man in space for surveillance, target selection, repairs and positive command and control. The Soviets are obviously way ahead of us, so we need to be operational in space to protect our interests there.

The second reason for an operational command is our growing dependence on space systems. Over 70 percent of our long haul communications are handled by satellites. As CINCAD, I can perform my missile warning mission only because I have the Satellite Early Warning System. Our military weather satellites provide key data to all Services. The Global Positioning System will let us navigate worldwide with unprecedented accuracy.

With this growing military dependence on space has come an ever increasing national security resource commitment — from $2 billion in FY 73 to over $8 billion in FY 83. Such commitment dictates that we provide the most effective management possible.

Another reason for having a Space Command is when the Shuttle provides routine access to space, it will require initiative, imagination and the proper organization to take full advantage of man's presence there.

Finally, as you know, President Reagan announced on July 4, 1982, at the Shuttle landing in California, that one of the primary goals of the U.S. space program is to strengthen the security of the United States.

All of these converging factors led the Air Force leadership to conclude that now was the logical and necessary time to establish an operational Space Command.

Why Colorado Springs?

But why in Colorado Springs? I could rephrase the question to ask, why not Colorado Springs? Because we have been in the operational space world for 20 years.

Our original charter was to do the space surveillance job—to detect, track and catalog all objects in space. The pertinent point is that we are the one agency to provide a space catalog for the entire free world. It is the very foundation for operations in space. It provides the blueprint of what is up there and presently, there are about 4,800 objects in space. And it requires about 30,000 observations a day to maintain that catalog.

Part of the surveillance function is space object identification where we determine the mission and status of satellites by using our wide and narrow-band radar sensors. Also, we must predict, confirm and report space objects which can survive reentry into the earth's atmosphere because a reentry space object can look like an incoming warhead.

We also have the responsibility for protecting our U.S. space assets. For instance, we warn satellite owner/operators (such as the Department of Defense [DOD], the National Aeronautics and Space Administration [NASA] and the National Oceanic and Atmospheric Administration [NOAA]) whenever one of their satellites is flying through potentially dangerous areas.

In addition to protecting against hostile threats, we provide collision avoidance information prior to the launch of space mission by comparing the programmed flight path against the space catalog.

We also serve as the laser clearinghouse for all DOD and DOD contractor facilities. Whenever anyone wants to radiate into space, they must contact us. After checking the catalog, we give them a time window. Again, the objective is to preclude inadvertent damage to satellites.

We also support the Shuttle by providing collision avoidance information, before launch and while the Shuttle is on orbit. During STS-4, you probably heard on television that the Shuttle would come within eight miles of a Soviet Satellite; we provided that information to NASA. We keep a precise position on the orbiter in case we might be of use in a contingency and we predict and confirm where the fuel tank lands in the Indian Ocean.

This protection role will increase greatly in future years with countermeasures, hardening and when more satellites have a maneuver capability.

Our third chartered responsibility is the possible future task of satellite negation. Both the national and DOD space policies direct, within such limits imposed by international law, the continued development of an operational ASAT capability to deter threats to friendly space systems and to preserve our right of self-defense.

As most of you know, our ASAT will consist of a miniature vehicle launched from an F-15 stationed on each coast at Langley and McChord Air Force bases (AFBs). As CINCAD, I will have operational control over the ASAT; Tactical Air Command (TAC) will provide day-to-day management of the interceptor; and Missile and Space Command will operate the Mission Control Center in Cheyenne Mountain.

These three operational space tasks—satellite surveillance, satellite protection and satellite negation—are conducted in the Space Defense Operations Center (SPADOC) in Cheyenne Mountain. This one-of-a-kind space command post is a fusion center where intelligence meets operations. It is the one place where the status of all national security and civil satellites is maintained.

Thus, with this operational space charter, and with our already strong nucleus of space expertise and facilities in Cheyenne Mountain, the logical and economical location for the operational Space Command is Colorado Springs.

Now, what will the command look like? It will have a built-in experienced space staff. It will have a pool of space expertise in the Chidlaw Building in Colorado Springs. It will have extensive facilities in Cheyenne Mountain to support the Space Mission—The Missile Control Center, The Space Defense Operations Center, and it will have the communications and computer capabilities in Cheyenne Mountain to carry out the space mission.

The Space Command has already been given responsibility for two operational satellite systems—the Satellite Early Warning System (SEWS) and the Defense Meteorological Satellite Program (DMSP). Today, as CINCAD, I have operational control of the SEWS while the Strategic Air Command (SAC) provides day-to-day management. Tomorrow, as CINCAD, I will still have operational control, but then as Commander, Space Command, I will provide day-to-day management. In operating DMSP, the DOD's Weather Satellite System, Space Command will assume day-to-day management of the mission control facility at Offutt as well as the command readout stations at Loring and Fairchild Air Force bases (AFBs).

As more substance, we will operate the Consolidated Space Operations Center being built nine miles east of Peterson AFB in Colorado Springs. One side of the CSOC will be a Satellite Control Center which will be a clone to the Satellite Control Facility at Sunnyvale and the other side will be a Shuttle Control Center which will be a clone to the Johnson Space Center. Imagine the close operational linkage when you combine the center in Cheyenne Mountain (SPADOC) responsible for assessing the threat and advising on countermeasures with the facility (COSC) responsible for carrying out the defensive measures—that has to promote efficiency.

Further resources or substance include 6,000 Air Force personnel, contractors, four bases (Peterson, Thule, Sondrestrom and Clear, Alaska) and 24 sensor sites.

The first Space Wing will be established at Peterson to manage those sensors. The wing, headed by a general officer, will be responsible for training, standardization and evaluation.

One month after establishment of the Space Command, the Air Force established a Space Technology Center at Albuquerque reporting to LTGen. Richard Henry, Commander of the Space Division, with the center having responsibility for the unique space technical disciplines. From my perspective, we now will have a very coherent processing of space systems—from basic technology at Albuquerque through research, development, acquisition and launch by the Space Division and on-orbit control, management and protection by the Space Command. We will have, through the dual-hat role of my Vice Commander, Gen. Henry, an autonomy, an integrated cradle to grave concept that has been proven by time to work. Space Command adds another dimension to what has gone before.

So the Space Command will be an organization with a very important operational mission and with the required facilities and personnel to carry out that mission. It will have initial substance.

Space Command Functions

The last question is, what will the Space Command do? In general, we now have an operational command to manage, control and protect operational space assets. We will promote a much closer relationship between the research and development community and the operational world. We will provide a focus for centralized planning, consolidated requirements and an operational advocate and honest broker for USAF space systems. We will provide the operational pull to go along with the technology push which has been the dominant factor in the space world since its inception.

In general, that is what the Command will do. In particular, we will be the regular crew chief for things which either are not being done or which we feel need greater emphasis. In forming this major air command, we are putting together the hundreds of details in such a way that we do not miss a beat—there are a lot of balls in the air.

On September 1, we established the Space Command and on October 1, we established the Space Technology Center. On January 7, we formalized the Space Wing and activated the Communications Division. On April 1, we will take over Peterson from SAC as well as start to transfer the worldwide space and missile warning sensors and the two satellite systems. By October 1, 1983, we will have completed the transfer and will be in full operation. Our challenge will be to properly sequence the events and make things happen when they should.

Space Command will develop a space strategy for the first time. For instance, we need to look at space systems in terms of launch vehicle mix, sparing and on-orbit force structure. As a basis for the strategy, we will promote a comprehensive documentation of the Soviet space threat.

Next, we are going to strengthen the softest link in the development cycle of space systems, the statement of operational needs. Initially, the Space Command will consolidate, integrate and collate the requirements of all Air Force users and assure that they are reflected in the acquisition process. When Space Command becomes unified, it will provide these services and support the requirements of the other CINCs.

To develop strategy, another important task is to realistically incorporate space assets in exercises. Last March, we did that for the first time in a Joint Chiefs of Staff (JCS) sponsored exercise. It was a successful exercise because it pointed out areas for improvement in our operational space planning. We have plans for land, sea and air contingencies, and now we will tie together space activities. We think Space Command can make a major contribution by developing realistic space scenarios that will help purge the system.

Position management is another important task. There is no single point for coordinating satellite positioning, and surprisingly enough, some of our orbits are getting crowded. For example, the synchronous equatorial orbit is a popular one. There are 204 objects there today, of which 73 are operating spacecraft, including those of Russia, Indonesia, NATO and our own. Space Command can develop separation criteria and be the focal point for coordination with civil and commercial satellite owners.

Survivability is the area where the operational space commander can make one of his greatest impacts. As space systems grow in cost and weight, traditionally the first requirement to be sacrificed is survivability. Space command will develop a sound survivability program which looks at all the vulnerable nodes—the launch site, the tracking, telemetry and commanding facilities, the command, control and communications (C^3) network and the satellite itself — then, working with the other Air Force commands and the R & D community, advocate the program through the acquisition process. A unified command is needed to work this problem on behalf of the CINCs and service staffs. Space knows no service or command boundaries. Everyone benefits.

We will also be working three important support tasks—space education, space medicine and the Programming Process (POM). Space education is a major challenge. In the future, our biggest limiting factor could be the lack of skilled personnel in the space career field. As a priority task, Space Command will oversee space training and career development. In fact, we have already begun working with Air Training Command and consolidating all training for space operations personnel at either Lowry AFB in Denver, or Peterson AFB in Colorado Springs.

We have recommended that we establish a major in Space Systems at the Air Force Academy and will urge that more academy graduates are assigned to the space operations field. Air Force Institute of Technology already offers a master's degree in Space Operations Management at Wright-Patterson, and we will be promoting the enrollment to be expanded to meet our future needs.

The next concern is space medicine. We are already investigating how we can promote a clearer focus. An important component of man's ability to operate in space will be our knowledge of the space environment and the quality of his life support systems. For example, we need to better understand the effects of increased space radiation, zero G's and extravehicular activity. Space Command will provide the operational advocacy for research efforts to understand the space environment and for design improvements to increase man's utility in space.

The third task is the POM or Programming Process. Putting together the Air Force POM is like working a jigsaw puzzle—all pieces must fit for it to make sense. Nowhere is this more true than with space systems. New starts for satellite programs, supporting C^3, survivability measures and launch vehicles all have to be in balance or the pieces will not fit. Space command and the Space Division will work closely in reviewing and commenting on the total Air Force Space Program. We will co-advocate space requirements and hopefully make the Air Staff job a lot easier.

The last line is that the establishment of Space Command is occurring at the right time, in the right place, with the right people and we are all excited about it. The Space Command is a giant step toward a space program that will meet the President's policy goal of strengthening national security.

Space Based Defense

George A. Keyworth, II

Over the past two years, the nation has been preoccupied with two overwhelming concerns: the economy and national security. Those two things, in various forms, dominate policy making here in Washington and consume the major part of the President's time.

It turns out that both issues share a strong common element—their dependence on technology. Whether we are talking about changes in basic industries and their effects on employment; the economic consequences of changes in balance of trade; shortages of math and science teachers in our schools; strategic modernization of our armed forces; or preserving stability and peace in the face of the adventurism of the Soviet Union, we quickly find ourselves talking about changes in technology as well.

As Science Advisor to President Reagan, I find great reassurance—in fact, I am more than a little humbled—in our President's instinctive feel for this essential element of modern society. He has a deep common sense understanding of what has made America strong both economically and militarily—and he is determined to build on those unique American qualities to forge our future. Unlike any other president in my memory, he not only knows that our country is distinguished from all others by the depth of our technical expertise—but he realizes that our future lies in taking advantage of that expertise and making sure it is maintained.

I want to talk today about that realization and its profound consequences for our national security. I have particularly looked forward to this chance to talk about defense to you members of the Armed Forces Communications and Electronics Association (AFCEA). I know you are deeply involved in military issues and committed to a strong and secure United States. You and your

**This article originally appeared in the January 1984 issue of SIGNAL.*

colleagues represent the people who are critically important to our long-term success in maintaining a strong defense posture.

Changing Times

These are rapidly changing times in defense—and those changes are going to intensify in the near future. What kinds of changes am I referring to? Primarily, we are shifting to a more balanced military deterrent, with increased emphasis on conventional weapons. This reflects a change in strategy that anticipates what we hope will be a substantial reduction in reliance on nuclear deterrents in the future.

Even today, 85 percent of our defense budget goes for conventional systems—and for good reason. Although public debate over the adequacy of our deterrence has focused people's attention on Soviet improvements in nuclear weapons systems, the Soviets have been quietly achieving comparable advances in their conventional military as well. Over the past decade, they have managed to encroach on our technical superiority there, as well as in strategic systems. Just to point to two examples, they now have a tank, the 1-80, that is a good match for our best, and we can no longer be certain of air superiority. Moreover, they are more than ready—and able—to put as many units in the field as they need to achieve a superior force. Our response has to be to go back to our strength—technology.

For example, Soviet armored forces are a serious threat in places like Eastern Europe. But, should we continue to try to match the Soviets, tank for tank—or, lacking sufficient numbers, be forced to rely on tactical nuclear weapons? No matter how sophisticated a weapon a tank can be made to be, it is going to be vulnerable to pinpoint attack from a distance. In an era of smart munitions and new tactical radars, we should be able to mount such an attack—and perhaps come up with a system so superior that it nullifies the Soviet armored leverage.

That is not a technical pipe dream. Let me give another example, a very real one, of the kind of non-nuclear weapons systems advance that has so much potential—what are called low-observable aircraft. Using highly sophisticated science and engineering, we are finding ways to cheat radar systems, to reduce the radar cross section of aircraft and increase the craft's ability to penetrate enemy defenses. To my mind, this general area is the most important new military technology since the intercontinental ballistic missile (ICBM) and nuclear submarines; it is going to require continued research and development to improve it and, of course, to explore countermeasures.

Technical Advantages

There are plenty of other ways in which technology can give us high leverage on conventional military operations. But, sadly, we still do not do a good enough job of incorporating technical advances, let alone new science, into our conventional weapons systems. We need a larger part of our scientific and technical community addressing these kinds of opportunities.

Now, this discussion of changes in deterrents and improvements in conventional defense leads me to the topic I want to focus on today. That is a discussion of the President's new defense initiatives. These come from his speech last March 23 [1983] when he called for the nation to develop a means to deter nuclear war by preventing it from occurring, rather than by continuing to threaten annihilation of an enemy who starts a nuclear conflict.

Not surprisingly, the President's proposal to undertake the most profound shift in defense strategy in decades produced strong reactions. One reason was because this nation does not expect its President to initiate policy; we have had too many years of bottom up processes where the interest groups, the agencies and the Congress all have their chances to neuter new policies before the President has a chance to consider them. We are not used to boldness, nor are our institutions. Until this event, I had never really appreciated how slowly the nation accommodates to change—and that the pace of change in Washington can be positively glacial.

Well, what did the President propose? Was it, as the ever constructive press labeled it, a Star Wars fantasy to orbit nuclear weapons in space? Was it an attempt to brush off traditional ballistic missile hard site defense systems as if they were something new? Of course not.

First of all, the President proposed no system at all. He proposed a concept, one he had long cherished and which only in recent years has technology made feasible. It was the concept of creating at long last a strategic defense that no longer relied on retaliation for its effectiveness. Rather than asking us to continue to rely on the threat of mutual assured destruction, a situation that becomes more unstable each year, he offered the nation the hope of a new stability based on, as he called it, rendering nuclear weapons obsolete. And, may I add, I do not know of a single one of us who has worked on nuclear weapons who has not wished fervently for just such obsolescence.

At the same time, the success of such an approach—and we have to assume that the Soviet Union's ongoing efforts along a similar development path have a good a chance to be successful—would inevitably lead to increased emphasis on conventional weapons.

The President's Speech

It has now been six months (in March 1983) since the President's speech about ballistic missile defense. A lot has happened in that time—much of it centered on a high-level panel, headed by James Fletcher, charged with recommending a research and development plan to carry out the President's proposal. And, a lot more is going to happen.

When the President's announcement was first made, I was disappointed, but not surprised, at the vehement reactions from some of the traditional arms control activists. I watched with some amazement while a couple of dozen of them marshaled largely irrelevant technical arguments against the President's proposal for mutual assured defense while at the same time they embraced mutual assured destruction as the preferred alternative. That was really logic turned upside down.

But, I was much more concerned about the reaction of the broader scientific and technical community, upon whom the success of this new defense strategy would ultimately ride. There was, not unexpectedly, strong skepticism, not on ideological grounds, but more of a prove-it-to-me kind. So, I am delighted to find now that, among those technical people who have been part of or observing this process of trying to define a research and development path, much of that initial skepticism has given way to cautious optimism that we may be able to reach the President's objective.

The basis for that optimism is the tremendous technical progress over the past decade and, in several areas, over the past year or so. Consider the kind of progress we have been seeing in information processing, which would be a critical technology for any fast response system. Ten years ago, it would have been pointless to talk about data transfer rates of, say, a gigahertz per second. Our high speed processing capabilities were relatively rudimentary. Then, we were not even thinking of anything as ambitious as a 16K random access memory (RAM). Now, we see that for the past year the phone company has been installing 256K RAMs in the field, and adventurous research groups are already talking about having a four megabit RAM available within a few years. These kinds of jumps in data processing can change a technology very quickly.

We have also seen truly remarkable advances in our ability to use satellites for navigation, for communications and for reconnaissance. At the same time, we know of defense means to protect these kinds of vulnerable, space deployed assets.

Recent advances that compensate for atmospheric dispersion in laser beams, in conjunction with adaptive optics—mirrors that can be pointed electromechanically like phased-array radars—could permit operation of high power space lasers, but on the ground. That could permit us easier operation and protection of complex, expensive components. We are also seeing good progress in using ultrashort laser pulses to create damage quickly by impulse rather than through the thermal effects of slower acting continuous beams.

There have also been recent advances in traditional non-nuclear terminal BMD technology that we believe may be extrapolated to the needs of a mid-course intercept system.

Strategic Defense System

So, it is these and other recent developments that have breathed vitality into the possibility of a workable strategic defense system. Now the Fletcher panel — the group putting together the long-term research and development plan—has concluded that we can now project the technology, even though it has not been demonstrated yet, to develop a defense system that could drastically reduce the threat of attack by nuclear weapons.

This is true not only for today's weapons, but those that could reasonably be expected to be developed to counter such a defense system. This would be a multitiered array, probably designed to respond first, to ballistic missiles in the boost phase, second, to midcourse vehicles, and third, to vehicles during reentry into the atmosphere.

Needless to say, the research and development program that is finally adopted will have a funding impact. Some of the money to conduct it will come from new funds, but some will also come from shifting funds out of existing research and development programs. Our goal now is to keep building momentum for the progrm and to build national support for the effort, in order to attract the best talent to a formidable but worthy goal.

My feeling is that it is likely to take five or six years of research and development to bring us to the point where we can make the critical decisions about developing and deploying actual systems. Although there will be understandable temptations—and pressures—to move quickly to near term deployment of whatever is the best technology we have available, it is important to give the research and development a reasonable amount of time to explore some of the less well developed technical options.

For example, I would hate to see particle beams ruled out prematurely simply because they are not as well developed as, say, chemical or excimer lasers. And other possibilities, such as free electron and bomb pumped lasers, are at even earlier stages and still need to have their basic technologies demonstrated before their possible roles could even be assessed.

Incidentially, I would make one thing very clear: I do not see a critical role in this defense initiative for nuclear weapons *per se*. First of all, I am not sure that the uses proposed for nuclear weapons in space could not be performed with non-nuclear technologies. More important, the American people are not likely to enthusiastically support the placement of nuclear weapons in space. But to return to my thought—because of the need to give these nascent technologies a chance to unfold, I see the 1980s as the key time for weighing the feasibility of a variety of approaches to strategic defense.

At the same time, we have to guard against the other understandable tendency of scientists and engineers to become so fascinated with the research process that we never get out of the laboratory stage. To prevent that, we have to keep pressure on the research and development program to move to closure. My own preference is to plan for a number of demonstrations of the evolving technology. By that, I do not mean deployment of a piece of a system, but rather some periodic visible proof of progress—near term demonstrating of militarily meaningful systems.

I can suggest an example. It is certainly possible, though by no means certain, that an eventual system could employ a high power ground based excimer laser. I have already suggested the reasons for wanting such a device. Based on the technical progress to date, we might set as a goal the demonstration of a multimegawatt pulsed laser whose beam is corrected for atmospheric distortions. This would be, maybe something powerful enough to act as an antisatellite weapon, maybe even strong enough to be effective as far away as geosynchronous altitudes. I think we could do that before the decade is out.

Now, such a demonstration would not demonstrate a workable antiballistic missile (ABM) system. But, quite frankly, if I were a Soviet planner, I would quickly put two and two together and realize that an important part of the technology for an ABM system was well in hand and that development was more a matter of time than breakthroughs at that point. Such a demonstration would pressure the Soviets to take our arms reductions proposals much more seriously than they do now. Although the strategic defense program's goal would still be eventual deployment of a working system, we should not overlook its potential beneficial impacts on arms reductions as it progresses.

Mobilization of Resources

Well, then, how are we going to mobilize the technical skills and resources we need to carry out this ambitious program? Or more to your interest, how and where does U.S. industry fit into this new defense initiative?

At this point, I would have to say that the situation is still quite fluid. But in any case, we intend to cast as broad a net as possible by establishing a competitive program open to any interested participants. We will respond to good ideas, credible proposals and good people. My strong hope is that we will attract creative organizations and bright young people into this defense initiative, perhaps people who might not have even considered defense work before.

When you consider the kind of talent that flocked to defense work in the 40s and 50s, you realize that today we really tap an inadequate amount of the nation's immense brainpower to work on its defense. I have some hopes that we may see that change.

I said earlier that I was finding a marked shift in attitudes among the technical people who have become involved in this planning stage. I think, as the program evolves, other people are going to start to realize that the President's proposal, and the plan to achieve it, offer a way to get off the runaway horse of mutual assured destruction. I see it not only as responding to the same kinds of sentiments and concerns that drive the nuclear freeze movement—that is, a realization of the increasing instability of nuclear deterrents—but also as a lot more realistic way to achieve an even more desirable end result.

After all, this plan does not stop at freezing nuclear weapons or even reducing them to some lower, but still deadly, retaliatory level. In precise terms, it offers a way to stop an attack after it begins but before it can reach us. More to the point, in practical terms it removes the preemptive attack option from the military planners. If that means our worst disputes may then force a resort to conventional weapons, then we will have moved a long way toward assuring human survival.

Part II: Military Communications Satellite Systems

Editor's Introduction

There are two general classes of military needs for satellite communications. One is characterized by high volume, point-to-point traffic, while the other by lower volume, mobile user traffic. The latter is often described as "tactical" because the preponderance of the air, sea and land mobile users are expected to be employed in tactical operations. However, the flexibility of modern military forces, particularly that of airpower, often blurs the distinction between "strategic" and "tactical" communications, as do modern military command and control structures. A critical subset of the users in both fixed and mobile communities require a very high degree of communications survivability. Thus, there emerged advocates for three military systems to satisfy the needs of the three classes of users described above. There were also those who advocated systems designed along military service lines or functional lines, such as intelligence collection and dissemination. The resource implications of multiple system approaches were very substantial as were the potential impacts on military operations. Therefore, in 1973, the Director, Defense Communications Agency (DCA), was assigned the role of systems architect for military communications satellite systems for the Department of Defense (DOD).

The first two articles in this part describe the key activities of the Military Satellite Office (MSO) at DCA. The overall framework in which user needs, which vary widely, can be satisfied is of great importance to the future of military communications satellites. The establishment of the MSO at DCA, with Dr. Fred Bond as its first leader, resulted in the first systematic overall examination of requirements and the first iteration of a military communications satellite architecture. Both have had great impact on subsequent development activities in this field. Read in conjunction with the article which follows by John Tuck (BGen., USA), one gets an excellent overview of the evolution of the DOD military satellite communications systems architecture.

The Defense Satellite Communications System (DSCS) began operation in 1966 and is today (1985) deploying its third generation of spacecraft. The DSCS meets the need for the high volume user and fits the traditional notion of a "strategic long haul" communications system of a fixed plant, point-to-point nature. However, both satellite design (spot beams) and earth stations design (light terminals) allow for rapid extension into contingency areas or for rapid restoral of damaged segments of the system as well as use by the ground mobile force (GMF) community. A description of the DSCS as it has evolved is contained in the article entitled "DSCS: An Evolving Operational System" by Paul McLellan whose experience with the program is unique. He emphasizes the new military capabilities of the third generation DSCS III.

A later article, "Defense Satellite Communications System: Another Milestone" by LTGen. Hilsman and LTC. Garcia, updates the McLellan article but, more importantly, emphasizes "the most important part of the system's capability" which the authors emphasize as the DSCS control segment. DSCS III is an extraordinarily flexible military system, and although the control subsystem of any communications satellite system is important, in the DSCS III case, it is vital.

While the DSCS program has had continuity and stability, tactical (mobile user) satellite programs have experienced the opposite. The early experimental programs (LES-6 and TACSAT-1) proved the worth of tactical satellite communications and the importance as well as the difficulties inherent in system control. The article "Operational Considerations for Tactical Satellite Communications Systems" by John Kuhn describes lessons learned from the operational use of those experimental assets. Efforts to justify a follow-on operational system (TACSAT-II) foundered on an enormous requirements list. There appear to be three factors which have tended to inhibit the growth of military mobile user satellite services. First, the multitude of potential users generally overdrives the requirements list, hence the size, complexity and cost of the proposed satellite. Next, although the cost of the earth terminal itself may be modest, the cost of modifying the weapon system platform is sufficiently costly to deter all but the most essential user. Finally, the limitations of the UHF frequency band have prevented the development of a highly robust antijamming capability.

The Navy, which was totally dependent upon the HF radio medium, has had the greatest success in implementing a mobile user satellite system. The Air Force, for certain critical users (principally the Strategic Air Command), has opted for placing various kinds of transponders on other host satellites, e.g., DSCS, FLTSAT, and SDS in the AFSATCOM program. The Navy program progressed from leased service on the Comsat General MARISAT as a temporary measure pending launch of FLTSAT, to a mature FLTSAT and more recently to use of satellites leased from Hughes — the LEASAT program. Jon Boyes, Vice Admiral, USN (Ret.), in his article "A Navy Satellite Communications System" provides important insights into Navy requirements for satellite communications. However, equally important in this seminal article is his emphasis on the need to design and manage the system (earth and space segments) as a whole and of the importance of the system antijamming properties.

Fjetland and Whittenberg in their article "Tactical Satellite Communications" provide us a look at the MILSATCOM architecture from the viewpoint of the tactical community user. A persuasive case is made for joint use on a demand access basis at SHF and EHF as well as the traditional UHF frequencies employed in earlier tactical satellites.

A recurring theme throughout these writings about both "strategic" and "tactical" satellite systems is the importance of system control and the need for antijamming capabilities. Marc Spellman in the article "ECM and Satellite Communication Network Control" states clearly the importance of system control and how, in the military case, it is compounded by the need to protect against electronic attack. He describes the fundamental control functions and strategies, both design and operating, for achieving a jam resistant system.

The Evolution of Military Satellite Communications Systems

Frederick E. Bond and William H. Curry, Jr.

Satellite communications systems offer a unique transmission medium which can be exploited for a number of diverse military applications. Examples of these applications include: worldwide coverage with distance insensitive cost; communication to isolated areas; rapid extension into new locations; support of high data rates; beyond line-of-sight service to moving platforms including aircraft and ships; and truck transmission for command and control in a concentrated contingency or combat area.

In a number of cases, the medium can provide vastly improved reliability and capacity compared to high frequency (HF) transmission via ionospheric reflection or scatter. In other cases, it can substitute for multichannel radio relay and tropospheric scatter systems with a significant reduction of the logistics burden imposed by equipment, personnel and real estate. Also, vulnerability to circuit disruption at intermediate relay points is minimized.

Because the potential of this new medium has become more widely recognized, the demand for satellite service is increasing rapidly in the Department of Defense (DOD) community. On the other hand, satellite communication systems are costly, and the portions of the radio frequency (RF) spectrum allocated for their use are limited and crowded. Furthermore, the number of satellites in synchronous equatorial orbit (a strongly favored orbit for many

This article originally appeared in the March 1976 issue of* ***SIGNAL.*

applications) is growing rapidly for military, commercial and foreign applications.

Earlier work in the development of DOD satellite communication systems was oriented toward the restricted goals of satisfying requirements of specific user communities. As the work continued, the need evolved for a more unified technical approach within DOD for planning, development, acquisition and operation of all satellite communications systems. Some of the factors driving this need included:

1. The desire for improved cost-effectiveness;
2. The desire for improved spectrum utilizations;
3. The concern that continued piecemeal acquisition might soon result in an investment so large that future changes, regardless of their merit, would not be feasible;
4. The recognition that priorities and requirements of the user communities are sensitive to the effects of significant military and political events occurring worldwide. Hence, the systems must be sufficiently flexible to provide efficient service through a wide range of scenarios including contingencies, crises, limited wars and nuclear conflicts;
5. The discovery that many user requirements are intermittent with a "low duty cycle." This affords an opportunity for efficient application of shared services within and among user communities;
6. The recognition that user operational concepts are dynamic. Sufficient flexibility must be provided to permit some latitude in system employment; and
7. The awareness that satellite system technology is dynamic. The entire DOD community must be prepared to exploit those new advances which have demonstrated military utility.

Current Military Systems

The system presently serving the military long-distance trunk requirements and providing limited contingency and mobile service is known as the Defense Satellite Communications Systems (DSCS). Its primary mission is to support the Worldwide Military Command and Control System (WWMCCS).

The initial version, DSCS I, started in June 1966 with the launch of small and relatively simple SHF satellites in near synchronous orbit. The next generation (DSCS II) design is a larger, more powerful satellite including both steerable limited coverage and earth coverage antennas. Two were first launched in November 1971, but both developed anomalies after only a few months in orbit and subsequently failed.

After considerable redesign effort, the next two were launched on December 13, 1973. They provide service to a variety of super high frequency (SHF) terminals with transmitter power ranging from 100 watts to about 10 kilowatts and antenna sizes ranging from 3 feet to 60 feet. Two more satellites were launched on May 20, 1975; the launch, however, was unsuccessful, as the third stage of the launch vehicle failed to boost the satellite out of the parking orbit.

Six DSCS II spacecraft are currently in production and an additional six programmed with the objective of establishing worldwide service. Eventually four active satellites with at least two on-orbit spares are planned. Initial development work for the third generation spacecraft (DSCS III) has started to assure continuation of DSCS service through the 1980s.

Military UHF Applications

Activity in satellite applications in the military ultra high frequency (UHF) band (225-400) began in 1967 with the launch of the small Lincoln Laboratory experimental satellite LES 5, demonstrating low data rate (75 bps) transmission to aircraft, ships, land vehicles, etc. The more powerful LES 6 and Tactical Satellite TACSAT 1 satellites, launched in September 1968 and February 1969, respectively, permitted the technical and operational feasibility experiments to continue on a broader scale and eventually went into use as an interim operational system.

TACSAT 1 ceased operation in December 1972. LES 6 is still performing with an effective radiated power reduction of about 8 dB from its initial value of 27 dBW.

Experience gained with these systems has been incorporated into the concept of the Fleet Satellite Communications System (FLTSATCOM). FLTSATCOM will provide a global UHF relay as the primary transmission medium for beyond line-of-sight ranges for Navy ships and aircraft.

In order to provide an interim capability until FLTSATs can be orbited, the Navy has leased UHF services from the COMSAT General Corporation—the so-called GAPSAT. The satellite also provides services to commercial maritime users in the 1.5 GHz band; this application is referred to as MARISAT.

Two FLTSATCOM satellites are planned, one for the Atlantic and one for the Eastern Pacific region. Additional details on this system are provided in other articles in this issue. At the time of writing, the launch dates to establish this service were not firm but would appear to be no later than early spring of this year.

The FLTSATCOM satellite will also provide low data rate UHF communication channels serving high priority Air Force needs. The Air Force portion of the system is known as the Air Force Satellite Communications System (AFSATCOM). In addition to the channels in the FLTSATCOM satellites, AFSATCOM terminals will use transponders on other host satellites to obtain additional coverage.

Another Air Force program, which is still in the research and development (R&D) stage, is the development of a survivable system to serve the most critical general war command and control functions. Activity under this program includes the Lincoln Laboratory satellites, LES 8 and 9, scheduled for launch early in 1976.

In January 1974, a Program Memorandum was approved by the Defense Department covering the need for a family of small satellite terminals in the UHF and SHF bands to support critical Ground Mobile Forces (GMF) operations involving Army, Marine Corps and Air Force units. Initially,

Defense Communications System and Special Purpose
(Diplomatic Telecommunications Service, Presidential Communications, etc.)

Wideband Data Relay (Surveillance, Intelligence)

WWMCCS (Worldwide Military Command and Control System) Command Netting

Nuclear Capable Forces (Army, Navy, Air Force)

Ground Mobile Forces (GMF) (Army, Marines, Air Force)

Fleet Operations (Navy, Marines)

Air Force Air Operations (SAC, TAC, MAC, etc.)

Table 1. User Communities

the requirements will be served by channels on GAPSAT, FLTSATCOM/AFSATCOM and DSCS satellites.

Commercial satellite service is employed by military circuits in cases where uniquely military characteristics of a satellite system are not required or as an alternate means of transmission to enhance overall network reliability and survivability. As an example, over 200 transatlantic and transpacific circuits are leased through international carriers who, in turn, lease channels on both submarine cables and the International Telecommunications Satellite Consortium (INTELSAT) system. Also, there are several leases of domestic satellite service in the United States and Canada, especially for wideband application.

USER COMMUNITIES	System Capacity	No. of Terminals	AJ	Physical Survivability
• Nuclear Capable Forces (NCF) (mobile terminals)	Low	Modest	Substantial	Yes
• WWMCCS Command Netting	Modest	Few	Substantial	Limited
• High-Capacity Comms (fixed and transportable) DCS Wideband Data Relay	Very Large	Modest	Selective (w/lower capacity)	No
• Non-NCF (mobile terminals) Ground Mobile Forces Fleet Operations Air Force Air Operations	Large	Many	Selective	No

Table 2. Comparison of Needs of User Communities

CURRENT COMMITMENTS

	DSCS-II (SHF)	AFSATCOM Packages (UHF)	AFSATCOM FltSat (UHF)	GAPSAT FLTSAT (UHF)
DCS/SPECIAL PURPOSE	X			
WIDEBAND DATA RELAY	X			
WWMCCS COMMAND NETTING	X	X	X	X
NUCLEAR CAPABLE FORCES		X	X	
GMF	X		X	X
FLT OPERATIONS	X			X
AF AIR OPERATIONS		X	X	

Figure 1.

Systems Design Factors

The first consideration in approaching system designs is an understanding of the requirements that must be satisfied. The Military Satellite Communications Systems Office (MSO) has an ongoing requirements analysis for this purpose, conducted under the auspices of the Joint Chiefs of Staff. The effort to date has resulted in the categorization of users as well as their requirements.

Users have been grouped into seven communities, as illustrated in Table 1. A loose definition of a community is a group of users sufficiently large and having enough common communication characteristics to consider an exclusive satellite communication system. Such characteristics might include: user mission; terminal constraints; type, rate and quality of service; survivability; message characteristics; and permissible type of control.

Table 2 is a summary and comparison of the basic needs of the communities in terms of total capacity, relative number of terminals involved and electronic and physical survivability. Generally speaking, airborne and shipborne mobile terminals are served by SATCOM systems in the military UHF band because of terminal cost and platform constraints.

The DCS (Defense Communications System) wideband data communities require the largest capacity, and Nuclear Capable Forces community, the smallest. With regard to antijamming and physical survivability, the Nuclear Capable Forces community and a portion of the WWMCCS Command Net require the most protection.

In developing the requirements analysis, the totals for each user community have been categorized in three time periods: Near Term—1975-80; Mid Term—1980-84; Far Term—1985 and beyond. Furthermore, their sensitivities to scenarios (ranging from day-to-day operations through general war) have been measured, as affected by sizing of force structures involved. Figure 1 relates the user communities to the satellite communication systems which already exist or are presently in development and acquisition.

	GP-SHF	SURVIVABLE UHF	SURVIVABLE EHF	GP-UHF (Some SHF)
DCS & SPECIAL PURPOSE	X			
WIDEBAND DATA RELAY	X			
WWMCCS COMMAND NETTING	X	X	X	X
NUCLEAR CAPABLE FORCES		X	X	
GMF	X			X
FLT OPERATIONS	X			X
AF AIR OPERATIONS				X

Figure 2. Next Generation MILSATCOM Architecture

Further work is necessary to provide a better understanding of the statistical parameters of user requirements (e.g., what services can be provided on a time-shared, vice continuous basis) and to couple operational concepts into the requirements data base.

Next Generation Systems

Current programs include large commitments for terminals. Since terminal life spans are on the order of 15 years or more, it is imperative that next generation space segments address terminal compatibility. Other pertinent factors include cost, satellite size and complexity, system control, frequency management and survivability. System cost is a major consideration and must include R&D, production, launch, operation and replenishment.

Replenishment is principally related to satellite life, a factor which unfortunately is not well calculated. One would intuitively correlate a simple satellite with long mission duration. Despite a systematic review of military and commercial satellites, no precise relation between satellite complexity and life span has yet been found. Some make a case for the small, simple design; others argue that larger designs allow economies of scale and onboard redundancy to provide graceful degradation.

Some trade-off considerations include:

1. Establishing a replenishment strategy for complex multifunction satellites when only one or a few functions fail;
2. The problem of controlling accesses to many small satellites and providing interconnectivity between users. Accordingly, there are conflicting factors (cost, control, risk) relating to the determination of an optimum satellite size and complexity to serve a given set of users.

Recent satellite problems and launch failures have dramatized the need for much greater sparing when acquiring the space segment. If budget limitations are a significant factor, it may be necessary to plan for much less capacity than actually desired in order to assure adequate availability.

Frequency congestion and allocation problems are an important consideration for future military satellite communications systems. The DSCS and potential follow-on systems will occupy the 7/8 GHz band with the 20/30 GHz band available for future expansion. There are many non-satellite terrestrial radio systems already using the 7/8 GHz band.

Frequency clearances are difficult to coordinate. Non-trivial clearance problems have arisen in attempting to operate DSCS earth terminals at overseas sites. There is also an issue regarding the use of mobile or contingency terminals in the 7/8 GHz band since present rules do not recognize such service.

Congestion at UHF is already large and growing. Fading and scintillation limit system capacity and availability. Frequency reuse is difficult because most terminals use simple antennas with broad beams, and satellite antennas are for earth coverage. Future UHF system design must provide reasonable flexibility with regard to frequency assignments, compatible with terrestrial systems.

Other candidate frequency bands for future military mobile use are under consideration, but no obvious solution is evident. Most frequencies below 15 GHz are already so crowded that new services, such as satellite communications, cannot be easily accommodated.

Consideration is also being given to more efficient use of present allocations. Shared use of scarce satellite capacity must replace dedicated assignments for low duty cycle user. Relatively low capacity sharing systems already exist. More efficient demand assignment and time sharing techniques, along with suitable control systems, are presently under investigation by a number of different government and commercial agencies.

Other new technologies applicable to future systems include the use of electronically steered, multibeam antennas. Such techniques can provide flexible coverage, uplink discrimination against sources of interference and the ability to rapidly reconfigure coverage areas. The ability to deploy a large spacecraft antenna (30 feet) demonstrated in the ATS 6 program opens the possibility for narrowbeams at the military UHF band. The capability to steer scarce satellite RF power where and when it is needed will give satellite communications users a degree of flexibility unmatched by any other medium.

The Space Transportation System, or Shuttle, will also have considerable impact on future communications spacecraft design. The Shuttle as presently planned will be able to lift comparatively heavy loads (by today's standards) to a low altitude parking orbit. From there, a booster must be employed to lift communications satellites to more suitable higher orbits. The Shuttle will essentially be volume, rather than weight, limited.

This means that for the first time, the spacecraft designer may be able to trade cost, reliability and capacity for weight. The ability to place a four or five thousand pound spacecraft in synchronous orbit could allow the cost, size and weight of earth terminals to be dramatically reduced. But, at the present time, there are many uncertainties about Shuttle design so that the specific design impact on the spacecraft is difficult to predict.

Future MILSATCOM Systems

The MSO has proposed an evolution of present systems into a three-tiered architecture as shown in Figure 2. The DSCS III spacecraft design, currently in Phase 1 development, is planned to replace the DSCS II and establish the General Purpose (GP) SHF system.

The FLTSAT, presently under development, will carry UHF users until an advanced design spacecraft is available in the next decade to establish the General Purpose UHF system. It is contemplated that some SHF capability would be included on this spacecraft. The present Air Force satellite (AFSAT) system would evolve into a survivable system, which may incorporate some of the features of the LES 8/9 experiment, serving primarily the nuclear capable forces.

The MSO presents this evolution as one, but certainly not the only, alternative. While all the options have not yet been fully analyzed, it is expected that the total system costs estimated for the amount of capacity postulated will not be reduced if other alternative architectures are selected. Therefore, a central issue to be faced is how to prioritize the user requirements in the event that the total capacity requires an exorbitant cost.

Most work remains to be done in areas such as:

1. The level of on-orbit spacecraft spares necessary for adequate system availability;
2. The role for smaller, less costly satellites;
3. The type and degree of survivability needed;
4. Providing greater capability to support small terminals;
5. Developing strong positions regarding frequency management issues at the 1979 World Administrative Radio Conference;
6. Improved utilization of higher (20/30 GHz) bands; and
7. Minimizing transition impact from present to future systems.

Summary and Conclusions

Military satellite communications requirements are increasing rapidly. Several practical limitations exist on achievable capacity and survivability. New technology will permit growth, but cost, frequency spectrum limitations and other factors will always tend to limit the capability.

The system architecture must necessarily include a continuing requirements analysis and evaluation of alternatives considering overall system costs. The evolution from present to follow-on systems will pose a number of transition problems with respect to both the terminal and space segments.

Finally, the need to provide adequate system availability may be the most important single influence.

Military Satellite Communications Systems Architecture

John S. Tuck

Abstract

By the early 1980s, the Department of Defense (DOD) will have a substantial military satellite communications (MILSATCOM) capability in the ultra high frequency (UHF) and super high frequency (SHF) bands. A pragmatic financial assessment of the investment in UHF/SHF equipment suggests that DOD may not move so quickly into the extremely high frequency (EHF) band, except to satisfy critical deficiencies in jam resistant service and in wideband data relay service. Equipment investment, projected deficiencies in service and the expected state of technology have been considered by the MILSATCOM Systems Office in its preparation of an architectural framework for MILSATCOM systems development through the early 1990s.

This paper describes the evolution of the present UHF and SHF MILSATCOM systems and reviews the architectural framework for the development of next generation systems. The need to lay a sound technical and preoperational foundation is emphasized for the next generation of MILSATCOM systems with enhanced military unique properties.

Reprinted from EASCON '79, Conference Record Electronics and Aerospace Systems, October 9-11, 1979, Arlington, VA, pp. 612-616.

Introduction

In October 1973, the Director, Defense Communications Agency (DCA), was assigned the responsibility of functioning as system architect for current and future Military Satellite Communications (MILSATCOM) systems of the Department of Defense (DOD).[1] To emphasize the system architect role (as opposed to military program management), a senior civilian position was created in the DCA — the Deputy Director for MILSATCOM Systems, who would lead the MILSATCOM Systems Office commonly known as the MSO.

The primary goal of the MSO is to establish a single coordinated approach to long-range planning and development, leading to systems that are balanced in terms of reliability, survivability, security, flexibility and affordability. One of the specific objectives of the MSO is to generate and annually update an architectural framework that is consistent with this goal. Before examining the current architectural efforts of the MSO, a review is necessary of the first MILSATCOM architecture and of changes resulting from guidance from the Office of the Secretary of Defense (OSD) and the Congress.

The 1976 Architecture

The first comprehensive MILSATCOM architecture was published by the MSO in 1976. (Footnote 2 provides an excellent description of the early evolution of MILSATCOM systems and of the preparation of the first architecture.) At that time, there were two operational MILSATCOM space segments: the Defense Satellite Communications System (DSCS) Phase II satellites operating at super high frequency (SHF) and the leased Gapfiller (GAPSAT) transponders operating at ultra high frequency (UHF) on the Maritime Satellites (MARISAT). In production for future use, there were two other space segments, both for UHF operation: the Fleet Satellite Communications (FLTSATCOM) satellites, which each contained a significant Air Force Satellite Communications (AFSATCOM) package, and several AFSATCOM packages to be integrated on other host spacecraft.

The target next generation architecture that was proposed in 1976 comprised three space segments: a general purpose SHF satellite system (GP-SHF), a general purpose UHF satellite system (GP-UHF) and a survivable satellite system (AFSAT/SURVSAT).

Since the functional specification for the DSCS III satellite was in preparation at the same time as the architecture, the design features of the GP-SHF satellite were incorporated in the DSCS III specification. The proposed GP-UHF satellite, designed to service mobile users not assigned to Single Integrated Operational Plan (SIOP) roles, would permit the continued use of UHF terminals that already were deployed or in production, but with increased resistance to jamming. Some SHF capability also was included. The proposed AFSAT/SURVSAT satellite was primarily a UHF satellite designed to serve the SIOP users.

During the latter half of 1976, the achitecture was reviewed by the Joint Chiefs of Staff, the military departments and the OSD. In January 1977, the

OSD issued guidance to the effect that the next generation MILSATCOM systems would consist of the DSCS (with the DSCS III satellite), a General Purpose Satellite Communications System (GPSCS) and a Strategic Satellite System (SSS).

The GPSCS (a derivative of GP-UHF) was to provide jam resistant communications service to the following: mobile ground, shipboard and airborne users; designated nuclear capable forces; and the National Command Authorities. The MSO was assigned responsibility for developing system alternatives for the GPSCS and the SSS, the latter to consist initially of packages aboard host satellites to provide dissemination of emergency action messages (EAMs). In addition, the SSS program was to continue with research and development (R&D) efforts to generate survivable satellite technology and evolutionary system alternatives that were threat responsive. Since the SSS initially would be only the packages, this 1977 guidance has been referred to as the "two and a half segment architecture."

With the DSCS-II program suffering from launch casualties — two satellites lost in 1975 and two more in 1978 — and the FLTSATCOM program experiencing extended delays, 1977 and 1978 were not banner years in the Congress for MILSATCOM programs. In 1977 and 1978, Congressional concerns with respect to the DSCS program centered around upgrading DSCS II as opposed to replacement by the DSCS III satellite. (And now in 1979, after gaining approval for the DSCS III, there is the need for Congressional approval for two additional DSCS II satellites to assure continuity of service in the early 1980s.)

The FLTSATCOM program was truncated to three satellites by the Congress in 1977, but approval for two additional satellites was granted in 1978 to assure continuity of service between FLTSATCOM and the Congressionally mandated Leased Satellite (LEASAT). The MSO played a lead role in defining LEASAT, using the methodology reported at EASCON 78.[3] It is evident that the Congress is an important element in the MILSATCOM architectural process. Other elements in that process are discussed next, followed then by an examination of the current architectural efforts.

The Architectural Process

The MSO's primary goal is mentioned in the introductory section of this paper; Figure 1 suggests how the MSO's activities interact in the evolving MILSATCOM architectural process. Only the highlights of this process and the significant items that led to the current architectural framework are described here.

All requirements for MILSATCOM service are submitted to the MSO for incorporation in the MILSATCOM User Requirements Data Base (URDB). The requirement originator (e.g., unified and specified commands, military departments, defense agencies) must provide a complete justification statement, based on the operational concept and technical and economic considerations that indicate that a MILSATCOM system should be used to fulfill the requirement rather than an alternative transmission medium. The MSO, in collaboration with the reponsible agent for the particular MILSATCOM system, analyzes the feasibility and operational impact of including the new requirement in the URDB. Annually, the MSO submits the URDB to the Joint Chiefs of Staff for overall review, validation and approval. In addition, the MSO uses the URDB extensively as a baseline for the planning and sizing of future systems.

The concepts for these future systems are developed by the MSO in collaboration with the military departments and defense agencies. To assist the MSO in this process, a MILSATCOM Architectural Issues Steering Group was formed in early 1978 comprised of senior DOD civilians in the Washington area. The steering group members agreed there were two pressing issues that should be examined in detail: antijam modulation systems for mobile users and available frequency bands for future MILSATCOM systems. The steering group recommended the formation of an ad hoc working group to study each issue.

These two working groups were composed of representatives from the military departments, defense agencies and Federal Contract Research centers, and each group met periodically in 1978 under MSO chairmanship. Although their efforts were primarily technical in nature, the working groups did concern themselves with most of the elements in the architectural process of Figure 1: the user requirements; the system operational and transition concepts; and the influencing factors — the projected threat to MILSATCOM systems, the available and projected technology, the current and planned investment, the estimated costs for future systems and the political climate.

The final reports[4, 5] of these working groups were published in February 1979. Concomitantly, the MSO engaged the Lincoln Laboratory of the Massachusetts Institute of Technology to conduct a technology assessment for future MILSATCOM systems in the EHF bands. The recent report of this technology assessment also addresses the aspect of rain attenuation at EHF and the related issues of system availability, duration and impact of rain caused outages and site diversity.[6] In the face of increasing divergence between the January 1977 OSD guidance and the actual evolution of MILSATCOM

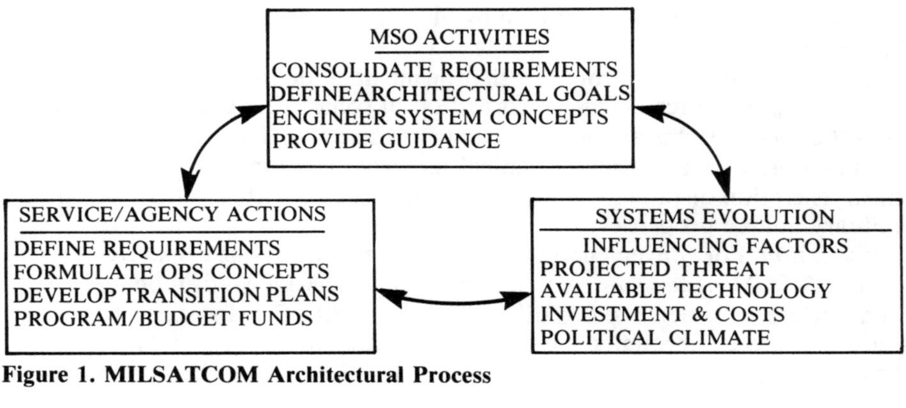

Figure 1. MILSATCOM Architectural Process

systems, the MSO turned to a renewed architectural outline. There were numerous MSO in-house studies conducted in addition to the activities of the two working groups, and a revised architecture took shape.

In November 1978, OSD redirected the GPSCS program in the light of Congressional guidance, the decision to use LEASAT satellites and the potential advantages of operation in the EHF bands. The GPSCS was redefined as a technology only effort to support the development of technology for future jam resistant satellite communications for the Services' tactical and mobile forces. The Services were directed to provide their requirements and specifications to the MSO for consideration and then transmittal to the Air Force for appropriate development and acquisition actions.

The stage is now set for a discussion of the MSO's "Framework for MILSATCOM Development." Published in late November 1978, the Framework is a succinct statement of plausible avenues for DOD satellite communications systems development. What follows here is an unclassified version of the Framework.

Framework for MILSATCOM Development

Overview: By the mid-80s the DOD will have substantial space segment capability for satellite communications embodied in the FLTSAT, LEASAT and DSCS II/III satellites and in the AFSATCOM packages. DOD investment in earth terminals and associated user input/output devices will exceed the substantial on-orbit satellite investment; thus terminal inventory will be a very significant factor in determining the pace and texture of new satellite acquisition. A pragmatic financial assessment of this situation, as well as the required "shakedown" time of the large earth terminal inventory, suggests slowing the rate of development, change and acquisition of space segments with completely new satellite designs, for the near term at least. A new system for the nuclear capable forces is probably an exception.

Despite a pause in major new system acquisition, certain critical, chronic deficiencies (e.g., antijam service for the Worldwide Military Command & Control System (WWMCCS) and ground mobile forces (GMF) and the need for additional capacity in several systems) must be alleviated. Pressing near term deficiencies shoud be rectified by block changes to existing systems (both terrestrial and spaceborne).

It appears that aside from the significant upgrade in communications for nuclear capable forces, plus the correction of some other critical deficiencies, no other new systems are necessary until the late 80s. The intervening period should be used to lay a sound technical and pre-operational foundation for the next generation of satellite communications systems with distinctly enhanced military unique properties.

Evolution: For convenience, the MSO has categorized the disparate users of MILSATCOM systems into three user groupings: wideband, mobile/tactical and nuclear capable. The evolution of the MILSATCOM systems to service these user groups through the early 90s is depicted in Figure 2 and is described below.

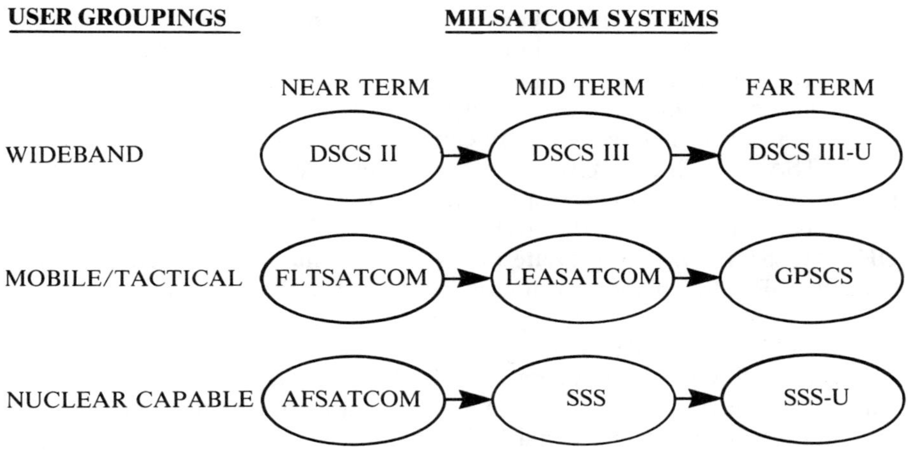

Figure 2. MILSATCOM Evolution

1. *Wideband Users/Systems:* With the exception of the FLTSATCOM uplink terminals for Fleet Broadcast, the investment in SHF terminals is associated with the DSCS user group. This group is referred to as "wideband" even though links through the satellite range from many Mbps to several Kbps, with antijam (AJ) links at a much lower rate. The DSCS is the most mature of the evolving MILSATCOM systems, but there are projected deficiencies in terms of capacity (for multi-Mbit/sec data relay and multichannel GMF links) and jam resistance (for data relay and for WWMCCS airborne, GMF to GMF and ship to ship links).

The near term satellite capability consists of the DSCS II, with a transition to the DSCS III in the early 80s. The DSCS III satellite will provide significant improvements in jam resistance, capacity, flexibility, in-orbit life and decentralized control.

Based on future threat projections and increasing requirements for service, the MSO anticipates an evolutionary upgrade to the DSCS III satellite in the late 80s. There is the potential for this block change upgrade (DSCS III U) to have EHF transponders and on-board processing.

• The EHF repeaters could service the multi-Mbit data relay users and the multichannel GMF community. The increased bandwidth available in this frequency band enables increased capacity and jam resistance for these user groups, and offloading these users to EHF provides needed capacity at SHF.

• The on-board processor could provide improved jam resistant service for the WWMCCS-airborne, GMF to GMF and ship to ship links.

These improvements to the DSCS III are of an evolutionary, low risk nature, and this block change approach is advocated, therefore, rather than initiation of an early DSCS IV program start.

2. *Mobile/Tactical Users/Systems:* Most of the investment in UHF terminals will be associated with this user group, including general Navy fleet communications, GMF single channel and manpack services and air opera-

tions. A UHF space segment capability will probably be required into the 1990s because of the sunk cost and because UHF terminals are relatively cheap and easy to install. The limitations of UHF, however, are well known: frequency congestion, lack of capacity and jam resistance.

The near term satellite capability consists of GAPFILLER and FLTSAT. As FLTSAT satellites expire in the early 80s, the LEASATs will be used through 1987 at least. A new system, GPSCS, is proposed for satellite communications for the General Purpose Forces as a follow-on or complement to UHF MILSATCOM to alleviate most of the military deficiencies of UHF systems. GPSCS, operating primarily with EHF uplinks would include onboard processing multifrequency crossbanding and (perhaps) EHF antenna nulling. (Although GPSCS is the current program title, the MSO prefers Tactical Satellite Communications [TACSATCOM] II as a more descriptive name. TACSATCOM I was the interim operational capability provided to tactical users by the experimental LES 6 and TACSAT I satellites in the early 70s.)

GPSCS would provide such military unique capabilities as jam resistant and cryptographically protected communications, command and telementry links, with near real time worldwide flexibility and autonomous control — features not characteristic of commercial or leased satellites.

3. *Nuclear Capable Users/Systems:* The nuclear capable community consists of a core of users requiring low data rate, highly survivable communications with a mix of terminal types including aircraft, ships, submarines and transportable earth terminals.

The present system, AFSATCOM, consists of the AFSATCOM packages on FLTSAT and other host spacecraft, and the inventory of ground command post and airborne terminals that communicate at 75 bps. Near term improvements to AFSATCOM will result from the addition of single channel transponders (SCTs) to the DSCS III and other host spacecraft; the SCTs one-way devices for emergency action messages (EAM) dissemination. Improvements in the critical features of jam resistance and survivability are essential for AFSATCOM's vital users, however, and the Strategic Satellite System (SSS) is proposed for the mid-80s. An attractive technical configuration for the SSS consists of the Strategic Forces Communications Satellite (SRATSAT) and proliferated SCTs. (This next generation SSS was approved by OSD in May 1979, but Congressional approval was uncertain when this paper was written.)

In the far term, SSS-U (a block change to STRATSAT) should achieve even greater effectiveness, serve a more comprehensive set of users and have a degree of backup interoperability with other space segments. For example, it would be desirable for SSS-U and GPSCS to be interoperable.

Current Activities

The architectural framework had a limited distribution within the DOD and has been well received by the DOD community. The MSO is presently preparing an "Expanded Framework" for wider distribution in the DOD and for inclusion in the Defense Documentaion Center so that contractors may have access to it.

The progressive improvement proposed in the architectural framework for the three major MILSATCOMM systems requires a vigorous research and development program to define, develop and test implicit advanced concepts and technology, preferably in an operational context. In the absence of systemic or architectural focus, the military departments have initiated independent development programs in various aspects of the future systems suggested here. In the interest of converging these efforts, the MSO is preparing a "Technical Program Plan for MILSATCOM Development." The plan will define a coordinated approach to the R&D required to insure availability of technology necessary to support the future systems. The MILSATCOM Architectural Working Group similar in composition to the two working groups that performed so well in 1978, is assisting the MSO in the preparation of the Plan. Advancing the state of maturity of EHF components and on-board processors is an essential consideration for the working group so that systems derived from such technology are affordable. The operational and control concepts and the transition strategies are prime considerations for each of the future systems.

In the Framework and in the Technical Program Plan, the MSO advocates a technology program that includes a developmental satellite payload and associated service unique terminals for joint test and evaluation in the mid-80s. The MSO considers the relationship between a processing satellite and the user terminals so inextricable that the technology must be demonstrated by operational users and judged for technical and operational utility in order to guide future production programs. It is hoped that this orderly process for the development of future MILSATCOM systems is acceptable to the OSD and the Congress; the MSO will continue efforts to influence the DOD planning, programming and budgeting system cycle to gain support for this orderly process.

Acknowledgement

The author expresses his appreciation to Paul Rosen and other members of the MSO staff for their assistance and encouragement during the preparation of this paper.

Footnotes

[1]U.S. Department of Defense, Directive Number 5105.44: Military Satellite Communications (MILSATCOM) Systems Organization. Washington: October 9, 1973.

[2]Frederick E. Bond and CDR William H. Curry, Jr., USN, "The Evolution of Military Satellite Communications Systems." *SIGNAL,* March 1976, pp. 39-44.

[3]John Edell, et al., "A Leased Military Satellite Communications System for the Early and Mid 1980s." IEEE EASCON Record, 1978, pp. 453-457.

[4]Defense Communications Agency, MSO Technical Report No. 79-1: "Final Report of the Working Group on Antijam Modulation Systems for Mobile Users (U)." February 1979.

[5]Defense Communications Agency, MSO Technical Report No. 79-2: "Final Report of the Working Group on "Frequency Selection For Future MILSATCOM Systems (U)." February 1979.

[6]D.J. Frediani, "Technology Assessment for Future MILSATCOM Systems: The EHF Bands." Massachusetts Institute of Technology Lincoln Laboratory Project Report DCA-5. Lexington, Massachusetts: April 12, 1979.

DSCS: An Evolving Operational System

Paul C. McLellan

The Defense Satellite Communications System (DSCS) is the key Department of Defense high capacity transcontinental transmission system. Operating at super high frequency (SHF), the space and associated ground segment provides critical connectivity for the National Command Authorities (NCA), Joint Chiefs of Staff (JCS), Unified and Specified Commands, Department of Defense Agencies, Department of State and, by resource sharing agreements, the United Kingdom and the North Atlantic Treaty Organization (NATO). Currently a six satellite constellation, consisting of four fully operating and two spare DSCS IIs, DSCS provides a multi-megabit wideband relay capability for 130 globally deployed earth terminals.

This article will focus on how the DSCS is evolving into an increasingly capable communications system during the 1980s. Primary drivers for system improvement are the need for additional channel capacity, electronic warfare resistance, digital technology efficiency and real time control. After touching on the program's history and management relationships, the following topics will be covered: system objectives, a description of the DSCS III satellite, developments in the earth terminal and control segments and a discussion of the program's current status and future plans.

This article originally appeared in the February 1981 issue of* ***SIGNAL.*

Phase I of the program, formerly known as the Initial Defense Communications Satellite Program (IDCSP), consisted of relatively simple 100 pound satellites with a single channel SHF transponder. The satellites also had a minimum telemetry system but no command system or batteries for eclipse operation. By 1968, 26 of these satellites had been placed into drifting near synchronous equatorial orbit by the U.S. Air Force's Titan IIIC launch vehicle to support some 35 ground and shipborne terminals. These satellites far exceeded their lifetime goals.

Following deployment of the Phase I system, military planners took advantage of reliability improvements in command and station keeping which were embodied in geostationary orbiting satellites. Moreover, improvements in SHF communications technology offered wider bandwidth channels, higher transmitter power and steerable antennas to focus radiated power into small geographic areas. Other improvements included rapid repositioning capability, modest nuclear survivability, secure commanding and telemetry encryption.

By 1968, the Defense Satellite Communications System—Phase II was approved to take advantage of the aforementioned state-of-the-art advances. Contract award for the design and development of this satellite was made in March 1969. Each satellite's communications payload included four channels coupled to either earth coverage or spot coverage antennas. The first launch of two satellites on a Titan IIIC occurred in November 1971. These satellites failed after operating over the Atlantic and Pacific oceans for 19 months and 10 months, respectively.

Phase I satellites continued to support communications until the second launch of two Phase II spacecraft was successfully accomplished in December 1973. Subsequent loss of four spacecraft due to launch vehicle failures delayed full operational capability until 1978.

The operational DSCS is being upgraded with more efficient digital communications subsystems and more reliable satellite ground terminals. To complement this ground segment, the space segment has two additional DSCS II spacecraft in production and a conservatively structured DSCS III development underway. Capacity derived from the upgraded DSCS is being used to support additional validated requirements including replacement of a few tropospheric scatter and line-of-sight microwave systems.

However, the DSCS is not designed to entirely supplant use of leased satellite communications. Rather, it serves to balance where possible, the transoceanic transmission media for the Defense Communications System (DCS) among military satellites, commercial satellites and commercial cable facilities. The diversity afforded by this routing provides for enhanced survivability during times of crisis and improved reliability and continuity in the day-to-day operation of the DCS.

Operational direction and management control of the DSCS is the responsibility of the Defense Communications Agency (DCA) through policy established by the Joint Chiefs of Staff. System developments are planned by DCA and programmed for budgeting by the military departments. DCA's focal point is the DSCS Program Management Office which is tri-Service manned. Spacecraft and airborne terminals are acquired by the Air Force, ground terminals and control equipment by the Army and shipborne terminals by the Navy.

Objectives

The DSCS is designed to accommodate unique and vital requirements which cannot be satisfied effectively by other means. Specifically, it has been designed and configured to satisfy the following operational objectives:

1. Support the Worldwide Military Command and Control System (WWMCCS) by providing communications service between the NCA/JCS and the Unified/Specified Commands, among the Unified/Specified Commands to the general war combat forces, from peripheral early warning sites and wideband users;
2. Provide a high capacity, reliable, independent communications capability to extend defense communications service in support of contingency and limited war operations and for the restoration of primary DCS transmission subsystems that may become inoperative due to natural causes, sabotage or direct enemy action;
3. Augment the DCS with a transmission subsystem capable of providing the wideband channels required to handle high quality secure voice, high speed data between automated command and control center facilities, high resolution graphics and imagery and rapid transmission of sensor data;
4. Provide DCS communication service to remote locations not adequately served by other means;
5. Provide an improved capability to support Navy ship-shore communications and other authorized users; and
6. Support the voice channel requirements of the Ground Mobile Forces (GMF).

Because DSCS II was not designed to meet the communications requirements imposed by all of the foregoing objectives, especially the heavy GMF burden, and the increased threat of electronic jamming, the DSCS III program was initiated in 1974. DSCS III offers significant improvement in resistance to electronic warfare, increased flexibility in meeting changing user requirements, enhanced nuclear survivability and longer design life than DSCS II. A detailed discussion of the capabilities of the DSCS III will follow, augmented by improvements taking place in the terrestrial and control segments.

DSCS III Space Segment

Unlike DSCS II, which has little on-board antijamming (AJ) capability, DSCS III has a 61 element receive Multiple Beam Antenna (MBA) to selectively null jammers with minimum disturbance to friendly users. During normal peacetime operations, the receive MBA coupled with the two 19 element transmit MBAs affords unmatched flexibility to optimize both up and downlinks to user locations unique to each satellite's orbital position. As user locations change, the MBAs can be rapidly reconfigured to meet these new demands.

Focusing on the Transponder Block Diagram in Figure 1, the flexibility of the DSCS III is evident. For example, the receive MBA can be connected to any or all of channels one through four. Likewise, the two transmit MBAs can be connected to the same four channels in whatever combination is required to optimize user needs.

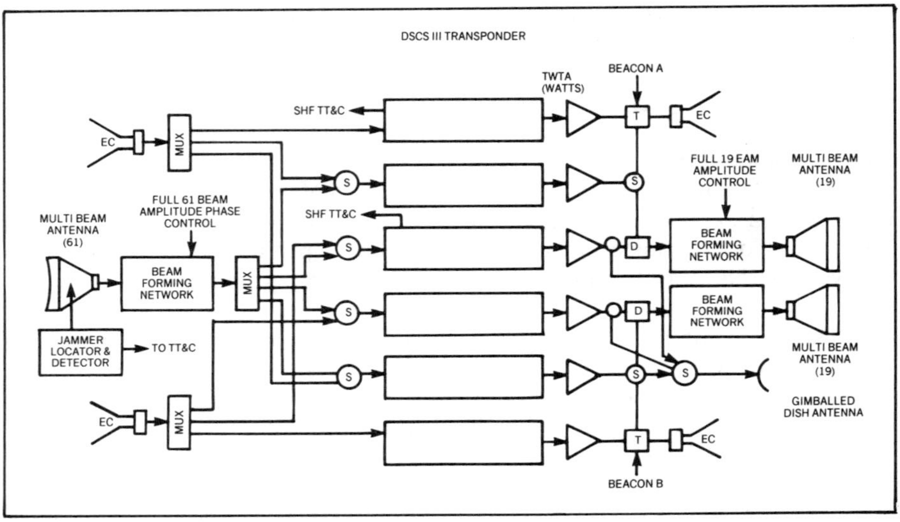

Figure 1. DSCS III Transponder

The Gimballed Dish Antenna provides the additional gain required to service the White House Communications Agency, JCS contingencies and GMF requirements of the Army, Marine Corps and Air Force. In order to fully capitalize on the flexibility provided by the antennas and multiple channelization of the satellite, significant control segment changes are mandated which will be covered after describing the remaining satellite features and the terrestrial segment.

In addition to providing satellite controllers with data on the status of critical components and command capability to alter satellite configuration, the Telemetry, Tracking and Command Subsystem is secure at both S-Band and SHF and antijam protected at SHF. For the first time, DCS ground terminals will have the capability to control the satellite at SHF. Satellite stabilization is achieved in three-axes via four skewed reaction wheels. The propulsion system is sized to accommodate the 10 year satellite life for both north/south stationkeeping and attitude control. Primary power is derived from either the 126 square-foot solar array or three nickel cadmium batteries. The satellite nuclear survivability meets Joint Chiefs of Staff guidelines. Launch of the approximately 2,400 pound satellite is accomplished on the Titan IIIC, Titan 34D/Inertial Upper Stage or Space Shuttle.

A secondary payload is carried by each DSCS III to disseminate Emergency Action Messages (EAM) from ground and airborne command posts to the force elements (FB 111 and B-52 type aircraft and missile launch control complexes). This Single Channel Transponder or SCT provides a high degree of jammer protection and security. Uplink signals are at either UHF or SHF while the downlink is at ultra high frequency (UHF).

Terrestrial Segment

Terminals employed worldwide in the DSCS include modified research and development (R&D) terminals (AN/TSC-54, AN/MSC-46), AN/FSC-78 and new AN/GSC-39 and Jam Resistant Secure Communications (JRSC) terminals. Operation and maintenance of these terminals may be performed by any one of the three Military Departments depending on location. The diverse types of terminals used range in size from the large, fixed AN/FSC-78 with 60 foot antennas, the shipborne AN/WSC-6 and its four foot antennas to the 33 inch AN/ASC-24 airborne terminal. Two of the newest terminals, the GSC-39 and WSC-6 are described in more detail below.

The AN/GSC-39 (fixed/transportable) terminal's antenna subsystem consists of a high efficiency 38 foot solid surface main reflector and a five horn monopulse feed system supported on a pedestal structure. The receive signal path is divided into wideband and narrowband segments. The wideband radio frequency (RF) segment provides signal gain over the full 7250 to 7750 MHz receive band; the narrowband down translates 40 MHz bandwidth slots to a 70 MHz intermediate frequency (IF) and 125 MHz slots to a 700 MHz IF center frequency.

Transmit and receive carriers are tuneable in 1 kHz increments over the full 500 MHz uplink and downlink frequency bands. An automatic carrier balance capability is provided which offers automatic leveling of each transmitted carrier with an accuracy of $\pm$ 0.2 dB. The terminal is designed with extensive and sophisticated fault location, monitors and alarms. It has been designed to accommodate up to nine transmit and 15 receive Frequency Division Multiple Access (FDMA) carriers.

The AN/WSC-6 shipboard terminal has two versions. The AN/WSC-6 (V)1 has a single four foot antenna, enclosed in a radome, and uses a narrowband BPSK modem. The (V)1 version is intended for use with Surveillance Towed Array System (SURTAS) projects. The AN/WSC-6 (V)2 has dual four foot antennas, radome enclosed, for full hemispherical coverage satellite communications. It uses a full duplex spread spectrum digital modem. Frequency band operation is 7.25 to 7.75 GHz receive and 7.9 to 8.4 GHz transmit.

The terrestrial and space segments are configured to operate in support of requirements validated by the Joint Chiefs of Staff. Actual system connectivity is a time variant function. However, an approved configuration for the near term is shown in Figure 2.

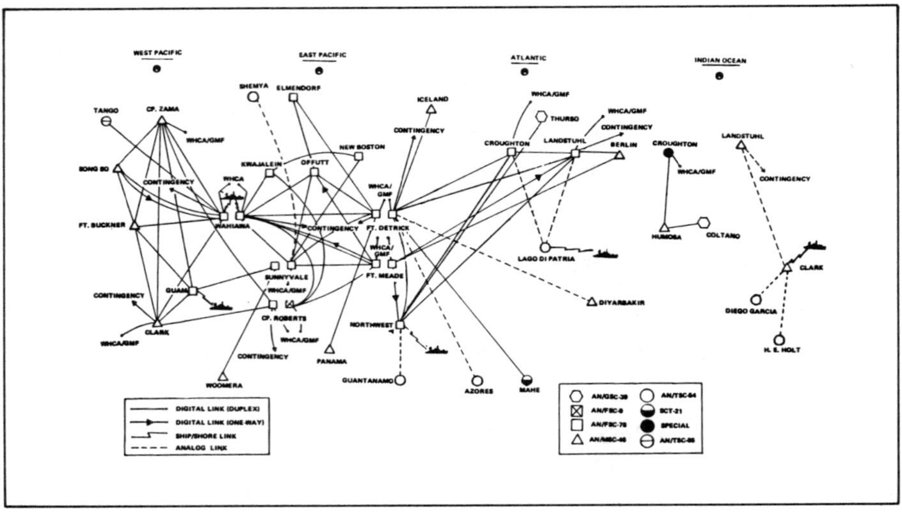

Figure 2. Typical Configuration

Control Segment

Today's control segment operates in a relatively static DSCS II environment whereby DCA exercises satellite communications control to establish and maintain circuits and links between authorized users. The Air Force Satellite Control Facility (AFSCF) does all spacecraft commanding, tracking and telemetry monitoring in coordination with DCA.

As alluded to in the satellite description, allocating the multitude of DSCS III satellite configurations presents a complex problem to ground controllers. For the first time, DSCS ground terminal operators will be able to alter the communications functions of the satellite by changing operating modes to meet user requirements at the satellites' operational locations. The AFSCF will continue satellite housekeeping and repositioning functions and will retain the capability to command communications functions. During an electronic warfare attack, reconfiguring the three Multiple Beam Antennas is a dynamic problem to resolve.

The number of users impacted in such an environment will vary according to each satellite location, but the total terminal population will expand from today's 100 plus to over 400 by 1986, requiring an automated control system. The DSCS Operations Control System (DOCS) is being developed to meet this requirement. Through computer assistance, the DOCS will manipulate composite network link equations involving antenna connectivity, effective isotropic radiated power (EIRP) and antenna gain-to-noise temperature (G/T) to provide satellite configurations most responsive to user requirements in peacetime through electronic warfare scenarios.

Such assistance will result in more effective use of available resources to achieve an increased system capacity for whatever situation the ground controller faces. Because DOCS is scheduled to be operational about the same time as a fully operational DSCS III space segment, an interim system to control early DSCS III satellites, the DSCS III Operations Support System (DOSS), will be available starting in 1981. Experience gained from DOSS utilization with the Satellite Configuration Control Element (which provides SHF command and telemetry processing at selected communications earth terminals) will be transferred into the DOCS development.

Program Status/Future Plans

General Electric's Space Division has been on contract to the Air Force's Space Division for DSCS III Full Scale Development since February 1977, to deliver two flight satellites and one qualification model satellite. The qualification model is well into its test program, is meeting required performance and is expected to complete testing by this month. The first flight satellite is scheduled for launch by mid-1981, along with a TRW produced DSCS II satellite on the Titan IIIC launch vehicle. Extensive testing of the DSCS III will be accomplished to validate its performance in an operational environment.

The U.S. Army Operational Test and Evaluation Agency will conduct independent Initial Operational Test and Evaluation in conjunction with DCA and Space Division. The second DCS III is schedued for launch in 1982, with the last DSCS II followed by the refurbished qualification satellite in 1983. After the launch of the first DSCS III, a Defense Systems Acquisition Review Council (DSARC) is schedued to decide upon production of 11 more satellites to meet operational requirements into the 1990s.

Production approval for 1981 is the key to maintaining a high availability space segment.

DSCS III is also designed to be compatible with the existing NATO SHF satellite ground segment. NATO plans to replace current NATO III satellites with a NATO IV to be launched in 1987. The NATO International Competitive Bid for NATO IV will specify that either of two options will satisfy NATO requirements—either a DSCS III or a NATO tailored option type satellite.

The architecture for upgrading the DSCS III and a follow-on system is currently being studied. Improvements such as on-board processing, crosslinks and utilization of the 20-30 GHz frequency band are being considered to further enhance antijam performance and provide for greater expansion of user requirements. Either modified DSCS III or NASA satellites could be used for initial test of 20-30 GHz satellite hardware before proceeding into another development. If approved, current planning would support launch of such a demonstration satellite in the 1988—90 time frame.

The AN/GSC-39 ground terminals produced by Comtech Laboratories under management of the U.S. Army's Satellite Communications Agency (SATCOMA) are being deployed. The first terminal was shipped to Thurso, Scotland, in November 1980. An additional 20 terminals are expected to be delivered by late 1983. The AN/GSC-39 will eventually replace all of the AN/

MSC-46s which have been in operation since the late 60s. SATCOMA is also managing the acquisition of two other satellite terminals to be used with the DSCS III satellites.

First, six AN/TSC-86s produced by RCA, which will replace the 80 AN/TSC-54 JCS contingency terminals also in the field since the 60s and second, the GMF family of terminals: AN/TSC-85, AN/TSC-93 and AN/TSC-94. A total of over 190 of these terminals should be in operation by 1986 to support the Ground Mobile Forces requirements of the Army, Air Force and Marine Corps. Funds are being planned for Army development of improved ground terminals to be compatible with the 20-30 GHz satellite.

The Navy's Electronic System Command manages the acquisition of all shipborne terminals and is expected to have 30 AN/WSC-6 terminals installed in the fleet by 1987. Finally, the Air Force's Electronics Systems Division is providing airborne terminals for the Advanced Airborne National Command Post.

Other key subsystems which are a part of the DSCS include: the 32 Harris JRSC terminals, electronic counter countermeasures equipment built by Magnavox, Harris PSK modems, Linkabit coders/decoders, TRW PCM and digital modems by Martin Marietta. About two-thirds of the currently deployed earth terminals have been converted from analog to digital modulation. The DSCS should be completely digital by late 1983.

Conclusions

The DSCS program continues to evolve to meet expanding user requirements and threat scenarios. Antijam SHF satellite communications will become a reality subsequent to the launch of the first DSCS III in 1981. DSCS III satellite production must be initiated in 1981 to ensure that the space segment is maintained with high availability.

Defense Satellite Communications System: Another Milestone

William J. Hilsman and Alfred R. Garcia, Jr.

It was 7 A.M., April 30, 1983, a quiet Saturday morning at Sunnyvale Air Force Station (AFS), California, when the cutover from DSCS II to DSCS III began at the new Defense Satellite Communications System (DSCS) Operations Center there. A small team of experienced satellite network controllers from the Defense Communications Agency (DCA) working with Air Force control personnel assigned to the 1999th Communications Squadron started the carefully orchestrated and rehearsed cutover.

At that time, 15 DSCS earth terminals located at such far-flung places as Kwajalein, New Hampshire, Panama and Alaska were directed to acquire DSCS III, the first of the Phase III satellites to be placed in operational service. The satellite was in geosynchronous orbit 22,300 miles above a point on the equator, 130°W longitude. A new era in military satellite communications had arrived.

Six hours later, the cutover was completed on schedule with no problems and all parameters nominal. Experienced controllers declared this cutover one of the smoothest in memory—a tribute to the skill of the DCA and Air Force controllers and professionalism of the Army, Navy and Air Force earth terminal operators.

With the advent of DSCS III, we are now better able to meet the unique and vital user communications requirements that are essential to national defense. Some of the DSCS users include the National Command Authorities, the Joint Chiefs of Staff (JCS), the various commanders in chief, several major command and control systems, the military department and special agencies.

Characteristically, these users are located almost anywhere in the world or can deploy to almost any point in the world, frequently with little advance

This article originally appeared in the September 1983 issue of* ***SIGNAL.*

notice. In addition, user high speed data requirements, isolated locations or lack of alternative paths for communications will dictate use of the DSCS for communications support. DSCS provides a completely U.S. government controlled communications system that introduces reliability and availability to our field commands and their subordinate elements.

DSCS Evolution

DSCS III is the latest satellite in the super high frequency (SHF) DSCS evolution. In 1966, we had the Initial Defense Communications Satellite Program (IDCSP) which was a network of 26 subsynchronous non-stationary satellites acquired by tracking antennas and used as point-to-point relays. With IDCSP we learned about satellite communications and operations; the system became very important to the National Command Authority during the Vietnam conflict.

The DSCS II was the beginning of our geostationary network of four operational satellites providing worldwide coverage: 12°W, Atlantic; 135°W, East Pacific; 175°E, West Pacific; and 60°E, Indian Ocean. Although the early DSCS II satellites have had a rocky life because of launch and premature failures, the system stabilized in recent years to provide a reliable, high capacity satellite transmission media. Thanks to the efforts of many people to make the space and earth terminal segments work properly, DSCS II became a byword for reliable communications.

Although the DSCS is generally thought to be only the satellite system, it is actually the synergistic product of three interrelated systems: the space segment, the earth terminal segment and the control segment. DSCS III is certainly the most obvious of the changes in the DSCS program, but it is only one of several major events occurring during the 1982-85 time period. In addition to our present terminals, a host of new earth terminals are being produced and fielded that will use the DSCS III satellite.

A new control system is being fielded that will significantly enhance the capability to control both DSCS III and, to an extent, DSCS II in a timely, efficient and effective manner. In addition, new spread spectrum modulation equipments, completion of the conversion to all digital operation and improved coding techniques are decreasing user vulnerabilities in threat environments and increasing system capabilities during the next few years.

As a "purple suit" program (jointly funded, manned and operated), each segment has been assigned to a military department for acquisition and life cycle management under the overall DSCS Program Management of DCA. The earth terminal and control segments, once fielded and in operation, are operated and maintained by the three military departments under the operational direction and management control of DCA.

The Air Force's Space Division in Los Angeles is responsible for the DSCS satellites, while the U.S. Army's Satellite Communications Agency at Fort Monmouth, New Jersey, is responsible for the fixed and mobile earth terminals that access the DSCS and the satellite and network control systems that make the entire system operate efficiently and effectively. For Service unique systems, the Navy is responsible for shipborne terminals and the Air Force is responsible for airborne terminals.

DSCS Characteristics

The DSCS III is a remarkable satellite with built-in toughness against threats, flexibility in supporting various user requirements and responsiveness to changes in threat or mission. Supporting both Frequency Division Multiple Access (FDMA) and Spread Spectrum Multiple Access (SSMA) users, one of the satellite's major characteristics is its multibeam antenna (MBA).

The receive MBA consists of a selectable 61 element antenna which can form a highly directional gain antenna that favors desired uplink user locations and nulls out unused or unfavorable locations.

The transmit MBA, with 19 elements, has the same capability to "fine tune" downlink power gain. With the single steerable gimballed dish transmit antenna (GDA), it is also possible to maximize downlink power to service small strategic and tactical earth terminals located in a relatively small area. The high gain MBAs are what gives the DSCS III its special capabilities that enable DSCS support to a variety of users in a variety of environments.

Electronic counter countermeasures (ECCM) capabilities in the DSCS III will also help the operations community. In addition to the nulling of offending signals available by reconfiguring the receive MBA, the DSCS III satellite is equipped with jammer locations electronics that enable controllers to perform any of several countermeasure techniques that reduce or eliminate the effects of the jammers. Both the FDMA and SSMA networks will be aided by this ECCM capability.

The Army's Satellite Communications Agency at Fort Monmouth, New Jersey, is acquiring and fielding new generations of earth terminals so that we can finally begin to meet most of our growing user requirements and replace aging terminals. The latest acquistion effort is the state-of-the-art medium terminal (SAMT), a terminal equipped with a hardened version of the reliable OE-222 38-foot antenna, and an all new electronics package designed for unattended operation.

The design concept is to have component redundancy, automatic switching capability and remote fault alarm. The SAMT should reduce personnel requirements while increasing system availability. The SAMT will be available in both fixed and van type configurations. It is currently scheduled for initial deployment in 1985.

The ongoing deployment of a new generation of SSMA equipment under the Jam Resistant Secure Communications (JRSC) program will give our earth terminals the capability to transmit and receive critical traffic through a variety of threat scenarios. We are still learning about network operations, but the system is beginning to work. By mid-1984, we will connect the Tactical Warning and Attack Assessment (TW/AA) sensor sites, via JRSC equipped earth terminals, into the major operations centers. Reliable and survivable communications will be achieved.

Control Segment

The DSCS control segment is the most important part of the system's capability, because control is what makes the entire DSCS operate properly. Communications satellite network control (SNC), satellite communications payload control (CPC) and ECCM network control are the functions that give the DSCS its robust communications capabilities, flexibility and responsiveness.

Control will be performed by eight DSCS Operations Centers (DSCSOC), two for each satellite area, operated and maintained by the MilDeps and under the operational direction and management control of DCA. This direction comes from the DCA Operations Center in Washington, D.C., through the DCA PAC Area Control Operations Center (ACOC) at Wheeler Air Force Base, Hawaii, for the East Pacific and West Pacific satellites and through the DCA EUR ACOC at Vaihingen, Germany, for the Atlantic and Indian Ocean satellites.

The first DSCSOC was activated at Sunnyvale AFS, California, in the fall of 1982, while the second will be activated at Fort Detrick, Maryland, late this summer. Primary equipment to be located in the DSCSOCs include the DSCS Operational Support System (DOSS); the DSCS Automatic Spectrum Analyzer (DASA); and the Satellite Configuration Control Element (SCCE).

ECCM remote displays, smart video terminals replacing teletype orderwires and computer assists characterize the DSCSOC environment. The people who will operate the DSCSOC will be our best network and payload controllers, because they will have the tremendous responsibility of controlling critical communications among highest priority users in both peace and threat environments. Control is definitely the most challenging and dynamic function in the DSCS III.

So what does all this mean? How do we use these capabilities? Using two scenarios of typical requirements, we can depict how DSCS support is provided to both the fixed and tactical communities. A typical change to the configuration of a fixed terminal would begin with a JCS validated requirement with some sort of lead time, whether in days, weeks or months.

DCA operations elements would prepare to size the network and links to accommodate the requirement, including planning multiplex arrangements, spectral filters, power and bandwidth parameters and frequencies. Telecommunications Service Requests and other key documents would be prepared to cause the system to be ready for the requirement.

On the activation date, the new traffic would cutover to operation and service would be provided. Although this procedure works for 75 baud to 750 Kbps, some equipment moves might be required for wideband requirements in the megabit range. DSCS is very flexible and can meet small to large requirements in a timely manner.

Typical Scenario

A typical tactical scenario would begin with a crisis, emergency or exercise in a remote part of the world. An operational CINC would quickly develop a communications requirement (along with many other requirements needed to accomplish his mission) and notify the JCS. The JCS would approve the basic requirements and cause DCA to review the technical aspects of the CINCs' needs.

In the case of super high frequency (SHF) satellite service, Headquarters DCA operational and engineering elements would develop a support scenario, calculate any trade offs necessary and determine the optimal method to configure the spacecraft and earth terminal segments. As the operations plans were modified and put into operations orders, the data and time for equipment arrivals and circuit activation would be established.

A few hours before link activation, the DSCSOC satellite network controller would use the DOSS to develop the exact parameters for a satellite configuration change. This change would be placed in the SCCE for translation by the communications payload controller into a spacecraft command sequence.

After a double check of the sequence, the command would be transmitted to the spacecraft for the reconfiguration. Within minutes, the 19 element transmit MBA or the gimballed dish antenna would be reconfigured to increase the transmit gain to the area of interest and the 61 element receive MBA would be reconfigured to increase receive gain.

Soon the mobile earth terminals in the area of interest would contact their satellite controller for their access parameters, including power, frequency and look angles. As the links came up, connectivity into AUTODIN, AUTOVON, AUTOSEVOCOM or any of several other networks would be accessed through the servicing tech control.

Last minute problems would be worked through by NCOs and Department of Defense civilian employees to ensure reliable connectivity to meet operational requirements. The end result is communications for the field commanders, wherever needed.

In conclusion, the DSCS provides a tremendous communications resource available to support unique and vital communications requirements. The combination of DSCS III, improved ground equipments and a dynamic control capability will significantly enhance both strategic and tactical communications.

Only the DSCS can support the wide range of earth terminals that talk to it, ranging from tiny three-foot antennas to giant 60-foot antennas, data rates from 75 baud to greater than 10 Mbps and users almost anywhere in the world.

A Navy Satellite Communications System: The Need for the Next Generation Is Now!

Jon L. Boyes

In support of national policy, the forces of the U.S. Navy are required to maintain a continuously combat-ready status throughout the world. The command and control of these forces demands the most efficient and reliable communications system that our modern technology and industrial resources can provide.

Periodically, in crisis situations, the dispersed forces of the fleet must be concentrated to support national policy. In many cases, this concentration is accomplished to provide a first wave of action in areas offering low support capability to our forces and a major threat from potential adversaries. Situations of this type characteristically require the command structure of the Navy

This article originally appeared in the March 1976 issue of SIGNAL.

to exercise responsive control over the measured action of powerful forces if national policy is to be served. The participation of the highest levels of our National Command Structure is often necessary and direct communication with the Officer in Tactical Command is often required.

The desired control must, of course, be supported by rapid and reliable communications. Historically, this communications support has been provided by a worldwide network of Naval Communications Stations primarily dependent on frequencies from very low frequency (VLF) to high frequency (HF). In recent years, political changes in overseas areas have raised significant doubt as to the availability of this network in the future. The Navy is, therefore, keenly interested in communications options that would minimize the risk of the denial of command communications and offer significant reduction in dependence on overseas facilities.

Naval Communications Requirements

A number of efforts have been made to determine what communications are required to support a deployed Naval force. From these efforts, two distinct systems of requirements have repeatedly emerged. First, it is necessary to link the shore based command centers by circuits capable of passing a small amount of high precedence traffic as well as a large volume of lower priority information. Second, it is necessary to link command centers ashore with commanders afloat by circuits that represent the chain of command. These circuits must be capable of passing a small amount of high precedence traffic with great reliability even in the face of jamming by a determined enemy.

From the Navy view, any system of communications must meet both of these requirements. Further, to be acceptable, any candidate system must be affordable in terms of total system costs. A number of acceptable solutions to the problem of linking the shore based centers have been demonstrated. However, the linkage of the shore and afloat commanders to the fleet elements for tactical purposes remains a major problem.

Naval Tactical Communications Users

In general terms, Naval Communications depend on the successful connection of two disparate classes of users — the shore based commander and commanders in the operational ships of the fleet. Of these two classes of users, the requirements of the commanders ashore most resemble those of commercial communications. In this case, where communications is a primary mission, large antennas are attainable; high power transmitters can be employed; quiet receiver sites can be sought; logistics support can be maintained; and, as a result, high volume traffic can be exchanged with other similarly advantaged installations.

The second class of users, commanders in the ships of the force, are, by comparison, significantly disadvantaged. Due to ship construction and stability considerations, antennas must be small; transmitters must be light and relatively low powered and receivers must operate in high noise environments. Logistics support and manning are restricted by the space allocated for communications in a vehicle where the prime mission is combat. These problems are compounded to a significant degree by the dynamic motion of the platform. Rolling of 30 degrees is not uncommon in even the larger ship classes and antenna stabilization is a significant problem.

In spite of these problems, there is an urgent need to provide the afloat forces reliable communications circuits that support a small number of high precedence, critical impact messages. The forward deployments of the fleet units require that these critical linkages of shore and afloat commanders be accomplished even in the face of jamming by a determined enemy. Furthermore, the communications should be accomplished in a manner that minimizes the probability of position location of the deployed force.

When taken as an ensemble, the missions of the Navy and the characteristics of the user classes dictate a system design which permits:

1. Direct, National Command Authority-to-platform connectivity;
2. Flexible and dynamic allocation of communications (channel) resources;
3. Accessing and modulation techniques which do not constrain the operation of highly mobile combatants;
4. Suitable data rates, operational disciplines and system control to counter jamming; and
5. Low probability of position location by enemy forces, balanced with detection and location techniques by other means.

Promise of Satellite Communications

To fulfill its mission, the Fleet needs reliable, antijam communications beyond the horizon. Techniques in development employing surface wave high frequency (HF) modes and ultra high frequencies (UHF) with remotely piloted vehicles or lighter-than-air vehicles may offer some protection against jamming but offer little capability beyond a 100-200 nautical mile (NMI) range.

Only satellite communications (SATCOMs) offer the reliable, range-independent, quasi-hemispheric connectivity that can meet the combined tactical and strategic communications needs of today's fleet. Yet significant revisions are required in today's SATCOM system concepts before robust (reliable and antijam) performance can be obtained from limited size shipboard and airborne terminals.

The Navy recognized the military value of satellite systems at an early juncture. Our interest in position location and control of dispersed platforms prompted developments of the highly successful Navy Navigational Satellite System (TRANSIT) Program to provide accurate and timely position location information. Similarly, our interests in communicating with these forces led to a series of satellite development efforts. From early studies of the satellite, it became clear that, conceptually, satellite systems could provide:

1. Reduced dependence on overseas facilities;
2. Reliable, flexible interconnection of widely separated users;

3. Distance-independent quality communications;
4. Improved low probability of intercept (LPI) capability, relative to HF systems; and
5. Acceptable jamming resistance.

To exploit these capabilities, the Fleet Satellite Communications (FLSATCOM) Program was initiated. The system was structured to provide effective broadcast capabilities to the Fleet along with a modest number of channels capable of handling secure voice, data and record traffic to and from the Fleet. In view of the substantial development time required, the COMSAT General Corporation (GAPSAT) Program was pressed forward to provide an early, interim capability for Fleet Broadcast and data channel service during the transition and testing phase, leading to full Fleet Satellite Communications service.

FLTSAT Lessons Learned

As originally conceived, FLTSAT had a rather modest, strictly military goal: to provide an antijam, dependable means of disseminating the Fleet Broadcast. The on-board SHF-UHF processor was incorporated to provide uplink jamming protection while the UHF downlink provided economy through the use of relatively simple shipboard receiver designs. The system was expected (albeit optimistically) to largely pay for itself by reducing demand for the large HF installations now providing this service.

Parametric analysis of satellite weight, ERP (Effective Radiated Power), and other features in the early design studies suggested that a few simple repeater channels could be added to provide unprotected communications supplements to existent long haul HF. Based upon experiences with extant UHF satellites of the time (Lincoln Laboratory experimental satellite LES-5, 6, Tactical Satellite [TACSAT]), where the difficulties of managing multiple users (particularly in power control, intermodulation products and Radio Frequency Interference) were recognized, narrow (25 KHz) channels were chosen for the design.

FLTSAT's problems began as the design concept passed through the multilayered, multiservice management structure. At each step, the largest alternative configuration was selected and additional features imposed. Protests by the original designers that uncertainties in the trade-offs made additions risky were overruled and, as long as there was any doubt that it was impossible to build the satellite, additional features were added.

These decisions caused a number of problems to surface. Central to these problems were the frequency plan and the consequent problems with extremely low level intermodulation products which appear at the receiver frequencies at a level much in excess of the thermal noise. While critics have insisted that a good frequency plan would cure FLTSAT's ills, the current frequency plan is probably as good as can be found — given the constraints. It is, of course, the constraints due to trying to use the heavily occupied UHF band and of trying to combine compatible functions that are at the root of the problem.

In facing these problems in the FLTSATCOM Program, the Navy has learned one strong and overriding lesson. If a tactical satellite system is to be obtained, it must be developed by a single manager, directly responsible for the performance of the entire system (including earth and space segments). The controlling element in the design must be recognized by the manager as the tactical role of the operational users and any suggested design changes must be made only if this primary role is not jeopardized.

Conventional wisdom has led to Military SATCOM System concepts approved largely on the basis of the space segment cost alone. This philosophy may provide many bits for a few large terminal users from a single common satellite (which makes sense for common carriers), but this concept is antithetical to a military system where the ability to get a single message through when it is needed is more important than sheer volume.

Integrated multifunction satellites are single-point targets for destruction, jamming or simply, failure. Positioning, repositioning, replenishment and orbit choices are severely constrained by common use. Most importantly, the cost-effectivness of large common use, tactical military satellites has not been proven by experience and economic trade-offs based on throughput under jamming that include the cost of the many small terminals of a Naval SATCOM System have not been completed.

The present FLTSAT Program is an integral part of the Navy's SATCOM plans and is urgently needed to provide reduced dependence on overseas facilities, robust broadcast services and reliable ship-to-shore information exchange systems. However, the time to plan the next generation of tactical satellites that will satisfy the full spectrum of military operational requirements has arrived.

Electronically survivable communications must be integral to this planning. Our philosophy and management procedures in conceiving and procuring military SATCOM systems must be consistent with this end.

New Generation of Tactical Satellite Communications Systems

The Navy is carefully appraising its approach to future tactical satellite system development. Several studies have been conducted in this appraisal which indicate that alternative approaches to tactical satellite system design are possible.

The alternatve approaches assume that long haul, point-to-point communications (including some connectivity to major afloat commanders) are provided by Defense Communication System (DCS) supplied satellites, tailored to that purpose, and that communications for the tactical commanders will be provided by satellites specifically designed to accommodate their unique needs.

The tactical satellite design approaches will, of course, permit the necessary interfaces with the DCS and our Allies through appropriate gateway terminals but will be sized to satisfy the antijamming communication needs of our combatants under realistic and stressing wartime scenarios. This design philosophy should permit smaller and less costly satellite system designs than would result from a design concept that mixes user classes in a single ecumenical design.

Considerable promise for meeting the Navy's communications needs seems to lie with the Military Satellite Communications System Office Architectures (MSOA) of the Defense Communications Agency General Purpose UHF (GP-UHF) satellite that is to be designed to share the communications burden of small, mobile terminals. The appellation of GP-UHF is unfortunate since what is described is, in fact, a multiband tactical satellite incorporating a sound mix of modern techniques thus potentially avoiding the pitfalls of packing too much into a single frequency band and spacecraft.

Current and projected terminal inventories and lifetimes should be considered and an orderly transition from the present FLTSATCOM System planned, perhaps to the extent of including a few FLTSAT/GAPSAT-like channels. Since some significant revision of the current narrowband terminal design is mandatory to attain even modest jam resistance, properties of these terminals alone should not be allowed to drive the next generation system design.

To obtain the most cost-effective system, the total system cost (satellites and earth terminals) and system performance must be carefully examined as an entity in order to apportion the complexity and cost of earth and space segments. The total cost of an SHF system consisting of many medium and large terminals, and four satellites sized to support them, may be completely dominated by the cost of the terminals.

Additions to satellite complexity invested in increased effective radiated power (ERP), onboard signal processing and/or elaborate multiple beam or adaptive nulling antenna systems may be rewarded by increased performance at lower overall cost if the terminal cost can be reduced. A UHF system, on the other hand, might be dominated by satellite cost, and improved performance might come cheaper through terminal improvements.

Promising high technology satellite design techniques (on-board signal processing, adaptive jammer nulling, true independent-multiple narrow-beams) must, of course, be considered in a unified approach to system design. However, constraining these investigations to only one frequency band is not advisable nor is any inordinate delay in the detailed design and development efforts for the next generation tactical satellite system.

Finally, if an objective cost-benefit analysis is not pre-judged by policy, significant system improvements might be realized by dividing the total satellite weight into a number of smaller satellites. Fixed weights for telemetry, antennas, altitude control systems, on-board signal processors, etc., will have to be duplicated with a corresponding loss in total ERP.

However, other militarily important benefits may accrue, such as reduced attractiveness to a one-on-one threat (if the jammer must divide his attack between a number of satellites, the total ERP may be more than recovered in throughput); improved coverage through a wider range of orbits (ship and aircraft terminals must face pointing and doppler correction problems even with geostationary orbits); graceful degradation, growth through additional numbers and so forth. If enough small satellites are substituted for current large, common-use designs, outages of individual satellites during eclipse could be tolerated, and the weight saved by reduced battery requirements could be invested to obtain more ERP than the single large satellite.

In the economically constrained environment of present defense budgets, all cost and risk mitigating design concepts must be pursued and exploited.

Conclusion

The time has come to recognize that in addition to the general requirement of all military Services for satellite support of point-to-point communications, a critical requirement exists for a tactical satellite system. To be useful in a military environment, this system must offer reduced risk in initiation, be economical in deployment and be robust in the face of jamming.

Studies and technology demonstrations have shown that satellite systems meeting tactical requirements are feasible. These tactical satellites could, in combination with point-to-point multiuser satellites, offer a balanced approach to the military exploitation of satellite capability.

Exploitation of satellite communications primarily as a replacement for point-to-point terrestrial communications circuits should be abandoned in favor of emphasis on the unique properties of satellite systems to satisfy requirements which cannot be met by terrestrial systems. Planning and development must be concentrated on the delivery of flexible, reliable and antijam systems to provide mobile forces crisis and contingency access and allow interoperability with the point-to-point system.

Solve the philosophical (tactical rather than ecumenical satellites) and managerial problems, and the technical problems will be solved in a straightforward manner. The Navy's tridimensional force structure (surface, subsurface and air) and our Marines' Assault Landing Force requirements encompass and embody the totality of the tactical commanders' problems and communications needs. A tactical satellite system design that satisfies these military requirements is urgently needed by our combined military forces.

The time for action to develop the next generation of satellites is now. We must bring Military Satellite Communications to a new course that points directly toward reliable, antijam tactical communications. It could well be that signals warfare, of which antijam communications is a major element, in the future scheme of things can mean the difference between solid connectivity and reliable command and control, and failure to control, ergo, defeat.

We must assess all available techniques, pick the combination most likely to provide maximum return without incorporating the ultimate in realizable complexity and get underway as soon as possible. Only such positive action will provide the robust communications that are mandatory to assure that our Fleet will continue to serve as the U.S. first response in crises.

Operational Considerations for Tactical Satellite Communications Systems

John M. Kuhn

Seven years ago a new era in military tactical communications was ushered in with the launches of the Lincoln Laboratory's Lincoln experimental satellite (LES-6) and the Hughes Tactical Satellite (TACSAT-1) from Cape Kennedy. These experimental communications satellites were to serve as test beds to prove out new technologies and concepts in communications via geostationary orbiting space platforms.

In mid-1970, having completed their experimental mission, the U.S. Air Force released the LES-6 and TACSAT-1 ultra high frequency (UHF) communications transponders to the authority of the Joint Chiefs of Staff (JCS) for use as a limited capacity operational satellite communications (SATCOM) system serving the needs of a select group of high priority military users.

Thus, the world's first military command and control capability utilizing tactical UHF satellites was born. Although admittedly limited in its user

---

**This article originally appeared in the March 1976 issue of SIGNAL.*

population, the Tactical Satellite Communications Interim Operational Capability (TACSATCOM IOC) has been exceptionally beneficial in gaining some firsthand insight on inherent advantages and problems associated with an operational tactical satellite communications system.

At Headquarters, U.S. Air Force level, the Executive Manager of the TACSATCOM IOC delegated the responsibility for operational control of this system in its day-to-day employment to a major command level, the Air Force Communications Service (AFCS). The AFCS Tactical Relay and Operational Center (TROC) was activated at Kirtland AFB, New Mexico, with three mission functions:

1. To be the scheduling agency in providing satellite access times to the users in accordance with JCS approved access priorities;

2. To be a total system monitoring activity, identifying system anomalies and aiding any user as necessary in establishing and maintaining communications when difficulties should be encountered; and

3. To act as a relay for operational traffic between users who were not under the "footprint" of a common satellite (the TROC was specifically located so that it was in the mutual coverage area of both satellites).

Although the relay mission of the TROC ceased in December 1972 with the failure of the TACSAT 1 spacecraft (due to spin up because of despin motor failure), the efforts of this organization were still required for the first two functions. The most useful information gathered on methods and procedures to improve operational performance of tactical SATCOM were (and still are) derived from accomplishing the second function—monitoring and assistance.

The efficacy of tactical SATCOM has been demonstrated in the TACSATCOM IOC by its employment for communications in Project APOLLO, Presidential communications support, in numerous military exercises (both U.S. and NATO) and in Middle East crises. It is also used daily by airborne command posts, ground command posts, Naval capital ships and ground transportable terminals.

Satellite Communications has historically been viewed as the "microwave tower in the sky," negating the need for expensive multiple hop terrestrial links for point-to-point communications over long distances. In the tactical arena, this characteristic is enhanced by the fact that most users would no longer be tethered to a line-of-sight relay point for effective high reliability, high availability communications.

When communication beyond line-of-sight is required, communications by satellite relay is now preferred to the use of high frequency (HF) systems with their inherently lower availability and reliability due to propagation fades and skip zones. The fact that communications by satellite is insensitive to earth terminal locations (assuming, of course, the earth terminal can see the satellite) allows the military commander increased latitude in force deployment while still maintaining effective command and control and increases the spectrum of dynamic response options available to counter a given threat or condition.

This positive improvement from a geographical viewpoint in exercising command and control of forces and weapons from launch to target and dynamic target reassignment options—based on having a dependable and reliable command, control and communication (C^3) system—continues to exert a profound influence on military planning and strategy for the future. However, it is important to recognize that military strategists will commit themselves to satellite systems as the prime means of C^3 only on the basis of their demonstrated capabilities in a broad range of operational environments.

This commitment must be consistent with the satellite system's ability to rapidly respond to changing situations with an availability and reliability commensurate with mission requirements. The TACSATCOM IOC has provided useful experience in this area of tactical SATCOM and the means to refine employment procedures.

Encountering Difficulties

Despite the fact that the number of TACSATCOM users has been limited due to available satellite capacity, certain difficulties have been encountered in providing responsive service even to this limited user population. To fully appreciate the situation, several factors which influence the employment of SATCOM, and TACSATCOM in particular, require examination.

It is important to recognize that all system users, within the geographical limits of approximately one-third of earth's surface, share a single common communications point—the satellite. Therefore, a single user's conformance or non-conformance to technical operating parameters could have a profound effect upon the entire user population. In multigrid terrestrial networks, this is not necessarily the case. If a single link in a terrestrial network is disturbed, normally only a small portion of the total grid user population is affected, usually those at the end-points of the disturbed link and those passing through the link for which no alternate routing is possible.

Second, assuming sufficient power is used, the satellite transponder "sees" the radiation of any source in its frequency band emanating in its direction, if that location is visible to the satellite's antenna pattern. Thus, the satellite transponder is susceptible to either intentional or unintentional interference from a large number of possible sources. Their location cannot normally be pinpointed without employing highly sophisticated techniques (such as satellite antenna spot beam steering).

In addition, the intentional jamming of the SATCOM system by a hostile force, either within the theater of operations or from a sanctuary area, can be easily accomplished. Employment of highly sophisticated techniques, usually involving additional high system costs, are required to counter this threat. The SATCOM system's susceptibilities to interference/jamming must be properly evaluated in the system design against the criteria of mission accomplishment versus the cost of countering these threats.

Since countermeasures to these can account for a substantial portion of the total system cost, it is incumbent upon the potential users to scrub their C^3 support requirements vigorously. They must specify prior to proceeding with the system definition as accurately and with as much detail as possible, the minimum essential communications support required for mission success in the anticipated operational environment. In addition to these general SAT-

COM considerations, there are several specific TACSATCOM IOC characteristics which influenced its responsiveness and utility.

Evaluation

The TACSATCOM IOC assets, as indicated earlier, were designed primarily as test beds for evaluation of technology and techniques. Therefore, additional cost and complexity were not added to the satellites for use in countering the above mentioned susceptibilities in an operational environment, so as not to jeopardize approval and accomplishment of that system's primary research and development (R&D) mission. Therefore, as was possible and expected, when inadvertent jamming did occur, the only permissible response was adjust and decrease user traffic to account for the effects of these parasitic signals.

When such interference conditions did exist, the TROC, equipped with only fundamental test equipment, attempted to analyze the nature of the interference in the hope of identifying its source. Many times the interference vanished long before arduous manual analysis could establish any significant identifying features.

Thus many short-term interferences were necessarily tolerated since a response to the situation could not be implemented in a timely fashion. Long-term interference required the use of special signal analysis equipment and days of analysis by special teams to identify the source of interference prior to initiating corrective action.

Further, the large majority of user terminal equipments were research and development assets and had a capability to pass mission traffic only, either voice or record, in a simplex mode. Thus, if a terminal was operating out of tolerance (usually too high uplink power), there was normally no means for the TROC to communicate with the offending station other than through command channels via terrestrial systems.

Delays in complementing communication corrective procedures following this course of action have been as much as eight hours. Further, since there was no effective way for the TROC to communicate with the entire user population through the system, the user access schedule was developed days in advance based on projected needs. This effectively negated rapid system response to dynamic change in user requirements based on the scenario of a developing situation.

All of these shortcomings taken together rendered TACSATCOM I unattractive as a prime means of C^3 for military forces. Of course, these particular assets were not intended to be that in the first place.

The major lesson of TACSATCOM IOC is that future systems—such as Air Force Satellite Communications (AFSATCOM) System and its follow-on systems, which are intended to be prime means of C^3 for large numbers of military users and weapons systems—must, in addition to being that communications system, provide for the inherent ability to control the system; i.e., identify the electromagnetic situation; evaluate its effects on the system; determine a corrective or counter response to it; disseminate response instructions to the users or the satellite or both; and confirm that the desired result has been achieved—all in a time frame which can assure rapid, accurate system responsiveness to user needs.

Summary

We have only seen the iceberg's tip in exploiting the advantages of satellite communications in meeting the challenge of our future military C^3 requirements. Much has already been learned and much more remains to be learned on the operational aspects of satellite communications in the tactical environment.

The system design questions arising from the need to have meaningful system control are not easily resolved or solvable. They require early and close cooperation between users, developers and system managers to identify and quantify the issues and then set upon a course of action that will yield a Tactical SATCOM system design that fulfills mission requirements and yet is cost-effective.

Tactical Satellite Communications

Gerald L. Fjetland and Edward L. Whittenberg

The night seemed especially dark as the flight of F-15 Eagles streaked eastward, 1,200 miles off the coast of Virginia. The flight was about 300 miles from an unidentified aircraft that could be an Air Launched Cruise Missile (ALCM) carrier. At least that was the projected position 15 minutes ago when the flight passed beyond ultra high frequency (UHF) radio range from the mainland. The intercept must be made before the "bogie" gets within cruise missile range of the mainland, otherwise each missile would have to be located and intercepted.

Suddenly the vector indicator on the "heads up display" lights up. "Turn left to 85 degrees—35,000 feet."

Word from Cheyenne Mountain—the "bogie" has maneuvered. The North American Air Defense Command (NORAD) is able to continuously track the "bogie" through the use of a space based sensor system and is able to update the interceptors via a satellite communications system.

Advanced Technology

Does all this sound a little far-fetched? A taste of Star Wars or Buck Rogers? It should not. The technology is here today. Space based sensors exist, and communications satellites capable of operating with fighter aircraft have been operational for several years. Today, these capabilities exist in bits and pieces, but, by 1990, we can have an operational tactical satellite communications system capable of providing effective command and control of our forces anywhere on the earth's surface.

**This article originally appeared in the March 1981 issue of SIGNAL.*

Last September, at a command, control and communications (C^3) symposium sponsored jointly by Rome Air Development Center and the State University of New York, Dennis Leedom of the Tactical Systems Division, Assistant Chief of Staff for Studies and Analysis, Headquarters U.S. Air Force (USAF), made three observations:

1. "We have all of the elements of C^3 lying out on many different tables."
2. "We are just beginning to understand how the elements fit together."
3. "We ought to be moving the tables closer together."

A major objective of this paper is to outline an approach to moving some of those tables together.

MILSATCOM Architecture

The current Military Satellite Communications (MILSATCOM) architecture, as defined by the Defense Communications Agency/MILSATCOM Systems Office (DCA/MSO), consists of three segments based upon the unique requirements of three communities of users. The first segment serves the nuclear capable forces, who need the highest degree of physical and electromagnetic survivability, worldwide coverage and high availability but do not need high data rates. The second segment serves the wideband trunking community.

These are medium to high data rate users, including the Defense Communications Service (DCS), some elements of the Ground Mobile Forces (GMF) and special users such as the Diplomatic Telecommunications Service (DTS). The final segment of the DCA/MSO architecture serves the Tactical/Mobile Community. These users require low to medium data rates, typically from small, mobile or transportable terminals. These users include Navy ships, the GMF and Air Force tactical networks.

The current system serving the nuclear capable forces is called the Air Force Satellite Communications (AFSATCOM) System. The AFSATCOM Space Segment consists of dedicated Air Force UHF channels on the Fleet Satellite Communications (FLTSATCOM) Space Segment and transponders aboard other host satellites. AFSATCOM reached interim operational capability (IOC) in 1979, and terminals are being installed in Strategic Air Command (SAC) B-52s, FB-111s, Minuteman and Titan launch control centers, airborne and ground command posts and the Navy's Take Charge and Move Out (TACAMO) aircraft which provide connectivity to the ballistic missile submarines.

DSCS

The present system for the wideband trunking network is the Defense Satellite Communications System, Phase II (DSCS-II). DSCS-II became operational in 1973 and continues to operate, providing super high frequency (SHF) SATCOM service to both the Department of Defense (DOD) and the State Department.

The Fleet Satellite Communications (FLTSATCOM) system presently serves the tactical mobile community. Operational since 1979, FLTSATCOM provides one fleet broadcast channel and nine fleet relay channels, as well as one wideband and 12 narrowband channels for the Air Force. (These Air Force channels are part of the AFSATCOM system discussed previously). Except for the Fleet Broadcast uplink, which is at X-band (8 GHz) all links are in the 225-400 MHz UHF band. Either processing or non-processing transponders may be used with the fleet broadcast and some Air Force narrowband uplinks. Use of processing transponders provides some jam resistance.

Current DOD planning includes a near term follow-on system for each segment of the presently deployed architecture. The DCA/MSO architecture includes a Strategic Satellite System (SSS) to follow AFSATCOM, a DSCS-III to follow DSCS-II and a leased system referred to as LEASAT, to follow the FLTSATCOM system.

The Strategic Satellite System was originally conceived to have a space segment composed of Single Channel Transponders (SCT) on DSCS-III and other spacecraft, as well as a dedicated spacecraft called STRATSAT in supersynchronous orbits for physical survivability.

However, the Congress has deleted funding for the STRATSAT, so the space segment will consist entirely of SCTs. The SSS will provide improved jam resistance over the existing AFSATCOM by using advanced modulation techniques in combination with SHF links to and from command posts and missile launch control centers. The downlinks to bombers and uplinks from bombers will still be at UHF, but the advanced modulation will offer some improvement in jam resistance. For the near-term, it appears that the physical survivability questions will have to wait.

The Defense Satellite Communications System, Phase III (DSCS-III) will offer significant improvement over DSCS-II, including increased channelization, variable beam antennas for improved jam resistance, a jam resistant command subsystem and multiple narrow antenna beams which will enable communications with smaller terminals. DSCS-III will serve an expanded user population including the Defense Communications System (DCS), Ground Mobile Forces (GMF), Diplomatic Telecommunication Service (DTS), Rapid Deployment Force (RDF) and other special military users.

LEASAT

In 1977, Congress directed that the follow-on system to FLTSATCOM could be a leased system. The resulting system is known as LEASAT. LEASAT will serve primarily the same types of users as FLTSATCOM, i.e., the Navy, plus some Air Force and Ground Mobile Forces (GMF) users. LEASAT will have four types of communications channels with characteristics similar to the FLTSATCOM channels. LEASAT will have less capacity than FLTSATCOM, and since it will operate at UHF, it is no more jam resistant. Perhaps the most pressing problem is that it does not address the total set of tactical users.

For several years, the Services have been doing planning for a future tactical satellite communications system, presently known as TACSATCOM II. However, there has been a lack of coherent planning as each Service concentrates on its own needs. Also, very little consideration has been given to ensuring that DOD presents a saleable approach that Congress would support.

One of the problems in convincing the Congress to fund satellite communications systems is that the DOD has not adequately explained the role of satellite commmunications in the context of an overall DOD communications, command, control and intelligence (C^3I) capability.

Another problem is that since the Services have been focusing on their own separate pieces of the problem, the DOD has not adequately addressed the C^3I needs of joint forces. These include users such as crisis/contingency forces and special operations. Some of these users are highly mobile and will need to communicate using small, low power terminals.

If we are to have a TACSATCOM-II system in the 1990s that is both responsive to the needs of the DOD and acceptable to the Congress, we must avoid some of our past mistakes.

We should avoid Service unique transponders. These proliferate the need for unique control systems and complicate the interoperability situation. Instead, we should strive for as much functional commonality as possible. As we transition to extremely high frequency (EHF), we must stress commonality of frequencies, modulation, coding, encryption, frequency hopping schemes, demand access techniques and satellite on-board processing.

By achieving this commonality, the concept of a truly joint mission satellite becomes realizable. Mixed user group satellites are a must to effectively connote to Congress the concept of a unified DOD MILSATCOM architecture that is focused on cooperative exploitation of space.

The concept of joint mission satellites could be carried still further. We could evolve eventually to a single generic type of communications satellite. A common, Shuttle compatible spacecraft with joint payload electronics is practical and technically feasible. Regardless of orbital schemes required to achieve global coverage, a single spacecraft design is practical.

We must not unnecessarily constrain the implementation of EHF technology. The Space Shuttle will be operational before EHF technology can be deployed. For instance, a 15 to 20 meter multiple beam antenna that is constructed in sections at Shuttle altitude is technically feasible. Once the antenna has been put together and tested, the spacecraft can be boosted slowly to its proper orbital location to minimize structural stress.

The TACSATCOM II user community will be very diverse, including users that are presently served on all three segments of the architecture. TACSATCOM II should serve, in addition to the present users of FLTSATCOM and AFSATCOM, some elements of the Defense Communications System and Diplomatic Telecommunications Service, as well as the Rapid Deployment force, Ground Mobile Forces and special military users.

The Force Multiplier Effect

When we consider the important characteristics of command and control systems, six of these are a natural for modern satellite communications systems. The key word is modern—forget about the huge installations with 60 foot dish antennas peering into the sky. These six characteristics are availability, mobility, flexibility, timeliness, jam resistance, and survivability.

Today's MILSATCOM systems have orbital parameters that ensure communications availability anywhere in the world. The use of higher frequencies leads to smaller, more mobile earth terminals and thus increased operational flexibility. The combination of worldwide coverage and highly mobile earth terminals allows the timely establishment of communications with any "hot spot." The evolution of satellite communications to extremely high frequencies (EHF) allows the use of techniques to negate the effectiveness of enemy jamming attempts.

Finally, while many would state that satellites are highly vulnerable to direct physical attack, the location of communications satellites more than 22,000 miles out in space provides a measure of survivability not found in terrestrial communications systems. The combination of these features leads to communications capability that can assure effective command and control of our military forces any time, anywhere.

Japanese Gen. Yamashita was reported to have replied to a question regarding Americans as jungle fighters, "They are not jungle fighters—first, they shoot away the jungle, and then they fight." There is a measure of truth in that statement. In the past four decades, Americans have enjoyed the ability to use a massive superiority in fire power to achieve military objectives. Americans have been able to trade ordinance for lives on the battlefield. From carpet bombing by B-17s in Germany, to air strikes on sniper positions in Vietnam, the philosophy was to heavily strike an area occupied by the enemy even when specific targets could not be identified.

Recent developments may have dictated a change in that philosophy. Reconnaissance by fire is rapidly being replaced by numerous sophisticated sensor systems that locate and identify targets on the battlefield, day or night, in any kind of weather. Suddenly, the battlefield is alive with targets. At the same time, the historical logistic superiority no longer exists. The situation has changed from one of an overabundance of ammunition and few identified targets to one of an overabundance of targets and a minimum of ammunition.

NATO Outnumbered

In Europe, the North Atlantic Treaty Organization (NATO) forces are outnumbered by well equipped Warsaw Pact forces. The NATO forces have the added disadvantage of allowing the enemy to take the initiative. While this situation is serious, it is not helpless. History is replete with cases where outnumbered and outgunned forces have prevailed. These were cases where the limited assets were utilized in a more effective manner. For example, Rommel, outnumbered and equipped with inferior tanks, took the starch out of the British in North Africa until he was ordered to hold an untenable position. Rommel's successes were a direct result of superior intelligence gathering and highly effective command and control of his forces. Communication is the key element in the exploitation of intelligence information and the effective command and control of military forces.

Information is only useful if it is in the hands of the decision maker. Information that is extraneous to the decision being taken will only delay the decision and can be viewed as noise in the system. Command and control is a three step process:

1. Locate targets with sufficient accuracy to fall within engagement envelope of friendly forces;
2. Provide target information to the decision maker in a usable form; and
3. Provide the action decision (orders) to friendly forces before targets have left their engagement envelope.

This process requires effective communications between the sensor and a data processing facility; between the data processing facility and the decision maker (command element); and between the decision maker and his forces. Whether we are discussing Caesar's legions, the Afrika Corps, or today's sophisticated weapons systems, there is an absolute dependence upon effective communications in order to make the system work.

As weapons systems become more sophisticated, the rapidity of the communications becomes more important. As warfare becomes more mobile, dependence upon fixed or cumbersome assets must be reduced. Units such as the Rapid Deployment Force cannot depend upon prepositioned communications for command and control. With satellite communications, effective command and control of forward elements can be accomplished by a stateside battle staff until forward command elements can be established.

Satellite communications also allow for the location of data reduction and correlation centers at great distances from the sensor systems and command elements they support. Not only does the remote location increase the survivability of the center by removing it from the enemy's engagement envelope but it also increases the mobility of combatant elements by freeing them from the task of relocating a cumbersome data center.

Communications with long-range aircraft and naval units at sea can be significantly improved by means of satellite communications. Satellites are the only means of providing highly reliable, jam resistant, long-range communications over the world's oceans. Satellite communications among the elements of a naval battle group can prove to be decisive in the air defense of the unit.

A modern satellite communications system is also an effective means of maintaining contact with special military operations. The combination of small antenna size and narrow beam pattern achievable at the EHF frequency band allows communications with low probability of detection by enemy forces.

TACSATCOM II

One of the most pressing questions in the MILSATCOM community is, "What sort of a system will replace LEASAT when the lease expires in 1988?" The final answer to that question is by no means certain, but the most likely possibility is a system known today as TACSATCOM II. The next few paragraphs will describe the major characteristics of such a system.

An important communications need of the tactical forces is the need for jam resistance at EHF. Taking advantage of the wide bandwidths available for spectrum spreading, and using on-board processing, a well designed system can withstand a determined enemy jamming strategy. Thus, TACSAT-COM II will use EHF. UHF and SHF will also be included for backward compatibility with existing terminals.

Besides its potential for jam resistance, EHF has other advantages. The narrow antenna beamwidths which are obtained even with small antennas help to make an EHF system more covert. This covertness will allow communications to take place with platforms which previously had to maintain radio silence in order not to compromise their location.

Another advantage afforded at EHF is that it is possible with reasonably sized spacecraft antenna array to automatically steer antenna nulls in the direction of enemy jammers, thus reducing their effectiveness still further. To do this at UHF or even SHF would require large impractical antenna arrays in space. In order to make full use of the potential advantages at EHF, full on-board processing is needed.

This means that the TACSATCOM satellite will despread, demodulate and decode the signal then encode, modulate and spread the signal before retransmission. This has the effect of removing the jammer signals, allowing the satellite to devote all of its downlink power to retransmitting the desired signal instead of retransmitting an amplified version of the jammer signal to the user terminals.

The tactical user community must have the ability to communicate worldwide. In current MILSATCOM systems, if two terminals need to communicate and there is not a single satellite within their mutual field of view, the signal must be relayed by one or more other terminals. TACSATCOM II will eliminate the need for multiple hop relays of this nature by having satellite-to-satellite crosslinks. In this way, any terminal can commmunicate with any other terminal without going through an intervening terminal.

All of this sounds like a grand scheme, but it will not happen unless the DOD can convince the Congress to provide the resources with which to do it. If we hope to sell the Congress on this system, we must demonstrate that the DOD finally has its act together in the MILSATCOM arena. Thus it is imperative that the TACSATCOM II space segment be an integrated payload that all users share on a demand access basis. This approach has several implications.

First, the operational control of the satellite cannot be given to one or another of the Services. The operational control of the system should be integrated into the command structure that will be used during periods of conflict.

Second, the system must be both forward and backward compatible. The DOD presently has a large investment in UHF and SHF MILSATCOM terminals and is continuing to deploy such equipment. In the early years of deployment of a TACSATCOM II system, there will be only a few EHF terminals, and many UHF and SHF terminals. Therefore, the satellite, which might have at least a 10 year on-orbit lifetime, must be capable of evolving with the terminal population. A reasonable approach to satisfying this requirement is to build the satellite with large enough numbers of channels in all three frequency bands to serve the highest expected traffic requirements in that band. Then through power management, only needed channels are turned on at any give time. For instance, in the early years we would probably have all of the UHF and SHF channels turned on, and only a few EHF channels.

By the end of the satellite lifetime, we would have turned on more EHF channels to accommodate new users at that frequency and turned off some of the UHF and SHF channels as the terminal populations decreased in those frequency bands. This approach to the space segment allows us to grow from today's systems to the full capability we are striving for in bite size steps. That is, the movement to EHF will be evolutionary rather than revolutionary.

The final implication of this approach is that it should lead to a different thought process by the Services. The time has come to start treating satellite communications just like microwave links, high frequency (HF) and troop systems. In other words, MILSATCOM is simply another transmission medium in an overall C^3I capability, and should no longer be treated as something special.

We have described a system which will enable communications capabilities which were unheard of before now. These capabilities may sound like pie-in-the-sky wishful thinking. They are not. All of the major technologies required to implement such a system have been demonstrated or are currently under development.

New Technologies

The Lincoln Experimental Satellites 8 and 9 (LES 8/9) were used to demonstrate several new technologies, such as EHF uplinks and downlinks, satellite-to-satellite cross-links and full on-board signal processing. In 1978, satellite communications with fighter aircraft (RF-4C) was demonstrated, using an AFSATCOM terminal designed for a B-52.

With the advances in microprocessor technology since that terminal was designed, a terminal with even better capability could be designed which would still fit into the constraints of fighter aircraft. The ability to accurately point a directional antenna on an aircraft at a satellite has been demonstrated, and new low profile EHF array antennas are under development which will enable the use of this frequency band on high performance aircraft such as fighters.

Thus tactical satellite communications need no longer be limited to the UHF frequencies which have been used in the past. Power amplifiers for both satellite and terminal applications have also been demonstrated at these frequencies. For satellite-to-satellite crosslinks and aircraft-to-satellite links, the technology has been demonstrated to communicate at data rates of one gigabit per second using laser communications.

Finally, the means is at hand to construct large antennas on orbit using the Space Shuttle. Large multiple beam antennas on the spacecraft will enable communications with disadvantaged (small antenna, low power) terminals. In summary, all of the key pieces of technology have been demonstrated; what remains is to integrate them into a system.

We have before us a prime opportunity to convince the Congress that the Services can follow an integrated approach and be a team. If we do this, the results can be an enormous improvement in our overall C^3I capabilities.

Author's Note

Since the above article was published, much of what was postulated has come to pass. The Army, Navy and Air Force established a joint program for tactical satellite communications. Each of the services is developing terminals to work with the single integrated satellite payload. The success of this program is indicative of what can be accomplished by dedicated people working together as a team.

ECM and Satellite Communication Network Control

Marc I. Spellman

It can be effectively argued that the advent of the communications satellite has been the single most significant advancement in radio frequency (RF) communication systems in the last 30 years. The combination of communications capacity, availability and spatial coverage per dollar of invested capital is unachievable in any other network architecture.

The design features of the satellite system which provide these desirable characteristics, however, are the same ones that result in the requirement of a level of control not necessary in other communication network architectures. The single wideband repeater accessible by all network members requires that the transmission parameters of each and every network member be adequately controlled.

The reason for this is that a single uncontrolled network participant can conceptually be the cause for deleterious effects in terms of the capacity and availability of the communications of every other network participant.

In a military satellite communications system, both the necessity and complexity of the control function are compounded. This results from the fact that any well designed military communications system must be capable of addressing the threat of electronic countermeasures (ECM).

The design of the control system is impacted in several ways if it is to perform adequately in an ECM environment as well as in a benign environment. First, and most obvious is that the control links must be provided with antijam protection at least equivalent to the communication links. Second, since the communication links are provided with an antijam mode of opera-

---

This article originally appeared in the April 1984 issue of* ***SIGNAL.*

tion, which by its very nature will have different properties and characteristics, two modes of operation must now be controlled. Furthermore, a timely mechanism for transitioning between the two modes must be provided.

In this article, the satellite network control system is examined first. For both a non-military and military satellite communications network, the control system rationale and requirements are presented. The control system is then subdivided into its component subsystems and qualitative requirements are provided for the subsystems.

Following this, the strategies that are available to both the jammer and the communicator are discussed with respect to the operation and design of the control system.

To put all this into perspective, the article concludes with a typical flow of events that could successfully transition the network in response to a changing ECM environment.

Control System Rationale and Requirements

Briefly stated, the fundamental purpose for a satellite netowrk control system is to allocate satellite resources in some optimum manner. The satellite resources to be allocated are the downlink effective isotopic radiated power (EIRP) and bandwidth, and the allocations must be made to each satellite communications (SATCOM) terminal participating in the network. The allocations take the form of designations on the uplink signal parameters for each terminal.

From the network control point of view, an optimum allocation is one where the capacity available for reliable information transfer through the satellite is maximized for each user. Given that there is an upper system capacity based on the satellite transponder and earth terminal designs, it is the function of the network control system to determine allocations that result in a total operating capacity that approaches the maximum system capacity available from the equipment while maintaining an acceptable level of link availability. To do this, it must minimize wasted capacity and it must assure that the transmissions from network participants do not interfere and thus reduce system capacity.

How the control system both reduces wasted capacity and assures that network participants do not interfere with each other depends on the method by which the satellite resources are shared between the user SATCOM terminals. The method of sharing, which is generally called the multiple access technique, determines which parameters of the uplink transmissions must be controlled and, based on the particulars of the specific network, how accurately they must be controlled.

There are three principal multiple access techniques: Frequency Division Multiple Access (FDMA), Time Division Multiple Access (TDMA) and Code Division Multiple Access (CDMA). There is no intent in this article to compare the virtues or shortcomings of each, but since the control system requirements are related to the multiple access technique, a brief consideration of the three approaches is warranted.

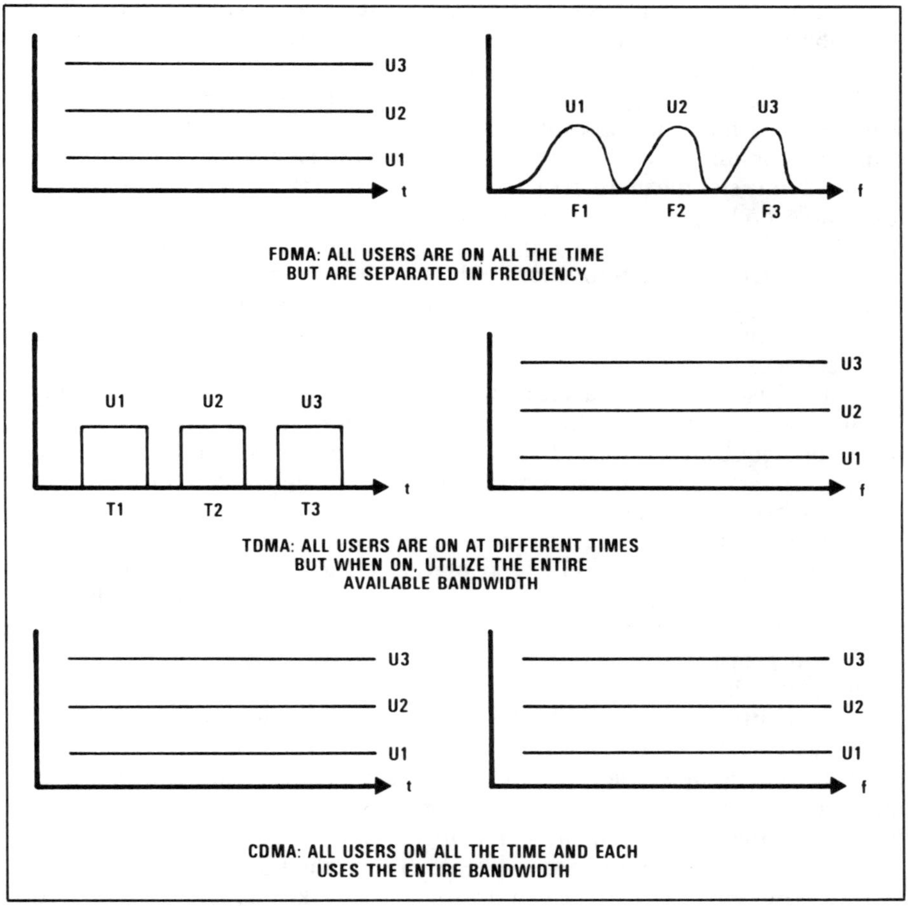

Figure 1. Multiplle Access Techniques

FDMA assigns a portion of the transponder bandwidth to each network participant. The channel spacings on the uplinks are determined by the instantaneous bandwidth of the uplink transmissions and the required frequency guard bands, as determined by the filtering characteristics in the earth terminals. From a control point of view, it is obvious that uplink frequency assignments must be adhered to in order to avoid interference between users.

In addition, since in an FDMA system all users are simultaneously present in the transponder, the uplink power for each user must be carefully controlled. This is necessary so that both potential intermodulation interference and power lost to intermods (due to non-linearities in the satellite transponder) are avoided, and that potential user capacity is not lost to excessive link margins.

Time Division Multiple Access systems assign a time slot in a master time frame to each earth terminal participating in the network. In the assigned time slot, the earth terminal has exclusive use of the entire transponder bandwidth and power. The earth terminal operates in a burst mode with a symbol rate consistent with the transponder bandwidth.

With respect to the control requirements, the time of transmission for each earth station must be carefully controlled since transmission overlapping in time will obliterate each other and transmissions not fully occupying an assigned time slot mean wasted capacity. However, the power control problem of FDMA no longer exists since only one user is present in the transponder at any time. The power control requirement is reduced to ensure that a constant minimum EIRP, enough to saturate the transponder, is always exceeded. The value is independent of the transmitting terminal characteristics, the information rate being transmitted from the terminal and the number of terminals in the network.

Code division multiple access systems separate user signals not in time or frequency but with an additional signal modulation. An orthogonal PN code is phase or frequency modulated on each of the user uplinks, allowing a receiver with a matched replica code to receive the desired signal without interference from the other transmissions in the network.

CDMA has the characteristic of multiple simultaneous users in the transponder, as does FDMA, so excess margin on any link equates to reduced capacity for other users. The power control problem with respect to intermodulation interference, however, is not as severe because of the interference reduction properties of the signal structure. Like TDMA, there is a timing accuracy requirement; however, in CDMA it is on a link by link basis (receiver and transmitter for each link), rather than on a networkwide basis.

Of the signal parameters mentioned above, the control of frequency and time are rather mundane. That is not to say either is an easy engineering problem, but it is the control and selection of earth terminal EIRP and bandwidth that is really at the heart of the network control function. These earth terminal parameters specifically determine the satellite allocations.

Based on the above, one can formulate two key types of quantitative requirements that can be placed on a satellite network control system. The first is to maintain no greater than some stated link margin on every link in the network. This is a legitimate requirement since, while link margin is fine for the particular link (because it assures a greater availability of the link), it is not fine for other network members or potential members.

Excessive link margin reduces the capacity available to all other users and potential users.

The second requirement is a maximum acceptable response time in which the control system must correct a deviation from the designated link margin on any link in the network. This requirement would seem naturally to be a function of both the polarity and magnitude of the deviation as well as on the particular link experiencing the deviation. For example, a negative deviation from a prescribed link margin means a statistically less available link and, therefore, a short response time would appear appropriate.

On the other hand, a positive deviation from link margin, which means a loss of potential network capacity, might require a response time inversely related to the amount of lost capacity. Thus for a given deviation (in dBs), longer response times might be acceptable for lower capacity links than for higher capacity links. Similarly, for a link of given capacity, the greater the deviation the shorter the required response time.

This fundamental issue of link availability versus network capacity takes on an entirely new dimension in a military satellite communications network. In this article, a military satellite communications network will be distinguished from a non-military network solely by its potential to be a target for electronic countermeasures. As a target for electronic countermeasures (ECM), the well designed military communications system must include, for each of its links, an antijam or electronic counter countermeasures (ECCM) mode of operation.

But such a mode, by its very nature, is a reduced capacity mode since it either uses power or bandwidth or both, far in excess of what is necessary for information transfer in order to defeat the potential jammer. Therefore, the trade-off between link availability and system capacity is greatly magnified. If one initiates an antijam mode of operation too early, there is a significant loss of capacity. If one initiates an antijam mode of operation too late, there is no link availability.

This substantiates the need for a timely mechanism for transitioning between modes. Complicating the situation further is the fact that with a jammer attacking the satellite transponder there is a definite question as to the capacity available to the network. Allocations must then be made in an environment far different than in the non-military network situation, which is characterized by a much more accurate data base.

Subsystem Definition

The control function can be divided into three separate subfunctions, each defining a related subsystem to perform an appropriate task. The three subfunctions are monitoring, allocation and communications and together they constitute the control process. Monitoring addresses the analysis of the downlink from the satellite. It is the information gathering process on which to base allocation decisions regarding the network participants.

The allocation subfunction determines the key uplink signal parameters of EIRP, bit rate and, if applicable, an antijam or clear operating mode for each of the SATCOM terminals in the network. The communication subfunction has the responsibility of communicating the allocation decisions made by the control system to each of the SATCOM terminals.

In attempting to develop some generic requirements for each of the three subsystems, the initial considerations go to apportioning the control system response time and maximum allowable link margin requirements to each of the subsystems. With respect to response time, a portion of the total budget must be ascribed to each of the three subsystems.

A finite time is necessary to process the received satellite downlink and determine the appropriate received signal characteristics. Similarly, time is necessary to complete the allocation process which is a calculation intensive task based upon a stored data base and the monitoring results. Finally, the communication subfunction needs a time allocation that must consider the limited data rate likely for control communications as well as the need to communicate possibly a different message to each of the earth terminals in the network.

The requirements for a maximum allowable link margin on each link in the network will impact the monitoring subsystem and the allocation subsystem. The monitoring subsystem is impacted because it has the requirement for gathering the data on which an estimate of existing link margin will be made. The allocation subsystem is impacted because it must include the calculation capability that makes the link margin estimate, as well as relate the required change in earth terminal uplink parameters, to the desired change in link margin.

More can be said about the requirements for each of the subsystems in the case of both a non-military and a military satellite network control system. For the monitoring subsystem, the basic requirement is the accuracy to which downlink signal power can be measured.

With respect to the measurement of downlink power, an output from the monitoring subsystem that can support a decision on whether or not the satellite transponder is saturated is an important monitoring subsystem requirement. When, as in the case of a military system, the potential for jamming is introduced, an immediate additional requirement should be placed on the monitoring subsystem; that is to support jammer identification and characterization.

The first step in that process is to be able to distinguish the effects of an equipment failure from the effects of a jammer. Once that can be accomplished, additional jammer characterization requirements, such as determining if the jammer is time varying or stationary in time, if the jammer is an uplink jammer or a downlink jammer and what frequency channels the jammer is attacking, are all legitimate requirements.

The allocation subsystem is also significantly more complex for the military network than for the non-military network. First, any time a jammer is introduced, the potential for a continually time varying network status is a real situation. The allocation subsystem must be able to react quickly and, even more important, must know when not to react at all. In addition, the jammer introduces a degree of uncertainty as to the available capacity left for the network, thus further complicating the allocation process.

Finally, with respect to the communications subsystem, the military SATCOM control system must provide a degree of antijam protection for the control link at least equivalent to that of the communications links that must be controlled. This will ensure a control capability at all times when the communications links are operable.

Strategies for Communicators and Jammers

The strategies for the communicator in a military satellite network center first around the alternatives available from which the network architecture can be selected and second around the various possible responses to the onset of jamming.

The prime consideration in selecting the network architecture is again, the fundamental trade-off of the availability of a particular link versus the potential capacity of the entire satellite network. It is apparent that there are two different operating modes for the military SATCOM network, a clear mode and an antijam mode.

The clear mode of operation is characterized by the desire to maximize the communications capacity of the network by reducing link margins to the greatest degree possible. This desire is predicated by the reality that most military satellite communication networks rarely, if ever, operate in an environment characterized by a determined jammer and therefore typically operate link non-miltary networks.

The anijam mode of operation, on the other hand, is characterized by an extremely significant reduction in the available capacity in order to assure a communications capability even in the face of the most determined jammer. This approach is justified since the scenarios that give rise to determined jammers are precisely those where link availability becomes most essential.

There are some alternative network structures as shown in Figure 2 and, therefore, strategies to effectively provide both operating modes. One approach is to use the same modem equipment to provide both the clear and the antijam operating modes and adjust operating parameters of the equipment to transition between modes as a result of control system directives. This, naturally, is not as easy as it sounds, since it puts the full transition burden on the

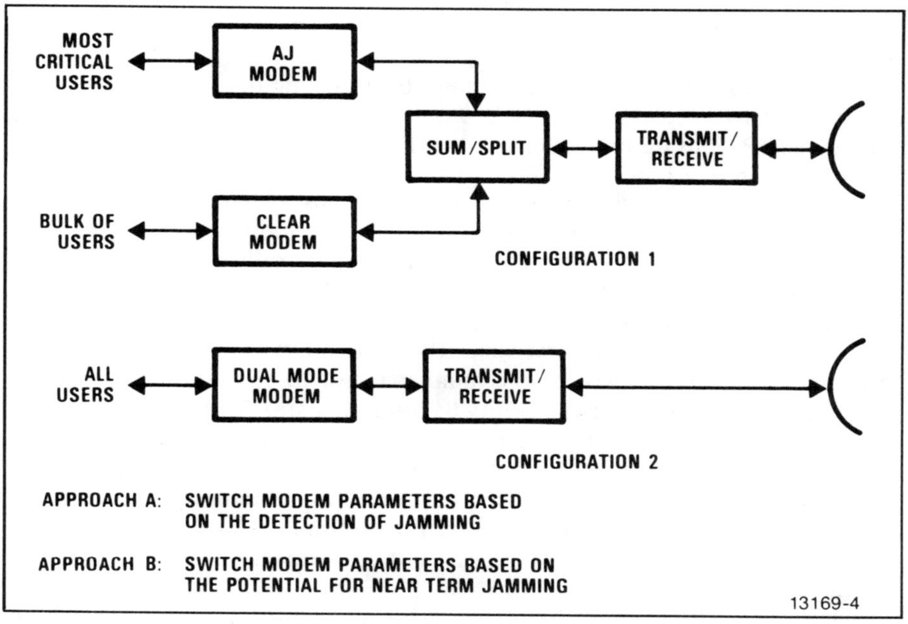

Figure 2. Alternate Architectures

control system; but it does have the advantage, if the transition can be done smoothly, of matching the right mode to the right scenario.

An alternate approach that greatly reduces the burden on the control system is to transition the network between modes, not as a result of a reaction to jamming as determined by the network control system, but rather as a reaction to the potential for future jamming as determined by an assessment of the political and military situation. The penalty paid here is the lost capacity either prior to the onset of jamming or if the jamming never occurs.

Still another approach is feasible if separate modem equipment is used to provide the clear and antijam operating modes. This would allow simultaneous operating of both the antijam and the clear modes of operation.

The approach would proceed with the few most critical channels assigned to the antijam mode of operation and all the other channels assigned to the clear mode of operation. The capacity of the network would only be slightly degraded due to self-interference between modes. The availability of the antijam links in the face of a jammer, while not as great as it would be if the bulk of the terminal EIRP was not allocated to the clear traffic, is still protected by the processing gain inherent in the antijam mode.

The transition process in this architecture does not involve the establishment of new links or the reassignment of users from one mode to another but merely the switching of the terminal EIRP from the clear modes transmission to the antijam mode transmission.

Independent of the architecture, there are some general responses that are available to the communicator once a jammer attacks the satellite. The possible responses include actions relative to the satellite, as well as actions relative to the earth terminals. For the satellite, if there is antijam spatial processing on board, its use to reduce the jammer power incident on the transponder is one available response.

A second response would be an attempt by adjusting the gain control of the transponder to keep the transponder out of saturation independent of the jammer EIRP. This naturally depends on the particular transponder design, as well as the jammer size.

At the earth terminals, there are three responses that are apparent: an increase in EIRP, a reduction in data rate or the initiation of the antijam operating mode. The appropriate response is naturally dependent on the specifics of the jamming attack.

From the jammer's viewpoint, there are alternate strategies available as well. In considering these strategies, the first point to realize is that the legitimate and realistic goal of the jammer is not to eliminate the ability to communicate but to reduce the ability to communicate. For both communicator and jammer, their abilities are not absolute but relative. The jammer must make the decision as to the strategy that minimizes the communicator's information transfer capability while requiring the commitment of a minimum amount of resources on his part. For example, all other things being equal, uplink jamming has the potential of taking down multiple links, while downlink jamming takes only one link.

One approach that might be effective for a jammer, depending upon the design architecture of the control process, would be to initiate an action that causes the communicator to transition his network. During the transition period, it is likely that communications must be interrupted. Once the network transitions, the jammer can cease his action, thus tempting the communicator to reconfigure to the higher capacity network state that existed prior to his initial action. If the communicator does, there will be another interruption in traffic after which the jammer can repeat the process.

This is an oversimplified example of potentially rewarding strategy open to the jammer. How rewarding the strategy can be made to be is a function of the requirements placed on the control system with respect to maximizing system capacity and the reaction time of the entire network to control systems directives.

Other alternatives available to the jammer include his transmission bandwidth, whether he attacks the control link or the communications links, and whether he attacks clear links only or he attacks antijam links as well. There are some general considerations with respect to each of these options.

The issue of the jammer bandwidth depends closely on whether antijam links, which nominally require the entire transponder bandwidth, or clear links are being attacked. In general, there is no rationale for a jammer to spread his energy across an entire transponder since he will then be no more effective against a clear link as he would be against an antijam link.

With respect to attacking the control function rather than the communications links, this is a strategy than can potentially achieve a jammer's objective of minimizing communications via a minimum commitment of resources. Communications will eventually break down without an adequate control function and the control function might very well be susceptible to a low power downlink jammer at the control terminal.

The issue of attacking antijam links as well as clear links depends on the degree to which the jammer is committed to minimizing the communication capability and his ability to commit resources. It is apparent that the jammer has already achieved a degree of success by forcing the communicator into an antijam operating mode, and it is also apparent that a greater degree of jamming resources is necessary to bring down the antijam links than was necessary to bring down the clear links.

Transitioning

The process of transitioning a satellite network in response to an ECM attack is naturally dependent on both the communications and control architectures of the individual network. Since it is extremely difficult to define a general approach to network transitioning, an example of a network transition process will be presented for a particular network with the following set of characteristcs:

1. The clear (no ECM) operating mode has all users operating on nonprotected communication links.

2. The protected communication links operate in the clear mode, without traffic, at a minimum EIRP in order to minimize interference with the clear links but sufficient to maintain lock at each receiver.

3. In the clear mode, the control link is a fully protected link operating both at a minimum data rate and minimum link margin between a single control terminal and each of the earth terminals that are included in the network.

4. All protected links include an automatic reacquisition capability, whereby the links can reacquire after a loss of signal of duration denoted as the reacquisition window.

5. The monitoring capability of the control system is located at the control terminal and is capable of identifying and characterizing jammers.

The transition from a non-ECM to an ECM environment for a network with the above characteristics would proceed as follows:

Step 1: An uplink jammer begins transmission.

Step 2: The clear links that are attacked lose the capability for information transfer and if the jammer is of sufficient magnitude, the antijam links will lose lock also.

Step 3: The monitoring function of the control system identifies the presence of jamming and must determine the appropriate response, depending on the assessment of the jammer magnitude.

If the jammer was of sufficient magnitude to force the antijam links into reacquisition, it is likely that a reduction in bandwidth and/or an increase in EIRP for the failed clear links will be of little value. The step to be taken then is to reallocate EIRP from clear links to antijam links.

If the antijam links remain locked, the initial response would apear to be a reduction in data rate and/or an increase in EIRP for the clear links.

For the remaining steps assume that the jammer was sufficient to knock the antijam links into reacquisition:

Step 4: The controller increases the EIRP for the protected control link to a level sufficient for all network participants to reacquire the control links.

Step 5: The controller now directs network participants to abandon the clear operating mode for the antijam operating mode, thus allocating terminal EIRP to the antijam links.

Step 6: The antijam communications links between network participants automatically reacquire (as long as the above steps are accomplished within the reacquistion window).

Step 7: The controller directs minimum operating data rates on all antijam links initially along with peak terminal EIRPs if the jammer has already driven the satellite into saturation.

Step 8: Based upon close coordination with the satellite controller and based also on the output from the monitoring process, the network controller can attempt data rate increases on the antijam links.

Conclusion

In summary, this article has pointed out the importance of network control for all satellite communications systems. In the case of a military satellite communications system, the importance is compounded by the threat of an ECM attack and the preventive measures typically taken by the MILSATCOM system.

The control system complexity for the military SATCOM system depends principally on the degree to which link margins are attempted to be reduced (network capacity increased). In one extreme where a system is designed solely with regard to link availability and, thus, with no emphasis on system capacity, the control system has little or no complexity. On the other hand, when a system is designed to adapt capacity depending on the current scenario, the network control system, and particularly the monitoring portion of it, can become extremely complex.

Part III: Other Satellite Communications Applications

Editor's Introduction

There are many examples of the use of satellites for specialized communications applications. A sampler of some, mostly taken from the pages of *SIGNAL* Magazine, is included here. Additionally, some articles which point to future directions or identify potential new applications or increased capabilities are included.

This part begins with an article by Wilbur L. Pritchard, a genuine pioneer in the field of satellite communications. "A Few Basics of Satellite Systems" serves to remind us of some fundamental truths. Drawing from those, he arrives at certain conclusions. Among them are a growing pressure for non-geostationary orbits and an irresistible trend towards higher frequencies, digital transmission, multiple beams and larger satellites.

Charles Taylor, in a panel presentation covering NASA's operational communications, describes the very complex NASA communications requirements for both earth orbital and deep space missions. He also forecasts the changes in the NASA tracking and data relay network that have begun to occur during the TDRSS era of the 1980s and beyond.

Today, we all take for granted the satellite weather pictures on the evening news broadcasts. Few realize the extent of the system required to deliver that information. David S. Johnson has been actively involved in weather satellites since 1958 when he participated in the development of TIROS-I. His article coauthored by E. Larry Heacock, "NOAA's Operational Satellite Service — C^3A," provides an overview of the environmental satellite system operated by NOAA, its management, the various applications of weather data and how command and control of the system is exercised.

Lt. M. M. Blume, USCG, relates how search and rescue activities can benefit from the wide area coverage capabilities of satellites in his article "Search and Rescue Satellite Aided Tracking (SARSAT)." The experimental system jointly entered into by Canada, France and the United States is described together with the U.S. Coast Guard's role. Although not described in the article, that joint endeavor, later joined by the Soviet Union is operational today. Recently extended until at least 1990, the SARSAT is credited with having contributed to saving 300 lives. Worldwide search and rescue and emergency location is clearly feasible using satellite technology.

Robert S. Magnant's "Satellite and Education: An Alternative" is an interesting summary of the development of communications satellites and the new communications capabilities that have emerged. The author sees one of the more important potential new applications in DOD as being its use by the education and training community. He describes the reasons for that belief and the benefits that would result. It is articulate spokesmen such as Magnant who cause us to look anew at experiments and to extrapolate revolutionary change to important activities which have made only modest changes in their methodology and technology over the years.

Greatly increased bandwidth for communications satellites will be required for them to remain competitive with optical fiber and to satisfy the growing demand for satellite transmission. Arthur R. Kraemer traces the development of communications at optical frequencies by means other than fiber in "Free Space Optical Communications." He concludes that free space optical communication is no longer a laboratory curiosity but a reality.

National security policy for telecommunications requires plans which fully exploit the capabilities of the commercial communications networks in case of national emergency. First enunciated in PD-53, now updated and revised in NSDD-97, this policy has resulted in a careful examination of the value of domestic communications satellites in restoral and reconstitution activities for the national security/emergency preparedness community. One aspect of this problem is dealt with in the paper "Commercial Satellite Interoperability Issues" delivered to the MILCON 84 conference sponsored by IEEE and AFCEA. Co-authored by Dr. John McLucas and the editor of this book, the paper describes the interoperability problems and lists various approaches to their solution. Emphasis is placed on cooperative and sustained efforts by the government and by industry.

Finally, the editor has included one of his writings, "The Future is Bright for Communications Satellites," to present one view of the impact of the growing importance of information in today's business world and the value of satellite communications in that environment. Also included is his tentative assessment of the competitive threat of terrestrial fiber optic transmission.

A Few Basics of Satellite Systems

Wilbur L. Pritchard

The use of satellites in many varieties of communication applications continues to expand at a rapid rate, and at the same time, the arguments between their use and conventional terrestrial means also continues to flourish. One might even think that the success of satellite communication would be assurance enough of its superiority, at least in those areas, but paradoxically much of the growth of satellite systems has been in areas where terrestrial means are capable of effective competition.

Intelsat has been growing at an average of 25 percent per year. Half its traffic is in the North Atlantic basin and probably half again of that, or 25 percent of the total, could easily be handled by submarine cables and microwave relay. Nonetheless, the advantage of being able to serve many small countries that previously had little or no international service makes its overall economic position sound. It is important to ask just what is this real economic advantage of satellites as compared to other means of communication.

An understanding of these economics is important, not only for commercial systems operators who depend on them for a return on their investment, but equally for the Department of Defense. It makes no difference whether one is looking for a maximum return on investment or the minimum present value of a cost, the system economics are decisive. Satellites have an indisputable advantage in two broad areas: one is in the creation of a new and large network in which there is little or no existing facility and the second is in the supplying of services to mobile platforms or to fixed platforms whose locations are not known *a priori*.

We can look briefly at both these categories in order to get an understanding of just how fundamental they are. A glance at Figure 1 shows the essence of the first advantage quite clearly. If there are "n" points to be connected by

This article originally appeared in the November/December 1979 issue of* ***SIGNAL.*

terrestrial means (whatever they may be), then not only is equipment required at all "n" nodes, but in addition n(n − 1)/2 links are necessary. On the other hand, in a space system equipment is required only at the nodes. There is also the cost of the satellite itself, but this latter cost is fixed. Even if the cost of equipment at each node is high compared to the nodal costs for terrestrial systems, there is only a very small cost which is proportional to the multiple interconnection. The term n(n − 1)/2 and the link costs in a terrestrial system, regardless of how small they are, will ultimately dominate the comparative total as "n" increases.

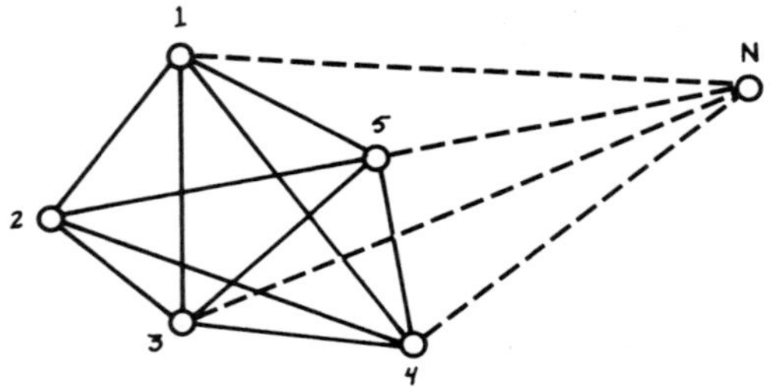

Figure 1a. General Network

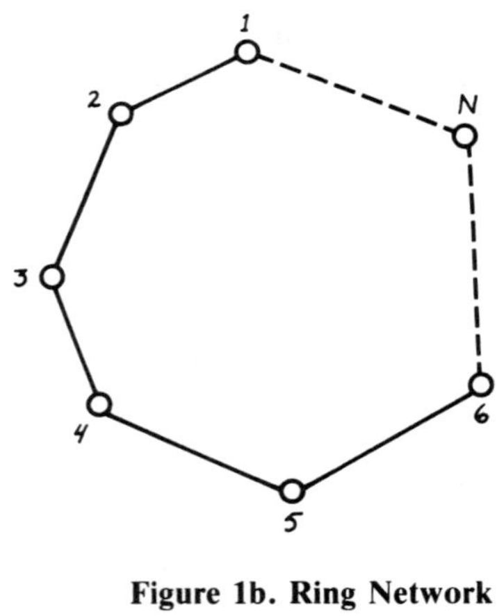

Figure 1b. Ring Network

If the network is arranged as a ring (Figure 1b), then only as many physical links as nodes are required, but the multiplexing system must in effect create the n(n − 1)/2 links. When this kind of network is geometrically and operationally possible and the traffic intensities are high, it can be competitive with satellites. These distinctions are important when one considers comparisons between satellites and other means — e.g., fiber optics. One often hears the claim that fiber optics will provide substantial competition because they are getting cheaper and cheaper. In order to clarify that argument, it is useful to point out that even if the fiber optics become free, the costs of the conduits will not, and therefore, for a large network with many points of the type in Figure 1a where there is no existing service, the satellite system will quickly become cheaper. This does not mean that fiber optics will not have a substantial application, but in my opinion, it will be in those areas where existing lines of less capacity are available and where one wishes to increase the service to very wide bands and where the geometry can be of the "ring" type such as in high density interurban traffic.

Communication with Terminals

The second advantage of satellites is in a sense beyond dispute, and that is their ability to communicate with mobile terminals and those fixed terminals whose locations are not known in advance. We are using the word "communication" here in its widest sense to include not only standard telecommunications between two points but also broadcast, such as the Fleet Broadcast of the Navy, and the inverse system of data reception from a large number of terminals. Today, we simply have no other satisfactory way of transmitting reasonable bandwidths over long ranges to mobile terminals.

If high data rates are required, the situation is even worse and we have no method at all other than satellites even for surprisingly short ranges. With these applications principally in mind, it is interesting to look at some fundamentally technical and economic problems and try to come to some conclusions that are thermodynamic in nature; conclusions, that is, that depend on the broadest kind of considerations and are independent of the technical details. If we are going to have systems dealing with mobile terminals in large numbers, then we can expect that these terminals will be small and it is the technology and economics of large systems with small terminals that is conspicuously the most interesting for the coming decades. Equation (1) tells the technical story "in a nutshell," if one notes that the area to be covered on the ground and the basic information rate, "desirables" in the

$$A_{cov}\left(\frac{C}{N_o}\right) = KP_T\frac{D^2}{T_S} \quad \text{Equation (1).}$$

A_{cov} = Area covered by antenna beam on the earth.

$\left(\frac{C}{N_o}\right)$ = Carrier to noise density ratio—a measure of the ultimate information capacity of the channel regardless of how much bandwidth is used.

P_T = Transmitted power.

T_S = System temperature at receive end of link.

D = Diameter of earth station antenna.

K = A constant comprising Boltzmann's constant, antenna efficiency, π and numerics.

system, are seen on the left side of the equation and the right side of the equation contains the physical size of the earth station antenna, the transmitted power in either the earth station or satellite and the corresponding system temperatures depending on which link is being considered.

The right-hand elements of Equation (1) correspond to the items that determine the cost of a terminal. The transmitted power and antenna size are obviously important to the cost and, to a lesser extent, on the downlink the receiver temperature also is, although one can expect the usual progress in technology, and adequately sensitive receivers for lower and lower costs. In broadcast applications and others where the downlink typically limits the performance, it is necessary to trade satellite transmitter power against station size, recognizing that the coverage is determined by the operational requirements.

Since there are many earth stations and only one satellite, the trade-off should be made on a cost basis. As the satellite power goes up, the cost of the earth station goes down and there is an optimum set of characteristics to minimize the total system cost. In some applications, such as Intelsat and some other commercial systems, this may not be so important since the space and earth station costs come out of separate pockets, but in the case of a U.S. military application, this kind of optimization is extremely important and unfortunately often not done. As an interesting "aside", note that to a first approximation, the satellite altitude, carrier frequency and bandwidth are not involved. Note that the confusing "figure of merit" (G/T) also does not appear. It is the physical size and not the gain of an antenna that is desisive.

An important point is that often the coverage area is fixed by operational requirements but that the total coverage area does not have to be all served simultaneously. In other words, it is possible to switch the power from one beam to another or equivalently to move one beam consecutively to different locations to reduce the area covered from the point of view of the radio frequency (RF) links while at the same time maintaining adequate operational flexibility. This can result in a tremendous improvement in performance at an increase in space segment costs reflected in larger antennas, beam switches and logic. Nonetheless, for the reasons mentioned, it will be worth the effort by reducing the total costs.

An analogous effect takes place in those systems in which the uplink limits performance. Rather interestingly, again it is the physical size of the earth station antenna which must be traded off against the receiver sensitivity in the spacecraft. In this case, however, it is more difficult to achieve really low system temperatures. There is not much improvement when spacecraft antenna "looks" at the earth with its 290K temperature and attempts to get improved system temperature by lowering receiver noise temperature. Nonetheless, the coverage factor is the same as on the downlink and again, the use of multiple narrow beams on receiving will considerably improve the performance of the system.

It is important to note that again the multiple beam or single switchable beam idea is dependent on the operational notion that all of the area does not have to be served simultaneously but can be served consecutively either in a programmed or random fashion. This is not always the case, and if it is not, there is no advantage whatsoever to the use of multiple beams. As we talk about multiple beams and high power, we come ineluctably to big satellites.

Higher Frequencies

There is little doubt that satellites, even for a single purpose, will grow in size by a factor of two or three in the next couple of decades, and if one considers the possiblity of multiple purpose satellites then this growth will be even greater. Big satellites are costly both for the satellites and for the launches, and thus the trend to narrow beams will be accompanied by the use of higher frequencies to allow small spacecraft antennas for the same coverage.

This trend to higher frequencies will be accelerated by regulatory necessities. Interference is becoming a worse problem every day and there is no doubt that satellite communication will have to go to 'k'' band and higher in order to avoid it. We are well familiar with the problem of atmospheric loss at the higher frequencies but see that problem only as one that will keep many engineers employed. It is a problem that is unavoidable but it will be solved.

All Digital Transmissions

Another trend that seems to us irresistible is that toward all digital transmissions. Satellite capacities are higher with single digital carriers and Time Division Multiple Access (TDMA) than they are with multiple carriers and Frequency Division Multiple Access (FDMA) because of the freedom from intermodulation. The previously mentioned multiple beam systems go extremely well with digital time division switching.

These satellite advantages will combine with the slow but steady change to digital in terrestrial systems to convert most of the straight telecommunications satellites to digital transmission. On the other hand, where no terrestrial interfaces are involved and where the terminals are very small, e.g., broadcast and mobile services, the analog GM systems will probably continue for a long time.

Other Orbits

The extraordinary success of the satellite communication program in geostationary orbit has led many to forget that although it is convenient, it is by no means necessary to use this orbit. In particular, there are many applications in which other orbits are convenient and not really much more expensive. For high capacity terminals where large antennas are used, such as in both military and civil fixed telecommunications systems, these antennas normally must be capable of tracking to accommodate changes in location and small errors.

Once a large antenna has this machinery, it really does not make much difference whether it tracks slowly for the aforementioned purposes or a little faster to follow a satellite in some non-stationary orbit. The tracking rates, even in very low orbits, are extremely slow. There is the disadvantage that more than one satellite is used. For instance, systems in the 12 hour 63.5° inclined synchronous orbit system used at least two satellites in orbit and require that the earth stations at some point during the day switch from one to the other. Again, this is a problem that is in the nuisance category and presents no real difficulty for ordinary communications. Both the United States and Soviet Union use this orbit successfully. A relevant point sometimes overlooked is that the link performance of a satellite is in no way dependent on its altitude.

Note again in Equation (1) that the satellite range does not appear. What happens is that as the satellite distance from a fixed terrestrial coverage area increases, the beam width narrows and the increase in gain exactly compensates for the increased inverse square law loss. We think increased geostationary orbit crowding may well make certain kinds of systems go to medium altitude polar and equatorial orbits, inclined orbits and the aforementioned non-rotating highly elliptical orbit inclined at 63.5°.

There will be increased pressure in this direction as the geostationary orbit becomes more and more crowded. This pressure should be on those systems that do not require the geostationary orbit, rather than on those systems for mobile terminals, data gathering and broadcast which almost have no choice but to use the geostationary orbit.

This will result in regulatory, political and economic competition that will develop slowly in the next decade or two and last a long time. A possible amelioration is through the use of large multipurpose platforms to economize on orbital space. The extent of this economy is not clear since two radio terminals even on the same platform still have possible interference although it might be easier to deal with. These platforms will offer substantial reductions in cost but the institutional and logistic problems connected with the multiple payloads will make them unsuitable for general use.

Conclusion

In conclusion, we see an increased use of satellites especially in the geostationary orbit, for mobile and multipoint applications, but increased competition with terrestrial means in more conventional applications and in regions where existing facilities exist.

We envision increased pressure for the more conventional satellite services to use non-geostationary orbits and we see the trend to higher frequencies, digital transmissions, multiple beams and larger satellites as irresistible. The combined regulatory pressures of frequency assignments and large bandwidths will force many satellite communications systems into the region between 10 and 20 GHz and even higher.

NASA Operational Communications

Charles A. Taylor

The National Aeronautics and Space Administration (NASA) has established a communications system which extends worldwide to meet the communication requirements of its space research and development program. This paper will discuss briefly the requirements placed on this system and the trends which we see for the next few years.

NASA space missions can be divided into two distinct groups, each of which has unique requirements for tracking and data acquisition support. The earth orbital group is by far the most diverse, with many satellities requiring simultaneous support from orbits ranging in inclination from equatorial to polar, and in altitudes from a few hundred kilometers on out to lunar distance.

These satellites produce large quantities of data which must be acquired, with a rapidly increasing requirement for real time data transmission back to the mission control centers and data processing facilities. There is, in addition, a requirement for a substantial flow of data between the control centers and the tracking stations for command and control of the spacecraft. The Spaceflight Tracking and Data Network (STDN) has been established to provide support to these space missions and has stations at 13 locations around the world. The network is supplemented by aircraft and a tracking ship.

The planetary or deep space missions operate at extreme distances from Earth and require continuous monitoring by the tracking stations. These missions are becoming more and more sophisticated, with rapidly increasing data rates and complex command requirements. The Deep Space Network (DSN) has stations co-located with the STDN in Australia, Spain and California.

The workload on the tracking networks has been heavy during the past decade, with 40 to 50 space missions being supported at any given time. At the

This article originally appeared in the August 1976 issue of **SIGNAL.*

present time, there are about 55 active payloads receiving tracking data acquisition support from the NASA networks, with about 80 separate missions to be supported at some time during the year.

The NASA space programs impose a severe and constantly changing set of requirements on the communications system. Analog and digital data of various bandwidths must be transmitted from the tracking stations; commands from the mission control centers must be transmitted to the stations for relay to the spacecraft; two-way voice and teletype communications must be provided between tracking stations, control centers and the many experimenters; and in the case of manned space missions, voice communications with the astronauts must be provided. It is imperative that these communications have high quality and high reliability.

With the success of missions such as Viking dependent on timely and accurate transmission of data and commands, we cannot tolerate failure of the communication system. It is our policy, wherever possible, to provide for diverse routing of the communication circuits in order to protect against natural disasters or major catastrophic failure of some element of the communication system. Of course, cost is of great concern to us in implementing the communication system and we are always conscious of the need to accomplish our mission at minimum cost.

NASCOM

All of the requirements for operational communications in NASA are met by a single communication system called NASCOM. This system is managed by the Goddard Space Flight Center (GSFC). The primary network switching center is at GSFC, with a West Coast switch located at the Jet Propulsion Laboratory (JPL) in Pasadena, California. Other switching centers, located in Canberra, Madrid and London, interconnect local circuits to the tracking facilities and experimenters with the trunk circuits to the United States.

The network makes extensive use of both submarine cable and satellite circuits to the foreign station locations, with increasing dependence on the satellites as requirements for higher data rate transmission increase. In the United States, we are beginning to utilize the domestic satellite services as these capabilities come into being, but most of the circuits at the present time are in the terrestrial facilities.

It has always been NASA policy to lease communication service to the maximum extent possible. Consequently, we own no circuits and very little communication equipment. This has resulted in a network which is very responsive to mission requirements, with circuits being called up or turned down as necessary to support the current workload.

Specific communications requirements have developed in lockstep with the space program itself, as described in Figure 1. The early space missions were relatively simple and were highly exploratory in nature, with limited requirements for transmission of data and commands in real time. Most of the data were recorded on magnetic tapes which were then shipped to data processing facilities for processing and later analysis.

- Past
    - Space Programs Highly Exploratory in Nature
    - Limited Requirement for Rear Time Data
        - Most Data Recorded on Tape
    - E.G., Early Mariners, Explorers
- Present
    - Space Programs Beginning to Exploit Previous Data
        - Requires Much More Data
        - Interactive with Control Centers
    - Much More Real Time Data Required
    - E.G., Atmosphere Explorer
- Future
    - Many More Specific Applications and Research Projects
        - Very High Data Rates
        - Highly Interactive with Control Center
    - Real Time Data
    - E.G., Spacelab
    - Tracking Stations Operating in "Bent Pipe" Mode

Figure 1. Evolution of Requirements

As the space program has evolved, however, the missions are requiring more and more real time data transmission. The spacecraft are becoming highly interactive with the mission control and experimenter facilities, and the communication system is being required to handle much higher data rates and tremendously increased volumes of data. This trend will continue in the future, and within the next few years the tracking and data acquisition networks will be operating in a "bent pipe" mode. That is, the tracking stations will be a part of a system which will not interpose time delay or intermediate processing in the flow of data between the spacecraft and the users.

Figure 2 shows the dramatic increase in the amount of data being transmitted by NASCOM. As stated earlier, most of the spacecraft data acquired in earlier years were recorded at the stations on magnetic tape, with a relatively small portion of the data actually passing through NASCOM. The steep slope of the image data curve in Figure 2 reflects the increasing importance of image data, such as that acquired by the Landsat spacecraft, as well as the continued improvement in the resolution of the sensors.

Much of the requirement for NASCOM transmission of the image data is resulting from the success of the Large Area Crop Inventory Experiment (LACIE) in the computer identification of crops. This kind of data is being used now on a semi-operational basis and will become more and more valuable as a crop yield forecasting tool. To be used in this fashion, however, the data

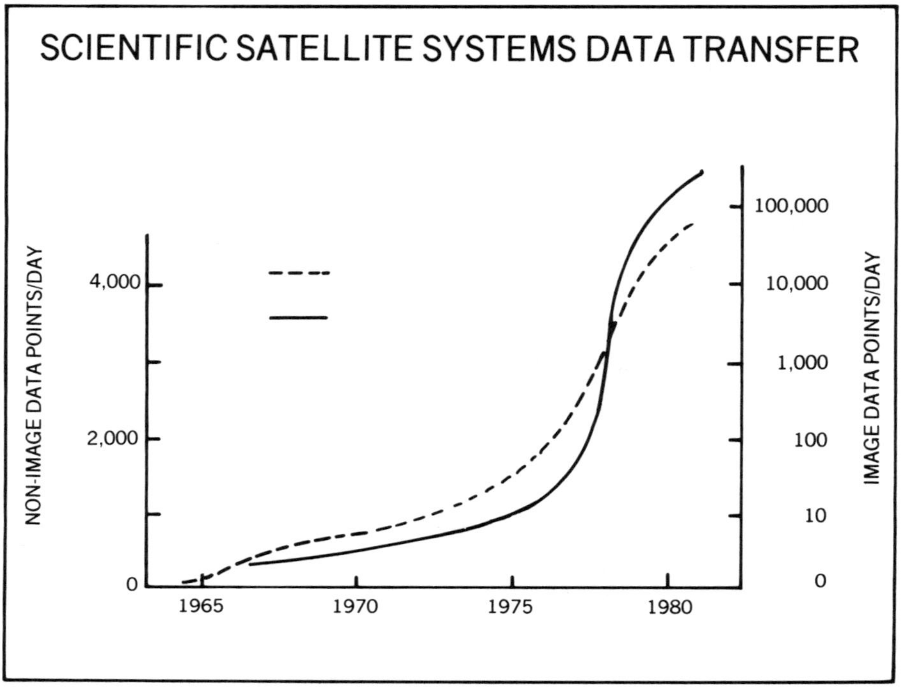

Figure 2. Scientific Satellite Systems Data Transfer

TYPICAL FUTURE PROGRAM DATA RATES

PROJECTS	LAUNCH DATE	MAX. DATE RATE
HEAD	1977	128 KBPS
MJS	1977	115 KBPS
LANDSAT-C	1978	15 MB/S 3.5 KHz
LANDSAT-D	1980	120-240 MB/S
SEASAT-A	1979	100 MB/S
SEASAT-B	1982	100 MB/S
SHUTTLE	1979	1.2 MB/S
SPACELAB	1980	50 MB/S

Figure 3. Typical Future Program Data Rates

must be processed and delivered to the users quickly, making it necessary to transmit the data through NASCOM rather than using magnetic tape.

Figure 3 illustrates the trend to much higher data rates, showing some typical data rates to be used by future space missions. The High Energy Astronomy Observatory (HEAD) and MJS are typical of newer scientific missions, and the Seasat and Landsat-D missions are typical of the coming space applications programs. The Spacelab, of course, will be a manned space program which will conduct both science and applications experiments.

In the early years of the space program, NASCOM data rates of 1200 bits per second (bps) were quite adequate, although requirements soon increased such that 7200 bps were needed. At the present time 56K bps circuits are being brought into use, and circuits with the capability will be in service to all tracking stations before the end of the year. In some instances, multiple 56K bps circuits are required, using inverse multiplexing to transmit the higher data rates. As the newer spacecraft such as those shown in Figure 3 are launched, it is expected that domestic communication satellite service will be used extensively, perhaps with earth terminals co-located with the NASA facilities. Several circuits have already been procured on the basis of proposals for co-location of the earth stations.

In order to obtain the bandwidths required for the future data rates, NASCOM will probably require full Domsat transponders within the next several years.

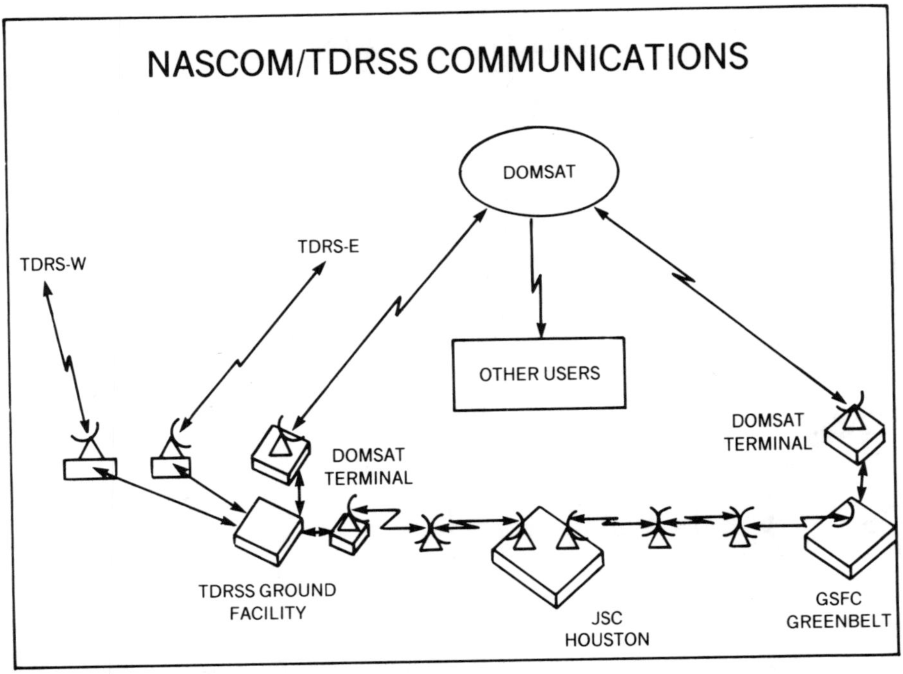

Figure 4. NASCOM/TDRSS Communications

The 1980 Network

About 1980 a new tracking and data acquisition system will come into being which will have a tremendous impact on both the space missions and the communication requirements. The Tracking and Data Relay Satellite (TDRS) system will support all NASA spacecraft in low earth orbit (under 8000 kilometers) and will eliminate the need for many of the present tracking stations. TDRS will consist of two satellites in synchronous orbit about 130° apart.

The research spacecraft in low orbit will transmit data to the TDRS satellites, which in turn will transmit the data to a single ground station in the United States. Commands to the spacecraft will be transmitted on the reverse path. Early in the development of this concept, a policy decision was made that the TDRS should be a service leased from the commercial sector. Proposals for the service are now being evaluated, with contract award anticipated for late 1976.

Figure 4 shows the communications which will be required with the TDRS. The primary communication circuits will be through a Domsat terminal co-located with the TDRS ground facility, one or more Domsat transponders and Domsat terminals at GSFC and at other user locations. A diverse route will be established, probably through terrestrial systems, for duplex 1.5 megabit per second circuits.

The tracking and data acquisition network as it will be in the 1980s is shown in Figure 5. The DSN facilities will be continued as they are today, but the

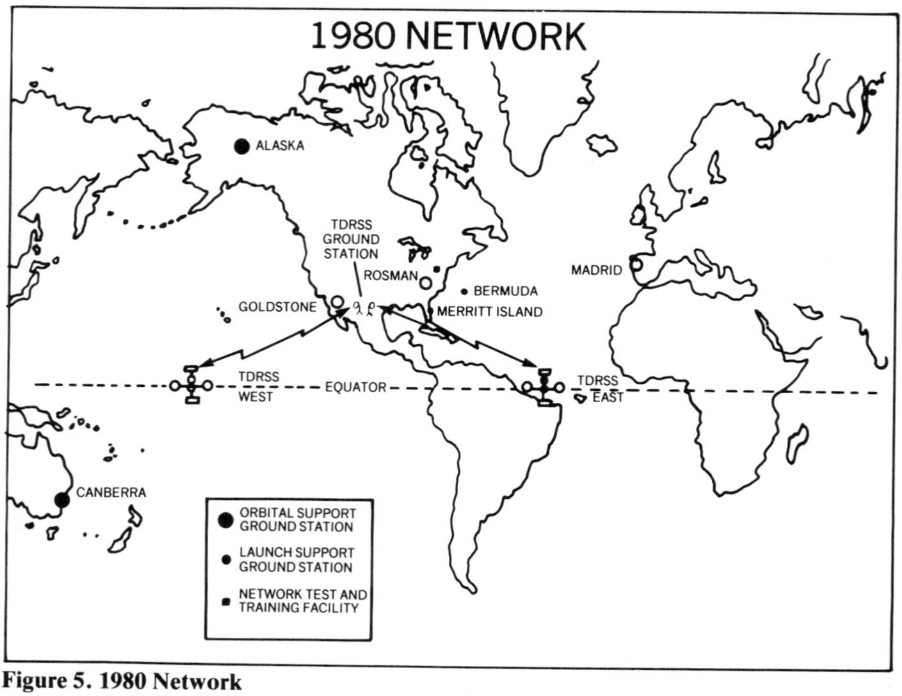

Figure 5. 1980 Network

1980 NETWORK DATA COMMUNICATIONS	
TDRS EARTH TERMINAL	1.5 MBPS (2) 50 MBPS
US STATIONS	
GOLDSTONE	50 MBPS
ROSMAN	50 MBPS
ALASKA	50 MBPS
OVERSEAS STATIONS	
MADRID	56 KBPS (2)
CANBERRA	56 KBPS (2)
LAUNCH SUPPORT STATIONS	
CAPE CANAVRAL	56 KBPS (2)
BERMUDA	7.2 KBPS (4)

Figure 6. 1980 Network Data Communcations

STDN will be much smaller. The STDN stations will be limited to those shown in Figure 5 to provide support to satellites with synchronous, highly elliptical or very high orbits.

The 1980 network will require very different communication capability than does the present network. Figure 6 shows the minimum capability requirements as we presently know them.

In summary, in order to support its space program, NASA has a requirement to transmit large quantities of data from place to place around the world in a highly reliable, cost-effective manner. This requirement is met by use of an integrated operational communication network made up of circuits and equipment leased from the commercial sector, with a capacity carefully tailored to meet mission requirements, and making maximum use of the new technology and capability developed by the industry.

NOAA's Operational Satellite Service: C^3A

David S. Johnson and E. Larry Heacock

What areas of the United States are threatened by severe thunderstorms or tornadoes during the next few hours? What is the best location for California's salmon and albacore fishing fleets to operate tomorrow? Is the latest tropical depression developing into a hurricane or dissipating? What routes should be taken by shipping fleets to minimize steaming time and save precious fuel? What winds aloft are being encountered by commercial airlines around the world? Will sufficient water be available for those U.S. farms which depend on irrigation next year? Answers to this list of questions and many more depend on data collected by one of the United States' unsung space heroes—the operational environmental satellites of the National Oceanic and Atmospheric Administration (NOAA).

All Americans are aware of the space exploits of the National Aeronautics and Space Administration (NASA) and most are aware that long-distance telephone, television and other communications are aided by satellites. Readers of *SIGNAL* are also aware that the Department of Defense (DOD) operates a number of satellites to ensure the nation's security and defense posture. Surprisingly few are conscious, however, that our daily lives are being increasingly affected by a different ongoing civilian space program which is operated by the National Oceanic and Atmospheric Administration.

This article originally appeared in the November/December 1979 issue of* ***SIGNAL.*

NOAA's National Environmental Satellite Service (NESS) is about to complete its second decade of providing services to the people of the United States and the world. Every issue of *SIGNAL* contains such common abbreviations as C^2, C^3 and C^3I. The NOAA satellite system can be thought of as C^3A where the abbreviation denotes command, control, communications and applications (of the data).

The timely, cost-effective and successful operation of this service is a tribute to one of the most smoothly functioning interagency relationships in the Federal Government. Under the provisions of the National Aeronautics and Space Act of 1958, NASA is responsible for and funds design and development of NOAA's spacecraft as well as supporting research and development in response to mission requirements specifications provided by NOAA/NESS. NASA also serves as NOAA's prime contractor for operational spacecraft and launchings under reimbursement from NOAA. NOAA, in turn, is responsible for C^3A once the spacecraft have been launched and checked out in orbit. Even this latter function, orbital check-out, is currently performed by NASA utilizing NESS facilities.

The present system consists of two basic types of satellites:

1. The Geostationary Operational Environmental Satellite (GOES); and
2. The polar orbit satellite series which carries the NOAA name, the latest series of which is the fourth-generation, television and infrared observation satellite (TIROS-N) series. Technical management of the industrial contracts for development and manufacture of both spacecraft series is the responsibility of project teams at NASA's Goddard Space Flight Center. The current production contract for the GOES spacecraft is with Hughes Aircraft Company, although satellites currently in orbit were produced by Ford Aerospace and Communications Corporation. All polar-orbiting spacecraft have been produced by RCA AE Division.

Once declared operational, command and control is performed by NESS's Satellite Operations Control Center (SOCC) located in Suitland, Maryland, through Command and Data Acquisition (CDA) stations located at Wallops Island, Virginia, for GOES and at Wallops and Gilmore Creek, Arkansas, for the NOAA satellites. Communications are via a mix of terrestrial leased lines, Domestic Communications Satellite System (Domsat) and the GOES satellites' own transponder system.

Applications of the System

The payloads of the satellites consist of remote sensing instruments for measurements of the earth's atmosphere, land and ocean surfaces, as well as the space environment; they also include a number of communications subsystems for reliably transmitting this sensor data to earth, for relaying data from remotely sited platforms to the central data facility and to broadcast data and analyses to weather stations around the globe.

Turning to the answers to the questions posed rhetorically in the introduction, the advent of the geostationary satellite has contributed heavily to the improved performance of the National Weather Service's (NWS) forecasting of severe storms since 1974. Once NOAA's NWS (which is the biggest single "customer" for NESS data) has forecast that conditions are conducive to severe storm development, one of NESS' two GOES satellites is programmed to shift from its normal two images of the earth per hour to four or more per hour. A computer derived "enlargement" of the endangered area is then passed via direct circuits to the NESS Satellite Field Services Station and the co-located NWS National Severe Storm Forecast Center (NSSFC) or National Hurricane Center where a close watch is kept on the development of these storms.

Since the launch of TIROS-I on April 1, 1960, satellites have played an extremely important role in tracking hurricanes and tropical storms. We are able to estimate quite reliably the strength of these storms, which is very important in making forecasts of where they are going to go and how severe the damage might be. The geostationary satellites now provide essentially continuous monitoring.

In support of California fishermen, the sea surface can be monitored in the infrared portion of the radiation spectrum to detect temperature differences, including those caused by the phenomenon called upwelling. It is known that the food that fish feed on (so-called phytoplankton) is concentrated in upwelling zones. The satellite senses differences in sea surface temperature, from which upwelling can be detected. This information is provided to the West Coast tuna and salmon fishing fleets. They sail directly to the location detected by satellite rather than searching in the blind. In this way, they often catch their fish much more rapidly and efficiently.

One of the more ingenious uses for satellite observations also is dependent upon infrared sensing. We can measure the location and approximate strength of the Gulf Stream, including large eddies or circular pools that develop along its edges. The Gulf Stream moves north-eastward with a speed of several miles per hour and very strongly affects commercial ship traffic. If a boat travels with the Stream in the region of the strongest current or on the correct side of the eddies, of course the vessel is going to move faster, get to its destination sooner and save fuel.

Conversely, if a ship is going in the other direction against the Stream, it will be slowed down dramatically, and operating costs will be increased significantly. Observations now are broadcast regularly to marine interests offshore, whereby they can know where the Stream and its eddies are. Shipping in high latitudes and the Great Lakes also is provided ice information derived from satellite images to safely extend the shipping season.

A new use of the geostationary satellites which employs their communications capability involves not only our GOES satellites, but also the Geostationary Meteorological Satellite (GMS—called Himawari) of Japan and the European METEOSAT—all of which have similar, compatible Data Collection Systems. Called Aircraft to Satellite Data-Relay (ASDAR), this application provides for special data systems to be installed on wide-body jet aircraft operated by several of the world's airlines, including at this time Pan Am, SAS, KLM, Singapore Airlines, Lufthansa and Quantas as well as the U.S. Air Force (USAF).

These black boxes use data generated by the aircraft's digital inertial guidance systems and other normal on-board sensing systems to generate enroute weather messages which are relayed by the geostationary satellites to the appropriate ground station operated by NESS, the Japan Meteorological Agency (JMA), or the European Space Agency (ESA), depending upon which satellite receives the data. It is then possible to get accurate measurements of the average winds, temperature and pressure height along the routes flown by these airplanes. Furthermore, because the data are relayed in near real time, ASDAR represents an important source of upper air data for medium-range weather forecasting.

In the western United States, satellite data are used to monitor snowpack by measuring the extent of the snow cover. This information, combined with the relatively sparse ground observations that are available to hydrologists, is used to give a much better estimate of the total amount of water stored up in the snowpack. Another role of the satellite is to assist in determining when this snowpack is going to start to melt, to aid in estimating the runoff. This is of use in two ways: in improving water management by controlling reservoir levels; and assisting in warning of flooding when there is rapid melting of snow.

Command, Control and Communications

The NESS satellite system normally requires two geostationary and two polar orbit spacecraft to provide the data base upon which its services are based. In addition to these four satellites, there are other spacecraft (particularly geostationary ones) which have suffered a degree of degradation requiring launch of replacements, but which are still capable of providing some back-up service. Thus the Saellite Operations Control Center (SOCC) must maintain a measure of control over these satellites as well. At times, the number of geostationary satellites being controlled has reached as high as five.

The polar orbiting TIROS-N series of satellites, unlike the geostationary spacecraft, are in communication with the ground stations only briefly on each orbit. Therefore, they require intensive command and control during overpasses to insure readout of their highly perishable data, to update programs in their on-board computers and to monitor and adjust on-board support subsystems, such as power supply and attitude control.

In order to manage this widely diversified program, the NESS Office of Operations (O/O) utilizes an impressive melange of communications media and techniques ranging from low bandwidth analog signals to high data rate digital communications.

The C^3 for each of the two GOES satellites is outlined in Figure 1. Basic satellite to ground communication is performed at the Wallops Island, Virginia, Command and Data Acquisition (CDA) Station, utilizing the S-band links allocated to the Meteorological Satellite Service (i.e., 1670-1700 MHz space to earth and 2025-2120 MHz earth to space). The process begins when a command is transmitted to the spacecraft to start the on-board primary imaging sensor (Visible Infrared Spin Scan Radiometer—VISSR). The sensor generates a 28 megabit/second data stream during the approximately 30 milliseconds that the spinning satellite "views" the earth on each of its 100 revolutions per minute.

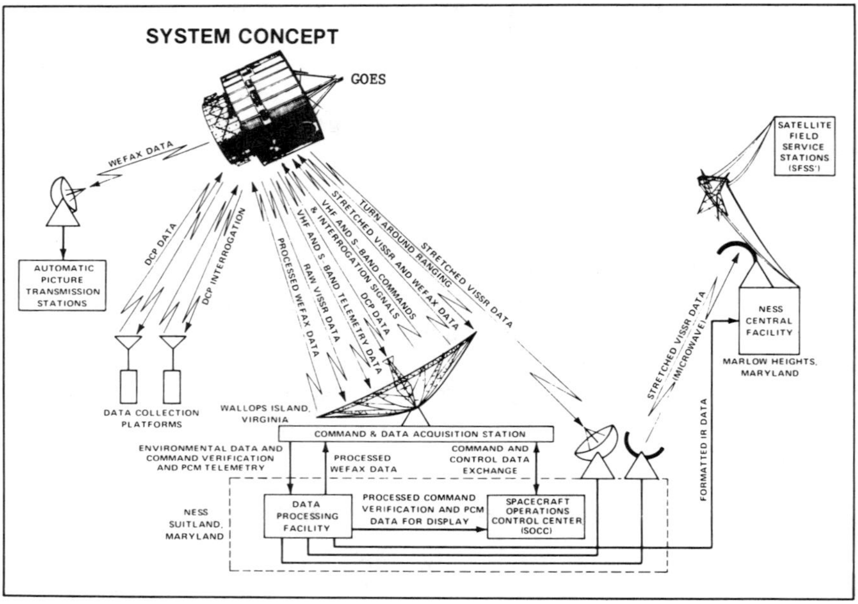

Figure 1. Command, Control and Communications System for One of the Geostationary Operational Environmental Satellites (GOES)

This burst of data is transmitted in real time to the ground where the data are buffered and retransmitted at 1.75 Mbs to the satellite during the 570 millisecond gaps between bursts of raw data. The spacecraft transponder then broadcasts this "stretched" data on the NESS facility at Suitland, Maryland (and to a number of users who have equipped themselves with appropriate receiving stations). At Suitland, selected portions of the data are diverted to the NOAA computer facility where quantitative processing takes place.

The full data stream also is forwarded via microwave to another NESS facility, where the images are broken down into several sectors, each having much lower total data content (varying resolution and geographic area such that the number of pixels is approximately equal in each output). The sectors are then transmitted at slow speed in analog facsimile format via leased telephone lines to NESS' Satellite Field Service Stations (SFSS) throughout the country. The sectors also are passed on to National Weather Service Forecast Offices, other government agencies and many users in the private sector.

Using the satellite data processed in Suitland, low resolution imagery also is formatted and sent (by telephone line) back to the CDA where it is transmitted to the geostationary satellites for broadcast to users having relatively unsophisticated receiving stations. This latter service is called WEFAX (for weather facsimile).

Telemetry plus data from the low data rate Space Environment Monitor (SEM) on board the GOES also are received at Wallops and sent to Suitland by land line where SOCC personnel monitor the spacecraft subsystems and generate appropriate commands which are sent to the CDA by land line to close the loop.

Figure 2 depicts the TIROS-N/NOAA C^3 operation. A significant difference in the modus operandi is necessitated by the fact that the two NOAA satellites are low earth orbiters (LEO) and hence communication with the spacecraft is only possible while it is within range of one of the two NESS CDA stations. Wideband data acquired at Wallops and Gilmore Creek by playing back on-board tape recorders is relayed to Suitland via domestic communications satellites of RCA American Communications, Inc. Spacecraft orbits beyond the range of Wallops and Gilmore Creek are received in Lannion, France, and relayed to Suitland via the eastern GOES spacecraft.

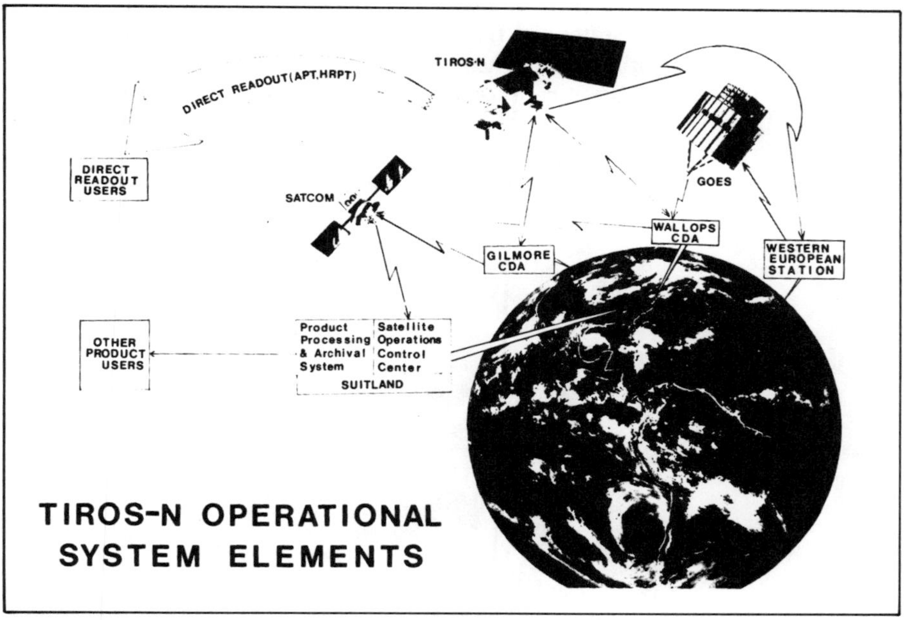

Figure 2. Schematic Representation of the Space and Terrestrial Communications Network used to Command, Control and Acquire Data from the TIROS-N Series of Polar-Orbiting Environmental Satellites

The data from the LEO, while it includes imagery, is much more heavily employed in quantitative applications such as the determination of sea surface and atmospheric temperatures. As a result, virtually all of the raw data remain in Suitland where they are processed in the NOAA IBM 360/195 computers to produce quantitative products used by NOAA's National Weather Service, the DOD and many other nations as part of the input to numerical weather forecasting.

Commands for the TIROS-N/NOAA spacecraft are generated in Suitland (at SOCC) and transmitted to the satellites via the CDAs during contact times. When necessary, they can be stored on-board the spacecraft for execution at a predetermined time. In addition to the worldwide data recorded on-board and acquired as described above, all of the sensor data are broadcast in real time from the satellite permitting reception by users equipped with suitable ground stations in all parts of the world. Indeed, the number of stations receiving data from the TIROS-N satellites' direct broadcast services is now nearing 1,000 in over 100 countries.

Summary

The NOAA satellite system is probably the least well-known of any U.S. space system. But it is the only civil program which is fully operational with a continuing commitment to provide necessary space remote sensing services to the people of the United States and the world. The necessary resources to fulfill this commitment constitute one of the most far-flung and comprehensive space systems outside the DOD yet represent a cost of less than one 5000th of the Federal Budget. As a civilian managed and operated system, it can concentrate its vital efforts on the "A" in C^3A: Applications of space remotely sensed environmental data for guarding, maintaining and improving the quality of life on planet Earth.

Search and Rescue Satellite Aided Tracking (SARSAT)

M. M. Blume

The Department of Communications of Canada (DOC), the Centre National d'Etudes Spatiales of France (CNES) and the U.S. National Aeronautics and Space Administration (NASA) are participating in an experimental search and rescue satellite aided-tracking project (SARSAT). The U.S. Coast Guard's participation in the project is governed by a Memorandum of Understanding (MOU) between NASA, the National Oceanic and Atmospheric Administration (NOAA), Department of Defense (DOD)(USAF) and the Department of Transportation (DOT)(USCG).

The objective is to achieve international cooperation in search and rescue by demonstrating that equipment carried on satellites in low altitude, high inclination orbits can greatly facilitate the detection and location of distress signals.

These signals are generated by Emergency Locator Transmitters (ELTs) and Emergency Position Indicating Radio Beacons (EPIRBs) carried on general aviation aircraft and some classes of marine vessels. The detection and location will be accomplished by relaying, via satellite, distress information to ground stations which will complete the information processing and transmit position location to rescue services.

**This article originally appeared in the January 1982 issue of SIGNAL.*

Current search and rescue (SAR) operations in the United States, Canada and France use the emission from low power radio beacons operating at the 121.5 and 243 MHz distress frequencies. These devices are carried on approximately 190,000 civil aircraft and 2,000 vessels as a result of existing U.S., Canadian and French legislation. This legislation requires civil aircraft and some classes of registered marine vessels to carry a device which may alert SAR forces to an emergency and assist SAR forces in locating the distressed aircraft or vessel. The emergency radio beacons are of two types: the ELT (used by the aviation community) and the EPIRB (used by the marine community).

Current SAR operations rely heavily upon ELTs and EPIRBs. ELT/EPIRBs are contributing to quicker finds of aircraft crashes, the saving of lives and reduction of flying hours necessary for search missions. Nevertheless, there are severe problems which limit the effectiveness of the current ELT/EPRIBs.

The main problem is the lack of an effectiveness monitoring system. On long overwater flights, International Civil Aviation Organization (ICAO) regulations require monitoring of 121.5 MHz. This affords minimal overwater coverage. However, no such regulations exist for domestic flights, and thus, even this minimal inflight monitoring cannot be counted on.

If an ELT/EPIRB signal is detected, direction finding equipment is usually not available to the person hearing the initially detected signal to determine the emergency location. Thus, the ELT/EPIRB's effectiveness in alerting SAR forces and locating an emergency is limited. Additionally, existing ELT/EPIRBs generate a large number of false alarms. Determining the validity and locating the alarms is difficult and time consuming. Unfortunately, existing ELT/EPIRBs exhibit characteristics that make solving these basic alerting and location problems by satellite techniques very difficult. The most confining of these characteristics are the relatively low radiated power, the poor spectral characteristics and operation in noisy communications bands.

Application of Satellite Technology

Canada and the United States have been investigating the application of satellite technology to the SAR problem. Experiments using the Orbiting Satellite Carrying Amateur Radio (AMSAT OSCAR) satellites during 1975 and 1976 demonstrated the validity of the proposed SARSAT system design. The U.S. Interagency Committee for Search and Rescue (ICSAR) study in 1976, recommended that a satellite aided search and rescue demonstration be implemented. Until 1976, the efforts by Canada and the United States had been conducted independently.

Now joint U.S./Canadian efforts have defined a mission to demonstrate a satellite aided SAR system which has been designated SARSAT. Meanwhile, France (CNES) developed the ARGOS system for environmental data collection and beacon location using a spaceborne instrument on NOAA TIROS-N spacecraft. This concept proved to be a viable applicable for distress detection and location purposes. During 1977, France (CNES) became a participant in the SARSAT joint demonstration project.

The SARSAT system will provide two modes of operation, a local or regional coverage mode and a full-orbit or global coverage mode.

In the regional coverage system, a spaceborne repeater will relay actual distress alerts in real time. These alerts will be transmitted from existing 121.5/243 MHz ELT/EPIRBs presently carried by aircraft and vessels and relayed to local user terminal (LUT) ground stations. The regional system will also include a spaceborne repeater for real time relay of experimental ELT/EPIRBs which will operate in the 406 MHz band.

The regional coverage system requires mutual visibility for real time distress alerting. The LUTs will compute ELT/EPIRB position information from the doppler profiles between the ELT/EPIRBs and an 850 km altitude, polar-orbiting satellite tracked by the ground system. This information will then be forwarded to a Mission Control Center (MCC) which will then disseminate distress alert information to the appropriate rescue coordination center (RCC).

Global Coverage

The full-orbit global coverage SARSAT system will operate only with the experimental 406 MHz ELT/EPIRBs. This system will provide global coverage by storing information, received from a 406 MHz ELT/EPIRB distress alert transmission, aboard the satellite. The onboard stored information is transmitted on command to a NOAA Control and Data Acquisition (CDA) ground station which forwards the distress alert information to an MCC for dissemination to RCCs.

For this mode, the spacecraft onboard instrumentation includes a 406 MHz receiver and a preprocessor which computes and time tags the doppler profile relating to the ELT/EPIRB position and stores this information. Recovery and storage of other distress alert data from the 406 MHz ELT/EPIRB transmission is also performed on board the spacecraft, as described later. This system will also work in real time if there is mutual visibility with the ground station.

This system will be implemented so that existing 121.5/243 MHz and experimental 406 MHz signals may be relayed, in real time, and located. In an actual emergency the existing ELT/EPIRB is activated and transmits a weak (approximately 100 mW) amplitude modulated signal. It is received and relayed when the NOAA spacecraft carrying the SAR instrument comes above the radio horizon into mutual visability of the beacon and LUT. For a system employing one spacecraft, average waiting times are less than six hours.

The spacecraft contains a receiver and a repeater for each of the 121.5 MHz, 243 MHz and 406 MHz emergency frequencies. The signals are doppler shifted due to motion between the receiver and transmitter. Each receiver translates its received emergency band to a band of frequencies between 35 KHz and 210 KHz. (See Figure 1). Each band is then phase modulated onto the 1544.5 MHz downlink carrier for transmission to the LUT.

An onboard receiver and preprocessor, for the experimental 406 MHz, will be carried in the spacecraft in addition to the three band repeater. The pre-

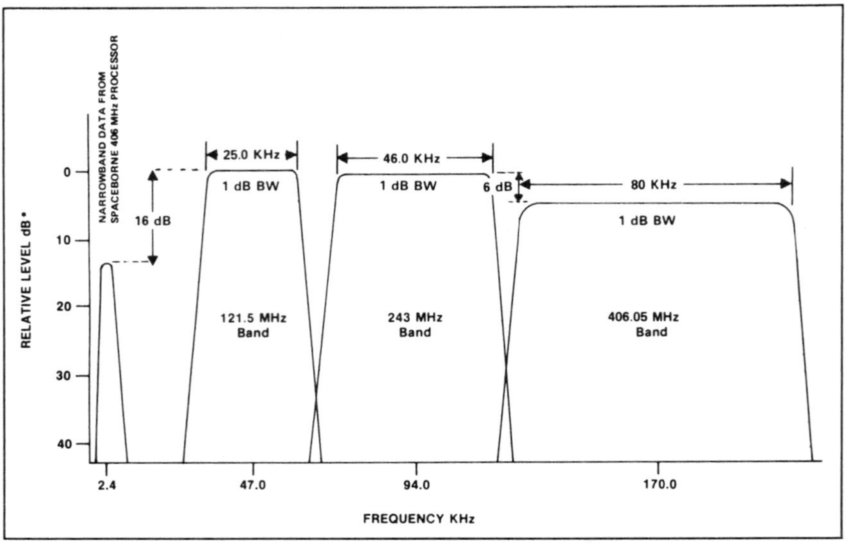

Figure 1. Baseband Spectrum on Real Time L-Band (1544.5 MHz) Downlink

processor measures and time tags the doppler information and handles the coded message data which the 406 MHz beacon will contain. The NOAA spacecraft telemetry system inserts the 406 MHz data into other data gathered by environmental sensors and stores it on one of the five NOAA spacecraft tape recorders.

When the NOAA spacecraft is in view of a NOAA Command and Data Acquisition (CDA) station, the data is dumped. Once on the ground, the data is routed to the NOAA data processing services section (DPSS) where the SAR instrument data is stripped and formatted and forwarded to the U.S. mission control center (MCC). There the position is determined, and the information is forwarded to the rescue coordination center (RCC).

Ground System and Data Exchange

The SARSAT ground system consists of four elements:

1. Local User Terminals (LUTs);
2. Mission Control Centers (MCCs);
3. NOAA ground system modified to support the SARSAT mission; and
4. Emergency Locator Transmitters (ELTs) and Emergency Position Indicating Radio Beacons (EPIRBs).

The LUT receives the phase modulated L-band downlink from the satellite. The downlink repeater data contains remodulated signals and data including: 121.5, 243 MHz and 406.05 MHz ELT/EPIRB band signals and processed 406.025 MHz ELT/EPIRB Experimental Test Unit data. Special processing will be used to search for the weak 121.5/243 MHz emissions.

Once a (121.5/243 MHz) signal is detected, its frequency is measured and the uplink doppler frequency is recovered. The time tagged doppler information, in conjunction with the spacecraft epheremis, is used to determine the ELT/EPIRB position. The LUT forwards, in near real time for regional mode, the alert and position of the distress.

The function of each mission center can be summarized as follows:

1. To monitor overall SARSAT system status in the area served by the MCC;

2. To provide operational input for SARSAT instrument mode and gain settings;

3. To receive stored 406 MHz incident data, stored instrument housekeeping data and ephemeris data, and to distribute the ephemeris data to the LUTs;

4. To collect a data base for evaluation of the system;

5. To evaluate long-term instrument health and trends from telemetry data;

6. To process 406 MHz stored data to determine locations and forward this to appropriate MCCs; and

7. To coordinate SAR incident data between USCG RCCs, USAF RCCs and international MCCs.

Mission Control Centers

Mission Control Centers will be in the United States, Canada and France. The U.S. MCC will be the designated contact point for the SARSAT Mission with the NOAA spacecraft control station and will coordinate spacecraft commands to the SARSAT instruments. The U.S. MCC will be operated and maintained by the U.S. Air Force. It will be co-located at Scott AFB, Illinois, with a LUT (also operated and maintained by USAF). The data flow between LUTs, MCCs and RCCs is illustrated in Figure 2.

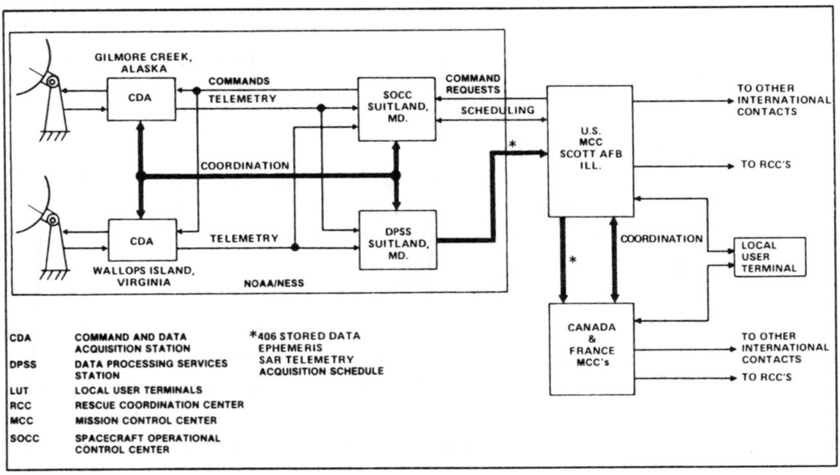

Figure 2. SARSAT Ground System

Operational control of the spacecraft will be through the existing NOAA/ National Environmental Satellite Service (NESS) ground system which will be modified to support the SARSAT mission. (See Figure 2).

Existing 121.5/243 MHz ELT/EPIRBs are designed for reception and audio demodulation with an amplitude modulation (AM) receiver. The electrical characteristics of ELTs and EPIRBs are identical. These devices often lack spectral purity causing widely different spectral shapes. This is a result of the phase incoherence in the carrier signal structure when the amplitude and sweep frequency modulation are applied.

Some ELT/EPIRBs exhibit good coherence properties with a well defined carrier component and a modulation spectrum symmetric about the carrier, but others have displaced carriers with unsymmetric sidebands due to superposition of phase and modulation spectral components.

Laboratory measurements have shown that phase incoherence makes the detection and tracking of these signals by satellite much more difficult. Consequently, the sophistication in the signal processing prohibits processing on board the satellite. This will allow only the regional coverage mode for existing devices. Furthermore, two satellite passes may be required to solve ambiguities in computed positions, of which the estimated accuracy is between 10 and 20 kilometers. Nevertheless, for the near future, these devices will continue to be carried by civil/military aircraft and certain classes of marine vessels, although wider applications are not envisioned.

ELT/EPIRBs (Experimental)

The 406 MHz beacon is designed to operate in a burst mode and with frequency diversity over the emergency band. It transmits every 50 seconds a relatively strong (approximately 5W) signal which lasts 440 ms. The signal is a carrier preamble followed by a digitally coded message. A low level 121.5 MHz signal may also be included to enable homing by equipped search aircraft. These experimental devices will be capable of message transmission in a 280 ms digital message. The data fields will be configured to provide the following information:

1. USER CLASS (air, marine, calibration, test, etc.);
2. Country Code (nationality);
3. Identification Code (a hull or tail number unique to the craft or distress);
4. Emergency Code (sunk/crashed, fire, medical, disabled, etc.);
5. Elapsed Time; and
6. Position (AUTO or MANUAL entry).

The characteristics and capabilities of the 406 MHz ELT/EPIRBs open the possibility of much broader application for future devices. Their characteristcs will preclude position ambiguity and provide an estimated accuracy on the order of 2-5 kilometers. Based upon anticipated performance and capability, it is assumed that usage will find much broader application in both aircraft and vessels. The accuracy may permit the use of these devices in coastal as well as conventional vessels.

Additional applications may include land based use particularly in remote environments (i.e., polar regions, mountainous terrain, etc.). The characteristics of these devices were chosen to be compatible with future geostationary satellite distress monitoring systems. (See Figure 3.) These characteristics will allow onboard spacecraft processing and, therefore, global (stored) mode of operation.

Parameter	Value
RF Signal	
•Frequency	406.025 ± .001 MHz
•Stability	1 part in 10^9 in 100 MS 3-5 Hz in 15 min. 2 KHz in 5 years
•Power Output	5 W + 1 dB; − 2 dB into 50 Ohms
•Spurious	50 dB below 5 W in 5 MHz Carrier harmonics 30 dB below 5 W
•Data Encoding	Bi-Phase L
•Modulation	Phase modulation 1.1 ± 0.1 radians peak
•Modulation Rise Time	The rise and fall times of the modulation waveforms must be less than 0.25 msec.
Digital Message	
•Repetition Rate	50 sec ± 5%
•Transmission Time	440 MS ± 1%
•CW Preamble	160 MS ± 1%
•Digital Message	280 MS ± 1%
•Bit Rate	400 Bits/sec ± 5 Bits/sec
•Bit Sync	All "Ones" (15 "ones")
•Frame Sync	000101111
•Continuous Emission	Not to exceed 45 secs

Figure 3. 406 MHz Experimental ELT/EPIRB Characteristics

Two coverages are available with SARSAT:

1. A regional mode for 121.5/243/406 MHz; and
2. A global mode for 406 MHz only.

In the regional mode, the LUT, ELT/EPIRB and spacecraft must maintain mutual visibility. This defines a geographic area of approximately 1000 Km radius around the LUT. The waiting time between periods of mutual visibility depends on the number of spacecraft, the orbit characteristics and the latitude of the LUTs. In a two satellite system, mutal visibility will occur 20 times a day at Kodiak, Alaska, and eight times a day at Pt. Reyes, California.

In the full-orbit mode, a 406 MHz ELT/EPIRB can be expected to be detected anywhere on the globe. The distress data is stored until the satellite is in sight of a NOAA ground station.

Capacity

The design goal for the 121.5/243 ELT/EPIRBs is a system capacity of up to 10 units although all parameters for a definitive analysis have not been determined. The 406 MHz system is designed as a random access system for high capacity. The design goal will be to receive and preprocess signals from as many as 90 active ELT/EPIRBs simultaneously.

Location accuracy will be highly dependent on:

1. Reference oscillator drift;
2. ELT/EPIRB oscillator drift;
3. ELT/EPIRB altitude uncertainty;
4. Ionospheric effects;
5. Radio Frequency (RF) link interference/noise;
6. Measurement precision; and
7. Ephemeris knowledge.

Performance analyses indicate position determinations of:

1. 10-20 kM (rms) 121.5/243 MHz system, and
2. 2-5 kM (rms) 406 MHz system. A design engineering phase will tentatively begin in the spring of 1982. This is contingent upon satellite launch. It will consist of a three month check-out and a 12 month user demonstration.

The objectives of the D & E are:

1. To test the operational performance of the systems in terms of regional and global coverage, probability of detecting and locating ELT/EPIRB signals, the ability to process multiple distress signals simultaneously and location accuracy;
2. To demonstrate and evaluate SARSAT utility in supporting actual SAR operations (reduction in notification and search time); and
3. To provide users experience with the system.

As part of the effort, the U.S. Coast Guard will operate and maintain two LUTs (Pt. Reyes, California, and Kodiak Is., Alaska). This provides coverage in the 11th, 12th, 13th and 17th Coast Guard districts of the Coast Guard Pacific Area Command. Each district has a Rescue Coordination Center (RCC). The Coast Guard LUTs will interface with the U.S. MCC (Scott AFB, Illinois). The MCC will separate the inland from marine alerts and transmit the marine alerts to the Coast Guard Pacific Area RCC or Coast Guard Atlantic Area RCC. The RCCs will initiate normal SAR procedures.

Satellites and Education: An Alternative

Robert S. Magnant

Because we live in a world that is telecommunications rich, where the forces of competition foster new information service offerings each day, we tend to forget that we are still in the formative stages of a new era. It is an age where technology is offering new means for meeting traditional requirements through the introduction and application of modern telecommunication systems as cost-effective alternatives.

While I will admit a bias for the potential of satellites up front, I am convinced that the reasons for this bias are historically well founded in the development of satellite telecommunications. It should not matter whether the satellites are used to link machines, individuals or institutions, or whether they are supporting military command and control functions or configured for education and training applications. A satellite should be perceived as an alternative technology that can be applied to a specific function or business need and not merely as a communicator's tool.

Satellite Developments

Within this country, a handful of communications entrepeneurs, using a new set of economics, have been offering commercial satellite solutions to business needs for almost 10 years. But without direct involvement in either aerospace or the telecommunications industry, you might believe that "communications by satellite" was a revolution of the 60s and let things go at that;

**This article originally appeared in the January 1984 issue of SIGNAL.*

communications satellites are as second nature to most Americans as moon landings or Shuttle launches.

The "live-via-satellite" caption that appears occasionally on television no longer gives any of us special cause for notice or excitement. Nevertheless, we are observing the birth of this magnificient era and the best results of this foremost application of space technology are yet to come. Some think that our Space Shuttle will do for the world of the future what railroads did for the development of this nation in the past.

If you do not recall what started the Space Age, let me refresh your mind to the events of just 25 years ago. On October 4, 1957, the Soviets launched Sputnik I, a 182 pound package of electronics that orbited the globe every 96 minutes. Because this was the earth's first artificial satellite, the United States suffered a serious blow to its global image as scientific leader of the world, and the public's image of our military superiority was shaken. It is extremely important to remember that, from the very beginning, satellites were inextricably linked to international politics and public policy.

This concept of using artificial satellites for communications was not a new idea; it has been around since 1945. Some may associate the name of Sir Arthur Clarke, the British author, with stories of science fiction and space only and not appreciate his most significant contribution. Although he may be most famous for authoring the movie "2001" with producer Stanley Kubrick, he is also a renowned scientist and responsible for conceiving of the geosynchronous satellite communications concept.

Clarke's proposal envisioned a system of three satellites, orbiting 22,000 miles above the equator, that could relay point-to-point or broadcast communications to any location on the globe. The geostationary orbit is that band in space above the equator, in which satellites circle at a speed equal to the earth's rotation, apparently motionless, balanced between gravity and the pull of deep space.

Because a synchronous satellite has approximately 40 percent of the earth's surface constantly in view, it can provide line-of-sight communications from any point to any point within that area. In its own unique way, the satellite has, in fact, added a new dimension to the world of telecommunications. Not too long ago, when someone asked Clarke why he had never patented his revolutionary description of a global communications system, he replied: "The idea of patenting the geostationary communications satellite concept never occurred to me, and my excuse for this is sheer lack of imagination."

That story should provide us with a humble reminder that we are all human, and that human communication and understanding are vital and must be at the very heart of a telecommunication system.

The meaning of satellite communication services to a potential user or to a specific community of interest is most important. The tendency has been for the technology to drive the market, but with a better understanding of the potential applications and the origins of today's commercial systems, the user's needs will become the dominant force that will shape future systems. But how did we get to where we are today? What makes us wake up one morning and suddenly find a booming commercial satellite industry? We know that policy played a major role; Sputnik definitely got things hopping.

But what factors have been shaping the development of satellite communications? The basic three that can be easily recognized are national policy, space technology and social or market pressure.

Policy, Technology and the Marketplace

The Communications Satellite Act of 1962 marked the establishment of a national satellite policy for the United States. The intent of the Act was to improve the global communications network to be responsive to public needs and the national objectives of this nation. This policy baseline that Congress selected for the launching of the satellite communications industry was not simple government regulation, but a set of special techniques intended to produce the results that a competitive economic structure would have provided. Remember that in 1962, the fundamental telecommunications industry was totally regulated for all intents and purposes.

The COMSAT Act created the Communications Satellite Corporation, a quasi-public entity established to develop a commercial communications system that was to be implemented and operated in conjunction with other nations around the world. From the beginning, it was recognized that foreign cooperation was essential if the global dimensions envisioned by Congress were to be achieved. Although COMSAT was initially viewed as an investment in technology, its operational achievements and its earnings potential developed much more rapidly than anyone originally anticipated.

The technology surrounding satellites played a major role in the process. During the early 60s, it still seemed doubtful whether rocketry could successfully achieve the orbital accuracy required for geosynchronous systems or whether small solar powered electronics could be used for the communications links in space as Clarke had previously perceived them. The random orbital technology of Sputnik required multiple satellites for continuous communications coverage and expensive earth station tracking systems. Nevertheless, by 1963, the technology necessary for synchronous systems had been perfected by Hughes Aircraft.

The broadcast properties that are inherent in satellite systems readily foster and accommodate new concepts for the acquisition and delivery of information. Now, with minimum construction, communications capacity can be brought to any location within a satellite's footprint or shared between multiple locations. In actuality, satellites have no terrestrial equivalent.

Societal pressures, the third factor that sparked this new industry, was directly related to market economics and shifts in regulatory policy. By the mid-60s, the bulk of the recent technical innovations in the telecommunications arena had taken place already. Terms like "competition" and "deregulation," however, were still new concepts that had yet to be introduced into this traditionally monopolistic marketplace. Initially, all the emphasis for satellites had been in the international arena.

The 1962 legislation had left the "domestic use of satellites" question cloaked in ambiguity, and for a time, it appeared that COMSAT would establish a monopoly in the domestic satellite business. In fact, Congress had not specifically precluded the use of COMSAT systems for domestic purposes, or the development of additional domestic systems. However, the established domestic communications carriers dwarfed COMSAT and the traditional international carriers and represented powerful financial interests. Because the satellite offered attractive alternatives with potential dollar savings, the domestic satellite question was raised by the emerging competitive sector of the telecommunications industry, rather than by the existing monopoly carriers, COMSAT or the government.

Specialized Markets

In 1965, the American Broadcasting Company (ABC), who was interested in obtaining cheaper transmission costs for the delivery of network programming to its affiliates, surfaced the issue to the Federal Communications Commission (FCC) in conjunction with Hughes Aircraft, who naturally was interested in selling more satellites. Until only recently, the networks always used sophisticated terrestrial systems exclusively, for passing commercial segments, sports feeds and specialized news around the nation. Traditionally, the networks' major interests had been programming, not transmission, and initial opposition to the ABC proposal from the vested interests was fierce. Since AT&T, COMSAT and some of the aerospace manufacturers believed that domestic communications satellites should be permitted only within their personal purview, the regulatory proceedings that were initiated to sort things out dragged on for almost eight years.

In 1972, as part of a major policy shift toward deregulation, the FCC established an "open skies" policy, determining that anyone financially and technically qualified could enter the domestic satellite industry. Many a corporate heart must have stirred when this opportunity of entering an established multibillion dollar market, by means of a new technology, was presented to them.

Canada had already set the pace for this hemisphere, and the world for that matter, in the use of satellites for domestic communications purposes. Its Anik satellites were already launched in 1972, but U.S. satellites (Western Union's WESTAR, RCA's SATCOM and AT&T's COMSTAR) soon followed. While several DOMSATs were in orbit by 1976, there was still much skepticism for the future of this infant industry. Only a few innovators were using the wide bandwidth of satellite systems for specialized applications, and satellite uses were being limited pretty much to the direct replacement of terrestrial telephone circuits in remote or dedicated applications.

But, in 1977, when the entertainment industry realized the satellites could create major changes in "the rules of the game" that was being played, things started happening. Speculating on markets where people would be willing to pay for special services, Ted Turner and other innovative broadcasters began to challenge the role of the traditional networks and their programming monopoly. In conjunction with a growing cable television industry, Turner and others knew that they could make alternate services available through specialized program feeds, reshaping the program content and the network controls. In conjunction with this, the FCC approved the use of small earth stations for the reception of feeds from sources like Home Box Office (HBO), public broadcasting, religious groups and sports networks.

Within two years, "specialized networks" were a reality. While projections for growth within that industry were dramatic, the actual earth station population for this community has long since surpassed the wildest expectations of the time. From a base of less than 100 earth stations in 1977, the number exceeds 50,000 terminals today. Through the satellite alternative, specialized networks now reach up to a third of the televisions within this country.

Defense Initiatives

As a direct result of the satellite's wideband capability, innovative communications carriers are now offering totally new services in addition to their traditional services. Business conferencing, for example, is being used by many firms as a supplement to, or as a substitute for, travel; newspapers are being delivered and printed remotely; consultants are routinely sharing ideas from the remote corners of the globe; and students and teachers are learning how to interact from thousands of miles apart. This economic technology is readily available and in use by many sectors today and there is an assortment of highly successful examples in actual operation. It is the general acceptance by users and institutions that appears to be lagging behind, and it is here that much of the research must be done and where the changes must take place before the full advantage of modern satellite technology can be realized.

As a direct result of some of the early National Aeronautics and Space Administration (NASA) experiments, satellites can already offer tremendous opportunities to the education and training community, particularly because of their ability to share and extend resources. At minimum, there are definitely areas of specialized education and training that are currently being neglected because of resource limitations that could and should be addressed through satellite communications.

One of the greatest possibilities for the satellite's broad applications may be within the Department of Defense's (DOD) education and training communities and I say this for three reasons:

1. The competition that exists between private institutions for student resources is not a factor within the DOD;
2. Common doctrine and instructional material must logically be shared across the services; and
3. Unlike the private sector, the cost of training within the military includes the movement and housing of the students.

A conference in June 1982 entitled "Education and Training by Satellite—A National Resource" highlighted the fact that all the facilities needed for remote classroom access and educational export are available now; nothing new needs to be invented. Interactive television can now be affordably arranged for discrete viewing by select groups. Because the composition of the audience being reached can be known in advance, users can adapt the information being exchanged accordingly, conducting discussions by satellite as if the collective group were at one location.

With such powerful communications tools available, compartmentalized thinking is no longer sufficient. The DOD spent approximately $10 billion last year for individual training and the manpower involved accounts for over 17 percent of the total military man years that are available to defense. Certainly, there are some resource sharing opportunities within those numbers that communications by satellite can provide.

The first educational experiments using satellites were in the mid-70s. They were conducted in conjunction with NASA's Advanced Technology Satellite program, but primarily in sparsely populated regions, specifically Alaska and the northwest portion of the nation. But it was the military, specifically the Army's Health Services Command, who took the first bold steps to transition commercial satellite communications into practical educational applications.

In the fall of 1978, the Brooke Army Medical Center at Fort Sam Houston began looking for alternative ways of providing graduate education for the medical professionals within its health care region. The problem was this: the education and training facilities were at the San Antonio based medical center while the doctors were at the community hospitals of Fort Hood, Fort Sill and Fort Polk, serving the military populations there. In a satellite environment, the solution was simple: bring the locations together electronically.

In less than a year, the traditional world of training and education had changed for the Health Services Command (HSC). Using Western Union's domestic satellite facilities, in conjunction with existing telephone and cable television systems, HSC became the first military organization to exploit the satellite's unique capabilities to meet educational needs. Leased uplink facilities in Dallas and Army owned receive-only earth stations at the remote locations comprise the network. Today, communications costs are approximately $700 an hour.

For more than three years now, the Brooke Army Medical Center and the Academy of Health Sciences have been broadcasting two hours of programming each day from studios at Fort Sam Houston. Real time video provides for visual presence and establishes the human connection with the lecturer or the patient as needed; voice interactivity provides for whatever information exchange may be required between the student and the instructor.

Through innovative programming, the network has grown like topsy, along with the rest of the satellite community. In addition to the medical treatment facilities at Fort Campbell, Kentucky, and Fort Leonard Wood, Missouri, some of the Veterans Administration hospitals, prominent medical centers like Dartmouth and the University of Vermont and several hundred other government and civilian institutions across the country have joined the HSC audience through existing earth stations and cable facilities. This highlights just one example where the hunger for sources of specialized expertise is clearly evident.

Imaginative Steps

What is needed today is the orchestration of this technology by both industry and DOD for its wider application and the general strengthening of this national resource. Like Arthur Clarke, we could be "limited by our imagination," but obviously the military's medical community does not suffer from that handicap either.

Gen. Charles Pixley, USA, the former Surgeon General of the Army, described the Health Service Command's initiative to Congress in this way: "To stretch our educational resources, we have taken some imaginative steps to apply the most modern communications technology available . . . a satellite system that could potentially link all of our fixed installations with each of our medical centers. . . . We are developing a tool that will make medical educational outreach readily available throughout the Army."

Gen. Donn Starry, USA, the former Commander of the Army's Training and Doctrine Command (TRADOC), echoed these sentiments in his closing remarks to the 1980 Commanders Conference. This three day event was conducted between the staffs of the TRADOC schools at 26 locations across the country through the use of commercial satellite communications. In Gen. Starry's words: "We've never ignored technology when it comes to weapons and other combat developments. . . . All we need is a little imagination and incentive to take the electronic road which appears to lead to such abundant rewards in training."

Although the defense community is sponsoring several telecommunications based training initiatives, it cannot be assumed that there is a focused defense initiative in this arena. Current efforts are still somewhat fragmented because common requirements are not genuinely understood. Nevertheless, the time may be appropriate to take steps to develop these communications tools on a national basis.

Discussions have been taking place between the military's five senior services colleges to apply telecommunications to joint war gaming efforts. Such an application would be a step in the right direction, but just the tip of the iceberg. The potential for telecommunication applications to education and training is enormous.

Common user broadcasting from and between the Services' primary centers of expertise and instruction could enhance the training process and have a force multiplier effect throughout the military. Senior Service colleges could set the pace for defense, applying the entire spectrum of communications by satellite to their educational experiences. What better way is there to introduce new ideas into the DOD than through the exposure of them to the military leaders of tomorrow, as an integral part of their War College training. Unfortunately, every time someone proposes a good idea, someone else brings up budgets. Whether we are teaching our children in school, our soldiers in the Service or our professionals in industry, we are developing our manpower, the nation's most precious resource. And for this reason, we can not afford not to invest in such systems; they are investments in the future.

Skeptics within the military should be cautioned that a competitive private sector is making the specialized applications of telecommunications very plausible. Trends in the commercial satellite market indicate that more powerful satellites and cheaper earth stations will continue to change the economics of their application. It is that advantage of multipoint distribution, which the satellite offers, that has driven the likes of HBO, United Press International or ABC to innovate. As more users learn how to apply these services, institutional structures will have to change.

Trends

Within the past two years, because of the growth in commercial applications, the FCC has authorized the construction of 25 new domestic satellites to replace space systems that were approaching the end of their operational lifetimes and to provide basic facilities for new entrants into the DOMSAT market. As we run out of orbital space, using technology and techniques developed in the 60s, new concepts, like space platforms, will have to be used to accommodate multiple systems.

But, what about today? Current systems can already provide advanced information services to specialized user communities like the military. Because of satellites, our capability to communicate and distribute information is global in scope and only limited by our imagination. As with the development of any system, it is people that make the difference: people with ideas who are out in the trenches, chipping away at pieces of the process.

From the individual perspective, progress may sometimes appear slow, particularly with no visible champion leading the way; the "Ted Turner of education and training" has yet to surface. Nevertheless, until "he" does, the brave innovators should take comfort in the vision of George Bernard Shaw, who perceived that "reasonable men adapt themselves to the world while the unreasonable ones persist in trying to adapt the world to themselves, concluding therefore that all progress depends upon unreasonable men."

Nothing could be more important to the future security of this nation than the best education that society and our institutional systems can deliver. Telecommunication systems will continue to evolve as an integral part of our institutions, offering training and educational alternatives that were essentially impossible to conceive of just a decade ago.

And satellites, particularly in a domestic role, will continue to influence and shape our traditional processes. As the technology reduced costs make specialized networks more practical and bring services tailored to specific applications closer to reality, the communications satellite may prove to be the most significant by-product of the space program to date and the tool that makes it possible to change from old ways of doing business.

Free-Space Optical Communications

Arthur R. Kraemer

Over the past ten years, free-space optical communication technology and system development have progressed to the point where operational use can seriously be considered. The purpose of this article is to summarize the progress made and to describe the special advantages that an optical communication system offers in an operational mode.

Free-space optical communications is defined for the purpose of this article to mean that the transmission medium does not confine the optical beam (as opposed to fiber optical communications where the optical beam is totally enclosed by the fiber). By this definition, free space then includes the vacuum of outer space, the atmosphere and water. Optical communication systems capable of operation in each of these media will be discussed. Particular emphasis will be placed on the principal advantages of optical communications, namely:

1. High data rates (Gbps);
2. Privacy, or low probability of intercept;
3. Antijamming properties; and
4. Nuclear environment operation.

The principal disadvantage of optical communication systems compared to Radio Frequency (RF) systems is the loss in the optical beam caused by scattering and absorption when the transmission medium is not the vacuum of outer space. However, there are methods to overcome this disadvantage for certain operations within the atmosphere. For the case of a water medium, optical communication systems operating in the blue/green portion of the spectrum can penetrate to depths of several hundreds of feet where penetration by conventional RF systems is virtually nil.

This article originally appeared in the October 1977 issue of* ***SIGNAL.*

Spaceborne Systems

There are several optical communication system configurations where one or both ends of the link are located in a satellite. A typical system configuration is shown in Figure 1. Here, low earth orbiting satellites, generating data from various locations, transmit the data to a synchronous satellite which in turn relays it to receivers located in the continental United States. Depending on the location of the low earth orbiting satellite during its data dump, a synchronous satellite to synchronous satellite crosslink may be employed.

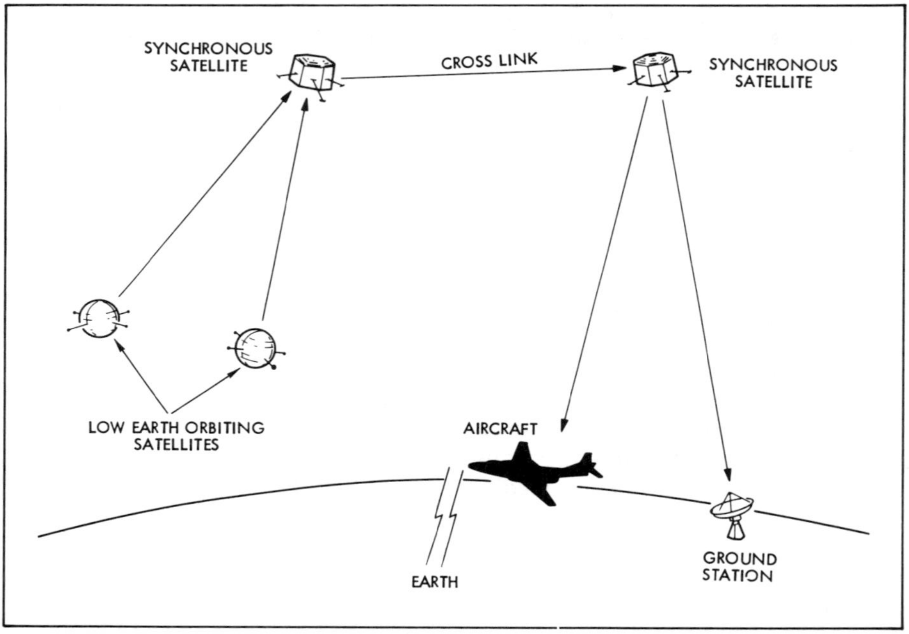

Figure 1. Spaceborne Optical Communication System Configuration

The final receiver can be located either on the ground or in an aircraft, thus providing a real time high data rate link to the point where the data is needed. There is no need for the low earth orbiting satellite to store data until it comes within range of a ground station and no need for a worldwide network of ground stations.

Direct Detection Systems

The technology required to provide a 1 Gbps data rate system for such a satellite configuration exists today. The Air Force's 405B Laser Communication Program, started in 1971, has progressed through system concept design, component development, brassboard and engineering feasibility model phases. Presently, the third iteration of the system is being designed for a planned spaceborne flight in 1981.

This system utilizes a frequency doubled neodymium: yttrium aluminum garnet (Nd:YAG) laser operating at 532 nm (green light) and a direct detection, as opposed to heterodyne detection, scheme.

Two different methods have been developed for providing the optical pumping required to drive the laser. The first uses a radiatively cooled potassium-rubidium (K-Rb) metal-arc lamp which draws 250 watts of prime power to provide a laser output power of 200 mw. The second uses the sun's rays directly to pump the laser, and requires no prime power to provide a laser output power of 400 mw.

In the configuration of Figure 1, the sun pumped laser is used on the synchronous satellites to provide the satellite-to-ground link and the synchronous to synchronous crosslink. The lamp pumped laser is used on the low earth orbiting satellite, where direct sunlight is not always available, and can also be used on the synchronous satellites to ensure that the link is maintained during the short eclipse periods. Both lasers are conductively cooled using heat pipes to space radiators, and do not require the flowing fluids and pumps normally associated with standard Nd:YAG lasers.

Several digital modulation formats are possible with this system, but the one that makes the most efficient use of the transmitted power is Pulse Quaternary Modulation (PQM). The concept is illustrated in Figure 2. The mode locked laser output consists of a series of 300 picosecond, polarized pulses at a rate of 500 Mpps.

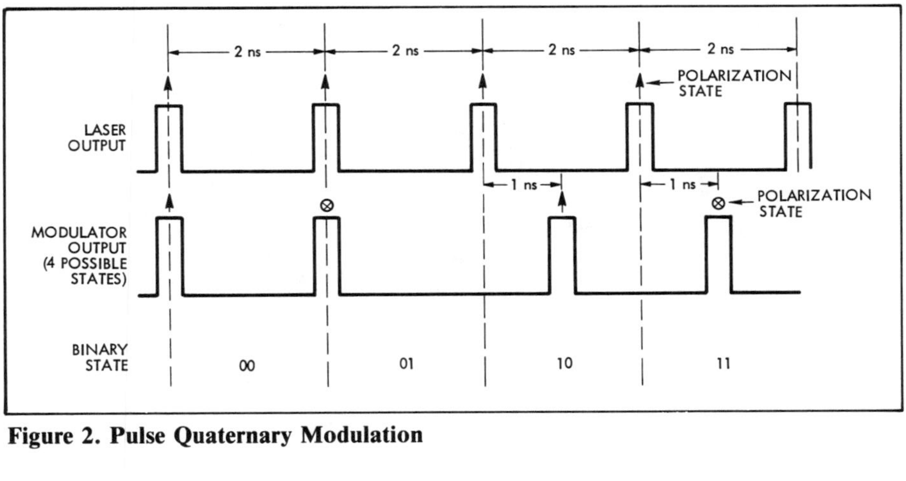

Figure 2. Pulse Quaternary Modulation

The modulator consists of electrooptic crystals which rotate the polarization of the incoming laser pulses, polarization beam splitters and optical delay lines. This combination is used to create four possible states, two time positions and two polarization orientations, for the pulse leaving the modulator. Since each pulse then contains two bits of information and the pulse rate is 500 Mpps, the information data rate is 1 Gbps.

One key feature developed for this system is the asynchronous multiplexer which allows the total 1 Gbps data rate to be made up from several sources of data. The synchronous orbiting data relay satellite can then simultaneously relay data from several low earth orbiting satellites without requiring the data streams from all of the satellites to be synchronized. Each satellite can have its own independent clock.

The receiver system utilizes a polarization beam splitter to direct the two possible polarization states to separate detectors. The detectors are dynamic crossed field photomultiplier tubes which are internally gated and have sufficient response times to detect the 300 picosecond pulses. Utilizing a phase locked loop to provide timing, the receiver decodes the two bits of information in each pulse by noting which detector received the pulse and in which time slot the pulse was received.

The transmitting optical "antenna" for this system is 19 cm in diameter while the spaceborne receiving optics is 61 cm in diameter. This combination allows for a 10 dB margin in transmitting the 1 Gbps data with a probability of bit error of 10^{-6}.

In addition to the small "antenna" sizes, optical communication systems utilize very narrow transmitter beamwidths and receiver fields-of-view. Thus, the gain of these small optical antennas is extremely high. This is the principal reason for the security and low probability of intercept provided by optical communication systems. However, these narrow beamwidths must be acquired, pointed and tracked to high accuracies.

The acquisition sequence utilized in the Air Force 1 Gbps system is shown in Figure 3. Here, the synchronous satellite contains the receiver and the low earth orbiting satellite contains the transmitter. (The sequence also applies for any of the other links shown in Figure 1.)

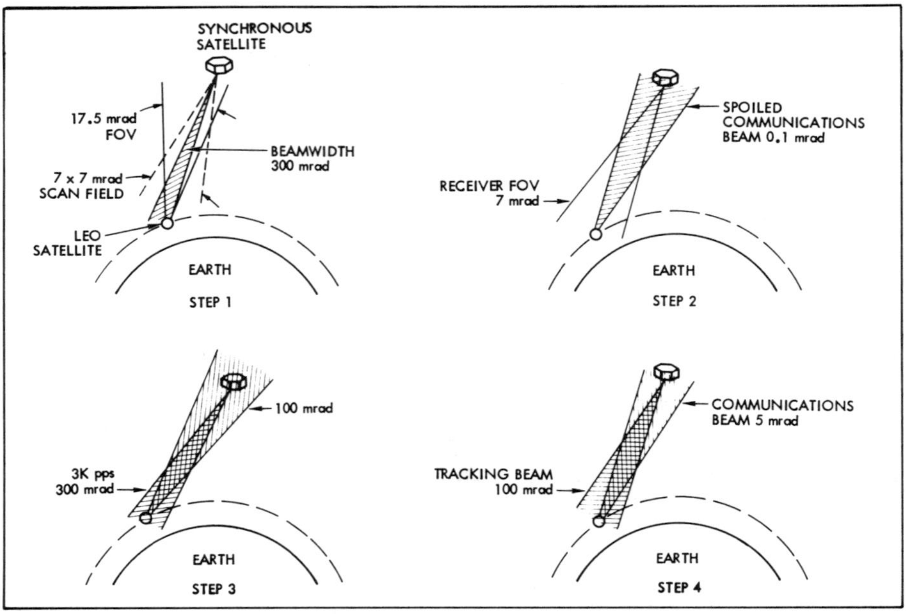

Figure 3. Acquisition Sequence

In step 1, the receiver initiates acquisition by scanning a 300 microradian (mrad) beacon laser beam over a 7 × 7 mrad area at a frame rate of 10 Hz. The beacon laser operates at 1064 nm (near infrared) at a pulse rate of 3 Kpps. The scan field angle corresponds to the synchronous satellite's attitude uncertainty, and with reasonable knowledge of the low earth orbiting satellite's ephemeris, the beacon laser will hit the low earth orbiting satellite 10 times per second.

Meanwhile, the low earth orbiting satellite points its 17.5 milliradian acquisition field-of-view (corresponding to its attitude uncertainty) in the direction of the synchronous satellite.

In step 2, the low earth satellite tracks the 10 Hz beacon laser and points the 532 nm communication beam (diverged at this point to 100 microradians) back along the path of the beacon.

In step 3, the synchronous satellite now tracks the diverged communication beam, stops scanning its beacon laser, and points it back along the path of the communication beam.

In step 4, the low earth orbiting satellite now receives a tracking beacon at 3 Kpps which allows sufficient tracking accuracy (better than 1 microradian) to narrow the communication beam to five microradians. When the synchronous satellite receives the narrowed communication beam, it also narrows the tracking beacon to 100 microradians.

At this point, each satellite is accurately tracking the other independent of its satellite attitude control system, and communication can start. This whole sequence has been repeatedly demonstrated, under simulated satellite motion and worst case vibrational conditions, to take place in less than five seconds.

The beacon laser not only serves as a tracking beam but also can transmit low data rate information. By modulating the intervals between the nominal 3 Kpps pulse rate (Pulse Interval Modulation), information at a 20 Kbps rate can be transmitted. In this manner, updated command and control information and new ephemeris data can be transmitted to the low earth orbiting satellite. This technique could also be used to provide secure, low data rate, command and control information to any satellite, even if it were not using the high data rate relay system.

The prime contractor for the Air Force 405B optical communications system is McDonnell-Douglas Astronautics Company. GTE Sylvania supplies both the lamp pumped and sun pumped lasers and the optical mechanical subsystem including the optical transmitters, gimbals and solar collector; Motorola provides the high speed asynchronous multiplexing and demultiplexing electronics; Varian supplies the dynamic crossed field photomultipliers; and ILC provides the metal-arc pump lamp for the laser.

Heterodyne Systems

Another optical communication system, utilizing CO_2 lasers and having a data rate of 300 Mbps, has been pursued under the sponsorship of NASA's Goddard Space Flight Center. A laboratory model of the spaceflight configuration was completed in 1975. However, due to changing NASA priorities, further development has been halted.

This system, operating at 10.6μm (infrared), uses optical heterodyne detection which requires that a CO_2 laser local oscillator be part of the receiver package. (Heterodyne operation is necessary at this wavelength because the available detectors do not provide the noiseless gain provided by photomultiplier tubes operating at shorter wavelengths. Therefore, to achieve system sensitivities close to the quantum noise limit, as achieved by direct detection at visible wavelengths, heterodyne detection must be used.)

Heterodyne detection has the advantage of being relatively insensitive to the background radiation. In fact, the CO_2 heterodyne system could operate with the sun directly within the receiver's field-of-view (assuming that the sun's energy directly focused by the receiving telescope does not destroy any internal components). However, in order to maintain a constant "If" in the detection amplifiers, the heterodyne system must compensate for the Doppler shifts in the optical frequency, which can be as high as 700 MHz for low earth orbiting to synchronous links.

The conductively cooled CO_2 transmitter laser developed for this system has an output power of 1 watt and requires approximately 60 watts of prime power. The local oscillator is a high pressure waveguide laser. The high pressure serves to broaden the bandwidth of the laser gain medium and allow the local oscillator to be tuned over a ± 400 MHz frequency range, thus tracking out the optical Doppler shifts and maintaining a constant heterodyne "If."

Although the CO_2 laser itself is inherently more efficient with respect to satellite prime power than a lamp pumped Nd:YAG laser, the reverse is true of the optical modulators. The 300 Mbps system uses an electrooptic crystal inside the laser cavity which, when electrically activated, couples a pulse of laser energy out. The technique is known as cavity coupling. The most suitable digital modulation format is nonreturn to zero (NRZ).

The modulator required 90 watts of prime power for a 300 Mbps data rate. The modulator drive power increases significantly with data rate. Although the cavity coupling technique has been experimentally demonstrated at 1 Gbps, an efficient unit suitable for spaceborne operation is not presently available.

The CO_2 heterodyne system is, however, readily adaptable to analog operation. By placing an electrooptic modulator within the laser cavity, the optical frequency of the laser can be rapidly tuned, resulting in an analog frequency modulation (FM) system.

The detector for the CO_2 heterodyne system is a mercury-cadmium-tellurium (HgCdTe) photodiode which must be cooled to approximately 100°K to achieve the desired system sensitivity. A radiative cooler weighing approximately 3.5 lb. has been developed to maintain the required detector temperature.

The transmitter and receiver antenna diameters are both 25 cm resulting in a transmitted beamwidth of approximately 80 microradians and a receiver field-of-view of 100 microradians. Acquisition and tracking are accomplished in a manner analogous to that shown in Figure 3. Since the transmitter beamwidth is wider than that of the direct detection system, the transmitter pointing accuracy need not be as high.

However, the receiver tracking accuracy must be higher because the alignment of the received optical wavefront must be maintained parallel to the local oscillator wavefront at the detector in order to maintain the optimum signal to noise ratio (SNR). In the direct detection system, the received optical energy need only reach the detector surface. The angle of arrival is of no importance.

The NASA CO_2 heterodyne system prime contractor was Hughes Aircraft Company. AIL supplied the detector system, and Arthur D. Little supplied the radiation cooler.

Open-Loop Pointing Systems

The Navy, under sponsorship of Naval Research Laboratory (NRL), conducted a study which showed that it is possible to point a 100 microradian optical beam from a single satellite, without benefit of a tracking beacon from the receiver satellite or from a ground terminal. The technique uses star trackers and computer processing to provide the pointing information to the gimbals and fine beam steerers of a system similar to the Air Force's direct detection system.

This method could be used to send information to any point on earth within view of the transmitting satellite where it is not convenient or feasible to have a beacon as part of ground terminal, such as in a submarine. The data rate would be far lower than 1 Gbps since a high peak power, low repetition rate (10 to 100 pps) laser would be utilized. The principal use would be to quickly dump information to remote locations.

Security Considerations

The narrow transmitter beamwidths and receiver field-of-view inherent in optical communication systems offer distinct advantages when the security aspects are considered. For example, the 5 microradian transmitter beamwidth of the U.S. Air Force's program covers an area on the ground 0.12 miles in diameter when transmitted from a synchronous satellite. (See Figure 4.)

The sidelobe power from an optical antenna is very low. As shown in Figure 4, beyond an area 0.8 mile in diameter virtually no sidelobe power exists. Similar dimensions for beam footprints exist for the other possible links shown in Figure 1. The tracking beacon beamwidths which may also contain data, range from 100 to 300 microradians. The corresponding areas for intercept have 2.5 to 7.5 mile diameters. These small beam footprints significantly reduce the probability of intercept even when compared to mm wave systems.

Since the receiver fields-of-view are also very narrow compared to RF or mm systems, an optical communication system is significantly less susceptible to jamming. For a typical field-of-view of 100 microradians, a jammer located further than 10 miles away from the transmitter would need a J/S ratio of

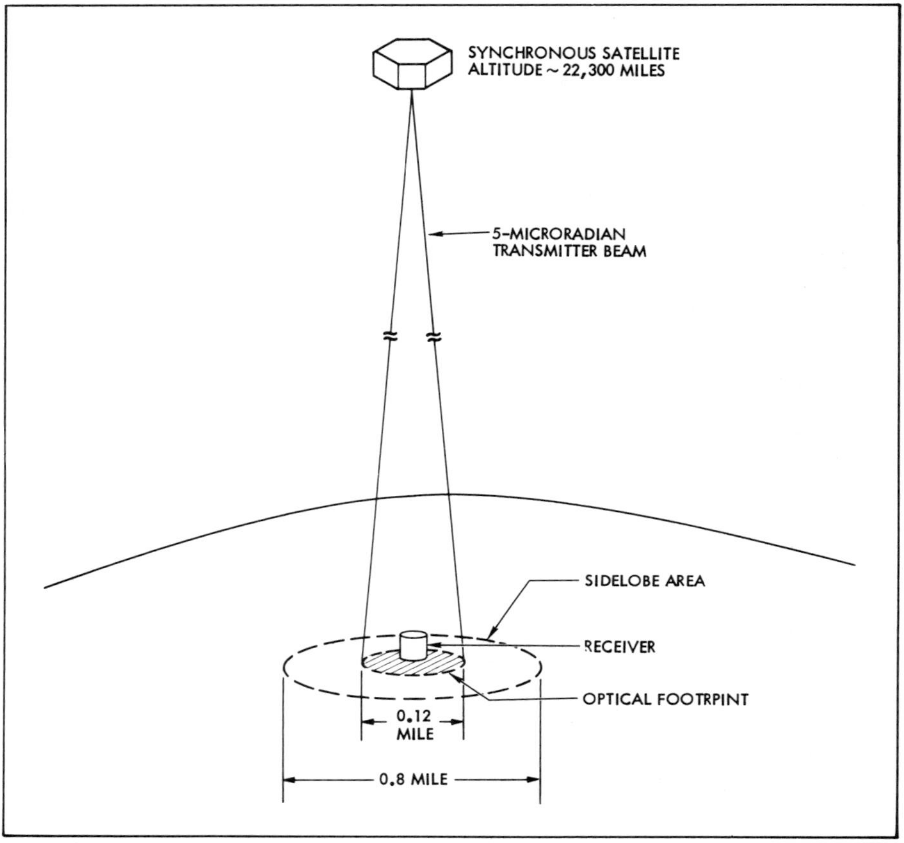

Figure 4. Optical Footprint on Earth for 5 Microradian Beamwidth

greater than 10^4. Although CO_2 lasers with power capable of providing this J/S ratio do exist, Nd:YAG lasers with this power capability do not exist.

Interruption of the communication link by ionospheric perturbation resulting from nuclear blasts is a problem faced by RF systems. However, for an electron density that a 60 GHz system is just beginning to penetrate but that causes an attenuation greater than 10^5 dB/Km, a CO_2 system operating at 10.6μm has an attenuation of about 1.5 dB/Km while a Nd:YAG system operating at 0.53μm, an attenuation of less than 0.1 dB/Km. Therefore, optical communication systems offer the capability of penetrating plasmas that cause blackout of RF systems.

Atmospheric Operation

The principal limitation of space-to-ground optical communication systems is weather. The optical beam will not penetrate thick clouds. Thin layers of cirrus clouds can be penetrated, however. Typically, an optical communication system operating in the visible portion of the spectrum will suffer only a 10 dB loss after passing through a kilometer of cirrus clouds.

The weather aspect thus places restrictions on the locations of ground stations. However, by having alternate ground stations properly located, a highly reliable space-to-ground optical communication system can be implemented. For example, based on studies of cloud cover, three ground stations properly located in the Southwest would provide a 99.9 percent probability that at least one was available. A three-ground-site configuration that included Washington, D.C., which has only a 58 percent probability of being available as one of the sites, would still have a 99 percent probability that at least one site was available.

The transmission of a clear atmosphere for an optical communication beam (532 nm) varies from 70 percent near zenith to 49 percent at 60 degrees from zenith. The low transmission at low zenith angles limits the suitability of a low earth orbiting to ground links. However, the period over which this type of satellite is within view of any ground station is short, and the information is best relayed to ground by using a synchronous satellite.

Atmospheric turbulence, which causes a laser beam to have "hot spots" and holes, has a negligible effect on a space-to-ground direct detection system. For receiver diameters much greater than 4 cm at $0.53\mu m$, the higher energies contained in the hot spots and the lower energies contained in the holes average out. For direct detection systems, the ground receiver aperture size can be increased from the 61 cm required for a spaceborne receiver to overcome losses caused by the intervening atmosphere since the weight of the ground receiver is not an important consideration. (The ground receiver planned for the Air Force's program has a 122 cm diameter.) However, for a CO_2 heterodyne system, the aperture diameter of the ground receiver cannot be larger than approximately 30 cm. This is because the heterodyne system must have an undistorted optical wavefront to maintain its sensitivity.

Due to atmospheric turbulence, undistorted wavefronts are limited to sizes of approximately 30 cm for a space-to-ground link at $10.6\mu m$. If the diameter of the heterodyne receiver were increased beyond this limit, the signal power would decrease because parts of the received beam would be 180 degrees out of phase with other parts, thus lowering the net received signal.

Ground Based Systems

The atmospheric attenuation caused by aerosol scattering and turbulence severely limits the range of horizontal optical communication systems. Since the scattering is more severe for shorter wavelengths, the longer wavelength CO_2 systems have better performance capabilities for horizontal links. The Army, under the sponsorship of the Army Electronics Command (ECOM), developed a 5 Mbps, CO_2 system capable of operation over a 5-mile path. The transmitter and receiver packages each weigh approximately 60 lbs. which makes the system relatively portable. The advantage of such a system is that data can be transmitted from point to point without being intercepted by anyone not directly in the narrow beam.

The scattering, which limits the range of direct line-of-sight optical communication links, can be utilized to complete over-the-horizon links since most of the scattering occurs in the forward direction. A study and experimental measurements performed by the Naval Electronics Laboratory Center has shown that an over-the-horizon link at 1.06μm is feasible.

Data rates of 2400 bps over distances of 50 to 100 Km are possible between two sea level points. The range can be extended to 200 to 400 Km by use of a remotely piloted vehicle (RPV) or aircraft at one end of the link. The range is dependent on the aerosol level present in the atmosphere, and the range for night performance is approximately twice that during the day due to the absence of the scattered sunlight background. The link would be relatively secure with the scattered power being down by 10 dB at 30 degrees off axis, 20 dB at 65 degrees and 30 dB at 110 degrees.

Underwater Systems

Optical communication systems operating in the blue/green portion of the spectrum can penetrate significant water depths. The penetration depth is highly dependent on the wavelength used and on the type of water. Figure 5 shows penetration depths for two typical systems operating at different wavelengths as a function of the water types. Further study, experimental system development and testing are required to fully establish the practicality of such a system.

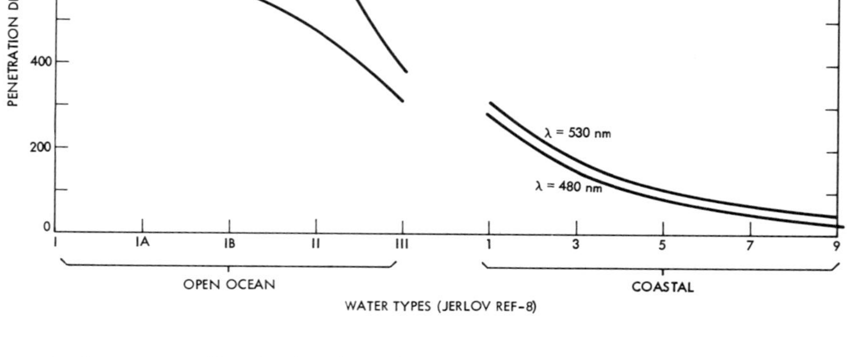

Figure 5. Penetration Depths for Blue/Green Optical Communication Systems

Summary

Optical communication systems operating in free-space are no longer laboratory curiosities but are realities. Several systems have been developed, and the Air Force is well underway with the development of a flight model of a spaceborne system. The distinct advantages of optical communication systems in the areas of data rate and security will soon be realized.

References

[1]Ross, Monte, "Optical Communications in Space," *Optical Engineering,* September/October 1974, pp 375-382.

[2]Kraemer, A. R., "Acquisition and Angle Tracking of Laser Communication Links," *Proceedings of Wescon,* San Francisco, California, August 1971.

[3]McElroy, J. H., McAvoy, N., Johnson, E. H., Degnan, J. J., Goodwin, F. E., Henderson, D. M., Nussmeier, T. A., Stokes, L. S., Peyton, B. J., and Flattau, T., "CO_2 Laser Communication Systems for Near-Earth Space Applications," *Proceedings of the IEEE,* Vol. 65, No. 2, February 1977, pp 221-251.

[4]McDonnell Douglas Astronautics Company—East, "Open Loop Pointing Experiment Study Final Report," Contract No. N00173-75-C-0559, Naval Research Laboratory, Washington, D.C., March 1976.

[5]Titterton, P. J., and Halus, M. N., "Privacy and Susceptibility to Countermeasures of Optical Communication Links," *Proceedings 1970 DOD Laser Conference,* San Diego, California, January 1970.

[6]Hughes Research Laboratories, "10.6μm Laser Communication Systems Final Report," Contract DAAB07-72-C-0051, ECOM, Fort Monmouth, New Jersey, February 1973.

[7]Naval Electronics Laboratory Center, "Extended Line-of-Sight Optical Communications Study," NELC TR 1988, June 1976.

[8]Jerlov, N. G., Optical Oceanography, Elsevier Publishing Co., New York, 1968, Chapter 10.

Commercial SATCOM Interoperability Issues

John L. McLucas and Lee M. Paschall

Abstract

The following is quoted from the Commercial Satellite Survivability Report to the National Security Telecommunications Advisory Committee (NSTAC).

> Satellite communications systems have a strong role to play in the day to day operations of our nation's telecommunications. They offer a means to quickly restore communications to isolated sections of the country. Under the most stressful conditions of a national emergency, satellite communications, in concert with other telecommunications systems, would aid the maintenance and recovery of the economic, political and social structure of the United States.

Government and industry have been working together to bring about the realization of this capability. We believe it can be achieved through a joint cooperative effort, with both government and industry fulfilling their roles to serve the national interest. This paper discusses how interoperability serves as a means to enhance the survivability of the national communications capability and meets the NSTAC charter.

Introduction

The purpose of this paper is to present an unbiased perspective on the feasibility and progress toward achieving commercial satellite communications interoperability — an initiative growing out of the U.S. Government's request to the telecommunications industry to assist in enhancing the survivability of commercial telecommunications systems which support national security and emergency preparedness needs.

The current ongoing effort to achieve interoperability has evolved as a response to Presidential Directive 53, National Security Telecommunications Policy, which subsequently has been replaced with new policy direction as outlined in National Security Decision Directive NSDD 97: "In 1981 the National Security Council (NSC) directed the National Communications System (NCS) to assess the vulnerability of national commercial telecommunications systems and develop initiatives to mitigate vulnerabilities so identified."

The NCS identified a need to establish a framework for joint government/industry planning. This activity resulted in Presidential Executive Order 12382 establishing the National Security Telecommunications Advisory Committee (NSTAC) to provide industry advice and expertise to the President on enhancing survivability of commercial telecommunications systems supporting national security and emergency preparedness (NSEP) requirements.

- The NSTAC is composed of chief executive officers (CEOs) from 30 telecommunications satellite manufacturing and information processing industries.
- They appointed an Industry Executive Subcommittee (IES) to conduct working level, issue oriented meetings through use of working groups: Resource Enhancements (REWG), Emergency Preparedness Procedures (EPPWG) and Funding and Regulatory (FRWG).

Through the recommendations of the NSTAC working groups, a National Coordinating Center has been established by Executive Order 12472, on April 3, 1984. Its purpose is to assist the NCS in supporting the President, NSC, Office of Science and Technology Policy and the Office of Management and Budget, in exercise of wartime and non-wartime emergency functions, in their planning and oversight responsibilities and in providing national security and emergency preparedness communications for the Federal Government under all circumstances.

Since the government is becoming more dependent on communication satellite services, the working groups identified a set of survivability initiatives and created the Commercial Satellite Survivability (CSS) Task Force to review these initiatives and develop a set of recommendations for consideration by NSTAC to be presented to the government. The overall goal of the CSS program is to develop an enhanced ability for commercial satellite communications to provide the telecommunications support necessary for national security in a national emergency environment. The CSS Task Force has recommended to the government a series of initiatives that would lead to enhanced communications survivability:

1. The government community must define current and future requirements for emergency communications through commercial satellite systems.

2. The NCS should define an architecture for emergency communications which clearly defines the role of commercial satellite systems.

3. The NSC must develop the necessary policy and funding process to support implementation.

4. A National Coordinating Mechanism (NCM) should be formed as a forum for the telecommunications industry to coordinate responses to the national security and emergency preparedness needs.

5. A Program Office should be established to implement the network architecture and the recommended survivability enhancements.

The CSS Task Force also identified and assessed six potential enhancements:

1. Contingency Plans and Emergency Procedures;
2. Communication Interoperability;
3. Control Survivability;
4. Communications Security;
5. TT&C Interoperability; and
6. Physical Security.

The consensus and recommendation of NSTAC participants was that the first two initiatives should receive top priority. Communications interoperability was established as a priority initiative to be implemented since it offered a significant potential to provide a means for rapid restoral and improved communications.

Feasibility of Commercial Satellite System Interoperability

Commercial communications satellite interoperability is an achievable goal. Interoperability between earth stations using any of the available C and Ku-band satellites opens a significant communication resource for trunking circuits. This resource can be available in emergency situations when communications have been disrupted by the destruction of key elements of various networks. Interoperability of equipment would provide the potential for restoring telecommunication service to a stricken area and connectivity of critical communications nodes/elements.

However, it can be achieved at several different levels: radio frequency (RF), intermediate frequency (IF), baseband and down to user end item equipment. Interoperability can be most readily achieved at the RF level. Interoperability at the RF requires the compatibility of polarization schemes and earth station hardware to the upconverter/downconverter level as shown in Figure 1. At C-band most satellite operators have designed satellites with 24 36 MHz transponders. A few satellites have 72 MHz transponders; however, the 72 MHz transponders can generally accommodate carriers normally operating on two 36 MHz transponders. There may be some exceptions where full data rate capacity cannot be restored.

There are two polarization schemes in use in the C-band: one uses the opposite sense of linear polarization on an individual channel for the uplink and the downlink. The other uses the same sense of linear polarization for the uplink and the downlink. Earth stations accessing satellites using the same or orthogonal polarizations are interoperable or can be made interoperable by rotating the antenna feed. However, earth stations accessing satellites using different polarization schemes are not interoperable unless each earth station is equipped with a four port feed. The four port feed allows the earth station operator to change the hardware configuration to use the other polarization scheme.

Interoperability between any two earth stations accessing any available satellite would require each earth station to be equipped with a four port feed, to have RF equipment which can be tuned to operate on any channel and to have a repositionable antenna. This would be expensive to implement in all earth stations; however, a subset of key earth stations could be provided with these capabilities. Futhermore, by cataloging an inventory of earth station equipment, a data base can be developed which shows which earth stations can interoperate with one another without equipment modification.

The situation in the Ku-band is more chaotic, since no standard frequency plan or polarization scheme has been developed. Thus, it would be harder to achieve interoperability between Ku-band earth stations for other than narrow-band transmissions. The attraction of the Ku-band is that earth stations are protected from terrestrial microwave interference, enabling small, on-premises earth stations to be erected anywhere. Thus, the deployment of small mobile earth stations in key locations would be advantageous in restoring telecommunications services after it was disrupted.

The next level of interoperability comes at the IF level. Interoperability at this level requires coordination between earth stations on a link by link basis. For example, each link must be using the same modem, which is tuned to the same carrier frequency and has the same code rates and transmission bit rates. Time Division Multiple Access (TDMA) modems must be of the same manufacturer make and model. They have to be operating on the same transponder with the same burst plan. Frequency modulation (FM) systems require the same carrier frequencies, preemphasis characteristics and deviations. Analog systems are perhaps the only systems where two modulators of different manufacturers can be made to talk to one another.

Interoperability at IF requires a much more thorough examination of the hardware involved. It is not solely a matter of repointing an antenna to a common satellite and tuning to a common channel. Compatible equipment must be present at each earth station, and coordination between earth stations on the use of the equipment must be present, much more so than for RF interoperability. The amount of coordination between earth stations can be minimized by defining a standard link for exigencies.

The last level of interoperability is at baseband. Usually, the satellite common carriers point of reference is the multiplexing equipment used to multiplex voice or data channels onto a single carrier. Certain multiplexer equipment used by common carriers, i.e., analog voice multiplexers, digital D3 multiplexers, can interoperate. However, other equipment such as ADPCM multiplexers, with digital speech interpolation and voice/data multiplexers, are manufacturer unique and need the same model with the same multiplexing configuration to recover the signal. As common carriers seek more efficient ways to transmit voice/data, the less likely there is to be interoperability between systems at random. However, prior planning and the implementation of a standard link can provide interoperability between transmission systems.

One could go further and speculate on the interoperability of Customer Premise Equipment (CPE). However, under stressed conditions, the first priority is to establish open voice communications between two points. Here the telephone in all its shapes and sizes comes into play. Fortunately, the proliferation of telephone equipment makes it more than likely that a rotary or Touch-Tone phone is at each end. Once voice communications is reestablished, a single voice circuit can be used as an orderwire to establish links on other circuits using other CPE.

Assessment of an Approach to Interoperability

A degree of interoperability exists today. Satellite transponders are essentially transparent to the signals that are translated through them. Therefore, the space segment does not present any major problem in achieving interoperability. The major consideration in developing interoperability is the configuration of the earth stations and their availability at critical nodes.

Some earth stations are currently interoperable, particularly those of any individual carrier/satellite operator. Many of these are compatible end to end, since most carriers attempt to establish some degree of standardization in both waveform and equipment. Interoperability also can be achieved readily at RF on an intercarrier basis. Intercarrier interoperability at IF or baseband is not readily achievable and any attempt to do it universally is neither a practical nor economically viable alternative. However, for future implementation or retrofit implementation on a selected basis, an interoperable standard link could be defined.

Achievement of Interoperability

Several initiatives must be undertaken to establish sufficient interoperable capability to provide an effective enhancement for commercial communications restoral in a crisis or national emergency. The following actions will form a basic foundation to establish an initial restoral capability through satellite communications station interoperability.

1. Catalog earth stations and determine those which can now interoperate at RF;
2. Identify those earth stations at key nodes which should be modified to provide a level of interoperability;
3. Determine the desired level(s) of interoperability standardization;
4. Define an acceptable standard for minimum essential communications interoperability; and
5. Determine what regulatory issues must be addressed in order to ensure that preestablished authorizations exist in the event of a crisis.

A standard communications link has been defined that should provide sufficient capacity for minimum essential communication in a national emergency. This standard link could be used to establish interconnectivity between any two earth stations in a time of crisis. It would provide a single TI using a D3 multiplexer. This link establishes a 24 circuit capacity consisting of voice or data (up to 56Kbps) between any two nodes. The commercial network survivability (CNS) task force is now exploring how the public switched network may be used during a time of crisis to ensure priority circuits remain functioning or can be quickly reestablished using this standard link.

The technical means of achieving interoperability is probably not as difficult as developing the policy, procedures and funding responsibilities required to implement an interoperable system capability. In addition to defining the technical means of achieving interoperability, the government must:

1. Define responsibilities and assignments;
2. Establish a government Program Office for implementation;
3. Define minimum essential requirements in terms of capacity and connectivity;
4. Define critical system nodes;
5. Establish priority assignments for various classes of users;
6. Develop protocols and procedures for implementation;
7. Define level and method of achieving interoperability;
8. Define a time phased architecture;
9. Establish necessary funding levels consistent with the architecture; and
10. Provide a mechanism for existing any proposed emergency system during peacetime.

Conclusions

A real start has been made through a combination of industry and government cooperative efforts to investigate commercial communications satellite interoperability and begin a program to establish a mechanism by which the government can use this resource during national crises to carry out essential functions. Near term actions have been identified which can provide initial interoperability. The cost associated with this effort may preclude the universal upgrade of all commercial earth stations, but the far term goal must be for all commercial satellite resources to be useable in a national emergency. In order to determine the cost exposures associated with a time phased interoperability upgrade, it is necessary for the government to identify the critical communications requirements and essential communication nodes in a prioritized hierarchy.

The next logical step, which already has been recommended and approved, is to extend the work of the task force to include the terrestrial tail circuits which extend to the major switching centers of both private and public switched networks.

The government must continue to support these efforts, through the implementation of regulatory and national policy actions and the establishment of an adequate funding mechanism/level, if a significant improvement in emergency communications is to be realized. The telecommunications industries also must continue to support these efforts by keeping knowledgeable and capable people committed to these national goals.

The Future is Bright for Communications Satellites

Lee M. Paschall

A shareholder said recently at the Radiation Systems, Inc., annual meeting that "Telecommunications is the most vital element in a stagnant economy." I believe that nearly all of us will agree with that comment. This article will seek to explain in part why that is so, what role satellites play now and will play in the future in what is generally regarded as the growth industry of the 1980s—telecommunications.

Fifteen years ago, Peter Drucker wrote that ". . . information, like electricity, is a form of energy. Electricity is the cheapest, most plentiful and most versatile energy for mechanical work. But information is energy for mind work. This is indeed the first era when energy for mind work has been available. Information through the ages has been all but completely lacking. At best, it has been expensive, late and quite unreliable. . . . The impact of cheap, reliable and universally available information will easily be as was the impact of electricity."

The imminent arrival of the information age has been forecast many times since and clearly has arrived for the one-way dissemination of information (and entertainment) by worldwide broadcast activities made possible by communications satellites of the INTELSAT system. Perhaps less dramatic, but of equal importance, is that the business information age—requiring two-way

**This article originally appeared in the February 1983 issue of SIGNAL.*

interactive information flow—is upon us. In no small measure, this has been made possible by the declining cost of information processing and the advent of the lower cost, high speed and nearly error free satellite transmission media available to businessmen today.

The more astute of those businessmen are beginning to make very imaginative use of that capability. One such example is a transportable earth station which follows a seismic exploration team into remote areas and electrically transmits the seismic data to the home office computers thus maximizing the cost-effective use of the seismic team. There are many other examples of present-day use just in my company's experience, but this article intends to focus on the future. Some trends are now clearly discernible and they pose both challenges and opportunities for user and supplier alike.

But before we examine them, what of the other major element of the information age—the home information marketplace? The direction and speed with which it will arrive is not yet clear. The alternatives are legion and growing in numbers. Will they be entertainment based or flow from the growing use of personal computers? Shopping from the home or working in the home? The list of questions to be answered is almost endless and the answers uncertain because public acceptance of information in the home, except for entertainment, is still an unknown. Numerous trials are underway to try to measure public acceptance, but the jury is still out.

It may well be that the home information market will follow the same pattern as did the business information market—forecast to arrive many times before it actually did so. When it does arrive, satellite communications will be a major player in the game because of its unique advantage in the broadcast mode of operation. In such a case, the return path is likely to be the ubiquitous telephone. Two-way cable and, of course, fiber optics as alternatives to satellites will certainly be major contenders in the densely populated metropolitan areas.

Trends which seem to be emerging that point to the future are:

- Management changes;
- Rapid growth in use of data;
- Large multifunction integrated networks;
- Need for improved decision making;
- Dispersal and automation of production facilities;
- Increased reliability requirements;
- Alternatives to the "last mile" problem; and
- Expansion of optical systems.

Organizational Changes

Several factors are causing management to look at the way companies are organized to compete in the information age where information system expense has become a significant cost of doing business. The first of these is the growing awareness of what the information systems are really costing and the search for ways to reduce or better manage those costs. More often than not, the telecommunications manager is concerned with the telephone system and the data processing manager with the data communication networks. Increasingly, one finds these two management functions overlapping and colliding with each other. Some have begun to bring both telecommunications and data processing functions under single management in an effort to try to reduce costs by combining telephone and data networks; others have tried out of frustration with our unending use of similar, esoteric jargon. But the real stimulus is being driven by efforts at defining and implementing office automation—the "office of the future."

Today's PBXs are beginning to look like computers and small computers can increasingly perform communications tasks. New functions such as "store-and-forward" voice, electronic mail, desk top computer access to central data bases, "smart" terminals which communicate and local area networks will further complicate any management distinction between telecommunications and data processing.

Another major factor impacting organizational structure is the growing diversity of services and suppliers in the competitive telecommunications structure emerging in the United States. More management resources will have to be applied by the user in the multiple vendor environment than when we lived in a regulated monopoly structure. Planning for, acquiring and managing networks obtained from several suppliers will require much more management attention and the skills required to do so are generally in short supply.

Growth of Data Communications

The use of data communications is growing very rapidly as information systems come into being. Voice communication still dominates the marketplace and continues to grow at six to eight percent annually. However, data communications represents a much higher annual growth rate—30 percent to 35 percent by most estimates. Moreover, data speeds are increasing—largely brought about by the availability of the high quality satellite transmission path. Fifty-six kilobit circuits are no longer a rarity and may soon be as widely used as 2.4, 4.8 or 9.6 kilobits are today. In the near future, 56 kilobit and 1.5 megabit circuits will be the norm in large networks.

Very large users have begun to see the utility and cost-effectiveness of large, integrated private networks which can combine many telecommunications functions. Dynamically controllable Time Division Multiple Access (TDMA) satellite systems enable voice, data, facsimile and video paths to be established and be rearranged in a single, wideband data stream as needed to satisfy traffic demands based upon patterns of the business or time of day. For these large users, the day of separate networks for data, voice or video is on the wane. Low speed controllable TDMA systems will offer similar service advantages to medium sized organizations which are heavy users of telecommunications.

The need for improvements in the quality and timeliness of decision making, particularly at the middle management level, is a very important part of improving white collar productivity. That need will cause a dramatic increase in the use of video teleconferencing at compressed data rates. Freeze frame at 56 kilobits and near full motion at 1.5 megabits are readily available today. Saving travel costs, or the time of high-level executives, has most often been cited as the benefit of video teleconferencing and, of course, this benefit is there. But the routine availability of video teleconferencing capability in ordinary conference rooms will produce a much greater impact. Moreover, the marginal cost attributable to video teleconferencing becomes much smaller in a multifunction integrated network based on controllable TDMA satellite transmission methods.

As American industry begins again to make major investments in modernizing its production facilities, there will undoubtedly be a trend towards dispersal of plant facilities outside of the major metropolitan areas. Energy conservation, labor costs and quality of life for the work force are just some of the factors which will further accelerate an already noticeable tendency. More automation will certainly appear, together with Computer Assisted Display Systems/Computer Aided Manufacturing (CADS/CAM) techniques to improve the quality and efficiency of manufacturing processes. Such a dispersed operation is almost an ideal situation to be served by communications satellites with on-premise earth stations linked together with digital data streams operating in a TDMA mode.

Increased Reliability

As large organizations replace manual processes with automated information systems, the reliability of computers and their interconnecting telecommunications becomes more and more critical. The more complex the system becomes, the less likely it is that a manual backup process can be made to work. The costs of even a modest interruption of service can far exceed the expense saved by buying the cheapest thing available. Quality and reliability will increasingly be dominating criteria for business as they have long been for the military.

Where the on-premises earth station is not appropriate, the last mile has been a longstanding problem. Technology and numerous entrepreneurs are beginning to offer alternatives to the universally available twisted pair with all of its weaknesses, particularly for high speed, error-free data. Solutions to this old problem are appearing in the form of digital termination systems, two-way coaxial cable, the advent of fiber optics and the expected arrival of digital devices which could greatly extend the utility of the venerable twisted pair itself. Highly reliable end-to-end high speed digital systems providing almost error-free service will become commonplace even where multiple vendors are involved in providing the service. The local facilities bottleneck will be broken for data streams in the major market areas.

No consideration of the future of communications satellites would be complete without considering the "threat" posed by fiber optic transmission. Fiber optics will clearly be a very cost-effective alternative to satellites over the longer distances which currently favor satellites. Fiber optic cable is, however, very expensive to install, particularly where right-of-ways must be acquired. Thus, their use for the mid-term will likely be limited to interconnecting areas of very high traffic demand.

To remain competitive by the end of this century, except for the unique applications of broadcast and rapid deployment, some technological advances will be required in communications satellites. Some are near at hand. Examples are: the National Aeronautics and Space Administration (NASA) 20/30 gigahertz program which will explore the much greater bandwidth available at those frequencies; cross links between satellites at optical frequencies; scanning beam and on-board switching/processing being planned for MILSTAR; greater throw weight to geosynchronous orbit from the Space Shuttle operating altitude; 10 year on-orbit lifetimes; improved linearity characteristics of transponders, etc.

Domestic communications satellite services have grown at an extraordinary rate during the last five years. One measure of that growth is the number of transponders in orbit. That number will have almost doubled during 1982 and, assuming Federal Communications Commission (FCC) approval of requests currently before it and successful financing by some new entrants, double again by the end of 1985. That growth has been largely due to the unique advantages of satellite transmission, generally agreed to be:

1. Distance insensitivity, making satellite transmission cost competitive for circuits over 500 miles in length, particularly for the rapidly growing high speed data market;
2. Very high quality transmission which produces the very low error rates required by high speed data users;
3. Generally much quicker response time required to install earth stations than that required to extend terrestrial facilities into remote areas; and
4. Area coverage which gives a unique advantage in broadcast applications.

Moreover, the trends which can be observed as we enter the information age demand the kind of transmission media at which communications satellites excel. Technological advances now underway will further enhance this young technology.

The future of satellite communication is indeed bright, even in densely populated, highly developed areas where optical fiber will be a cost-effective alternative. In less densely populated areas and in the developing world, communications satellites have no peer in providing the telecommunications infrastructure essential to the information age.

Part IV: Communications For Air Operations

Editor's Introduction

Today, there is much emphasis on making C^3 more capable and, in particular more survivable. President Reagan has placed improvements to strategic command and control at the forefront of his defense modernization program. Many have valid claims to have identified the problem of communications vulnerabilities. Although Gen. Richard Ellis regards connectivity studies by the Navy and JCS as particularly significant, none did more than he to bring the need for major improvements to strategic connectivity to its present high priority on the national agenda. In a pivotal speech to the Washington AFCEA chapter in late 1980, he concluded as follows: "Modernization of our strategic connectivity systems requires the same funding priority as that enjoyed by our newest strategic weapons systems." His speech "Upgrading Strategic Connectivity" makes a compelling case for that conclusion.

The Strategic Air Command (SAC) has particularly demanding requirements for communications as part of its command and control structure. Beginning with Gen. LeMay, all of its commanders have emphasized that point again and again. Gen. Dougherty, who delights in recalling that his military career began as a communicator (bugler in a National Guard cavalry outfit), is certainly among the most articulate of those SAC leaders. His article, "SAC Command, Control and Communications," vividly portrays the importance of communications to SAC and describes the technological and management improvements being made to further improve SAC's command and control process. It is not only an informative article but also a delight to read.

Every day of the year aircraft of the Military Airlift Command (MAC) fly into remote corners of the world. The communications linkage from many of those out-of-the way places is often tenuous at best. The global delivery of high priority cargo with an airlift resource always in short supply requires the effectiveness of each aircraft be maximized. Added to that imperative is the requirement to be able to rapidly reconfigure the force from normal route flying to massive support for contingency operations. Gen. P.K. Carlton's article, "C^2 in The Military Airlift Command," describes the command and control structure and the control philosophy employed by MAC to do that job.

Gen. Robert Dixon's article, "Tactical Air Force Command and Control," emphasizes the importance of the effective application of tactical airpower in the air land battle. The command and control system is characterized as a key force multiplier. Tactical air command and control and tactical air employment are regarded as inseparable processes, and they must be understood and practiced in the Joint Force context. Finally, C^2 systems must be designed to support commanders who must get "the last drop of effectiveness from available forces." Gen. Dixon was affectionately known in the Air Force for doing just that.

The late Gordon Gould deserves much credit for many of the improvements made in Air Force C^3 activities over the years. His 1968 article "Is The Past A Prologue?" deals with lessons learned in three wars. Always a "systems thinker," he was among the first to clearly articulate the requirement that equipment cannot be designed for purely tactical purposes. He states that there are at least three time frames to be considered in "tactical" situations and that today's worldwide command and control concept may require that tactical equipment be incorporated into the worldwide Defense Communications System (DCS). From those lessons, he concludes that "tactical" communications equipment must be designed so as to satisfy the technical standards required by the worldwide DCS.

MGen. Robert Edge, USAF (Ret.), in his article "Future Tactical Communications Requirements for the USAF" reminds us of a lesson far too often ignored. His thesis is that we must learn to balance the true communications needs against the tendency to exploit technology and to concentrate on those important system parameters which most need improvement.

The Air Force component commander must have a responsive and flexible command and control system in order to exploit the inherent flexibility of airpower. The Tactical Air Control System (TACS) is a network of facilities, equipment, people and procedures that enables the commander to plan, direct and control air operations. The TACS is well described in Gardner's article "Employment of Tactical Air Control Radars." He concludes that the TACS is the heart of the effective application of tactical air power and communications is its life's blood.

Paul Stoney was the best operating communications manager I ever met. His article "Communications in Transition" describes the problems facing a communications system operator with a rapidly growing data traffic demand to be served by an analog system of various age vintages. The fundamental and enduring theme is that telecommunications systems *evolve.* One cannot create a totally new system overnight that will solve all the newly emerging problems caused by the explosive growth of information systems.

Armies fight in the field; navies from their ships; and air forces from their airbases. This, in part, causes the differing management emphasis and styles amongst the military services when they address the "administrative" communications needs of their fixed based structure. The report by Col. William Bettis, USAF (Ret.), on the Air Force "Base Communications Mission Analysis" is included herein because it represented a pioneering effort to look at air base communications needs in terms of an information system. It represents the best analysis the editor has seen of the information transfer needs on modern air bases. In some of its contemplated solutions, it anticipated much of today's current emphasis on local area networks (LANs). Although somewhat lengthy, it deserves careful reading for the many insights it contains.

Finally, this part concludes with three articles ("Air Force C^3 Technologies," "Applications for Airborne Digital Communications" and "Military Applications of Laser Communications") which point to future development and equipping programs leading to further improvements in Air Force C^3 capabilities.

Upgrading Strategic Connectivity

Richard H. Ellis

Now that we know who our Commander in Chief will be for the next four years, I believe this is an appropriate time to discuss matters of far-reaching national importance with the military and industrial principals in the command, control, communications and intelligence (C^3I) community. Also, this is an appropriate audience because our nation relies heavily upon your collective output to provide the strategic connectivity necessary to carry out its deterrent and war fighting policies.

Today, the military capability of this nation is bound together by a network of high technology, automated command, control, communication and intelligence systems. They are essentially peacetime systems; however, . . . conceived, for the most part, in the late 1950s and put into operation in the 1960s. Many are located at fixed sites or depend on ground communications networks and, like all fixed terrestrial sites, they have become highly vulnerable to attack. As a result, communications and connectivity between our national leaders, the field commanders and the forces have become an area of increased attention and much concern in the strategic community.

The need to correct existing connectivity deficiencies and deploy new, survivable, two-way systems takes on even greater meaning and urgency when viewed in light of the growing Soviet threat. Fortunately, Soviet actions and other world events of the past two years, while alarming, have helped forge a strong national consensus and have resulted in an informed and more responsive Congress. In other words, a majority of Americans are now in agreement

This article originally appeared in the January 1981 issue of **SIGNAL.*

that if the United States is to remain an effective international power it must possess the capability to back national commitments with strong military forces and with the resolve to use them if need be.

The prerequisites for such military forces are people and hardware that are competitive with the forces they must fight. To achieve the required competitive edge, America's strategic forces will be deeply involved in across-the-board force modernization for the next decade. Such modernization, however, must be accompanied by a corresponding upgrade of supporting C³I systems if we are to take full advantage of advancing technology and to satisfy changing force management requirements.

Today, national policy for the prosecution of nuclear war differs from that of a generation ago. National guidance now requires the capability for a more measured and flexible use of forces throughout the spectrum of strategic nuclear operations. In the past, when the United States enjoyed clear-cut nuclear superiority over the Soviet Union, emphasis in the communications area was concentrated on force execution with little concern for what would be required to support escalation control or follow-on force management. But, as the strategic equation shifted over the years, there was a corresponding move towards providing the National Command Authorities (NCA) with greater flexibility and additional nuclear response options. This, in turn, significantly increased the need for more and better tools to support force management.

These changes in national policy mandate that our C³I system be survivable and remain operational through all phases of conflict. In other words, we must now possess the wherewithal to ensure continuous worldwide command and control of our forces under wartime stress with the ability to terminate hostilities at the appropriate time. That is the requirement. Unfortunately, Congress and the Defense Department (DOD) have experienced considerable difficulty in agreeing on the best way to achieve this much needed upgrade of our national strategic connectivity net. Meaningful progress in key areas has been disappointingly slow and, if national policy is to be converted into a realistic capabilities plan, much remains to be done.

Some of the problems and disagreements that have surfaced are understandable because of the complexity of the subject and the rapidly expanding technology involved. Nevertheless, the time has come to select a positive and decisive course of action for the remainder of the decade and then go out and do it. Like the follow-on to the B-52, we can study the strategic connectivity issue to death if we try hard enough.

Strategic Connectivity Concept

The overall C³I charter has been clearly spelled out in Presidential Directive 53, as well as other related documents. PD-53 states, in part:

> A survivable communications system is a necessary component of our deterrent posture for defense. In support of national security policy, the nation's telecommunications must provide for connectivity between the National Command Authority (NCA) and strate-

gic and other appropriate forces to support flexible execution of retaliatory strikes during and after an enemy nuclear attack.

With this Directive as a starting point, the first question that should be addressed is: What constitutes survivable and enduring strategic connectivity? Let me take a few moments and give you our definition. Although overlapping in a few areas, we envision strategic connectivity as being made up of seven basic elements.

The lead element, the attack detection network, is also the most demanding. Satellites and ground based radars in this network must provide detailed, unambiguous warning information and provide it quickly.

Next is attack characterization. After receiving detection inputs via survivable communications, the North American Air Defense Command (NORAD) Missile Warning Center must analyze the attack and provide the NCA with both warning and assessment information. Whether the amount of warning time provided by NORAD is adequate depends upon circumstances not entirely under U.S. control. For example, the proximity of Soviet submarines to Washington and coastal military installations greatly influence how much usable warning time is available.

Short Submarine Launched Ballistic Missile (SLBM) flight times also affect the length of time available for the third element of connectivity—the decision making process by the National Command Authorities. Here, we are talking about minutes, very few minutes, during which attack assessment information must be weighed, many possible response options evaluated and a correct decision made. Then, once a decision is reached, an Emergency Action Message must be formatted and disseminated to the forces in the field. The key links in this fourth element are the numerous components of the National Military Command System, including the nuclear CINCs' command posts—both static and airborne.

The fifth element is execution of the NCA decision. Here, responsiveness is critical. Unless all elements of our nuclear triad receive the message prior to impact of incoming Soviet reentry vehicles (RVs), we risk losing large segments of our retaliatory forces. In Strategic Air Command (SAC), systems such as the Primary Alerting System (PAS), the SAC Automated Command and Control System (SACCS—OR 465L), the Air Force Satellite Communications System (AFSATCOM) and the Emergency Rocket Communications System (ERCS), among others, currently provide timely connectivity to our forces.

Once initial execution is complete, we must have sufficient feedback data and intelligence information to evaluate not only the effects of our retaliatory attack on the enemy but also our own ability to continue the war. Satellites, reconnaissance aircraft and supporting communications will be needed to fulfill this sixth requirement of connectivity.

Finally, we need communications in the trans and post-attack period to reconstitute our surviving forces, generate follow-on sorties and retarget our forces, as required. This complicated cycle is then repeated until hostilities are terminated.

We believe this strategic connectivity concept provides the National Command Authorities (NCA) with a closed-loop capability to manage forces throughout a conflict. But to do this, the C^3I network must be survivable, enduring and two-way. Warning, intelligence and force status information must be continuously provided to the nuclear CINCs and the NCA and the decisions of the NCA must then be communicated down the chain to the forces in the field.

One-way communications simply cannot hack it in today's and, more importantly, tomorrow's environment. Let there be no mistake that SAC's requirement, and the nation's for that matter, is for a two-way system that provides for the continuous flow of information "up" and command and execution decisions "down." This is what effective force management and escalation control are all about.

Connectivity Problems and Solutions

There is no doubt in anyone's mind that we have a full plate before us as we go about transforming this concept of strategic connectivity into operational reality. There are problems in each of these seven areas. For example, in an attack, many of our fixed ground based sensors and communications terminals may be destroyed or become inoperative within the first half hour. We can expect massive outages of commercial circuits and power grids. Electromagnetic Pulse (EMP) and jamming will undoubtedly affect major portions of any remaining communications. The potential Soviet Anti-Satellite System (ASAT) capability is a likely problem area. Finally, the continuity of command and leadership at the highest levels, both civilian and military, must be understood and periodically exercised.

Having presented this somewhat gloomy assessment, I feel obligated to describe possible solutions. From my perspective, a program to correct these connectivity shortfalls can be divided into three basic phases.

The first phase has been underway for some time with important elements already completed. It involves a realistic evaluation of where we are today and corrections we should undertake in the immediate future. Several studies were conducted that took a hard look at our C^3 capabilities. The more significant of these was the Chief of Naval Operations' Connectivity Study and the one considered by many to be the most thorough top-to-bottom examination ever undertaken of our nation's connectivity architecture—The Joint Chiefs of Staff Connectivity Study.

The Connectivity Study, directed by the Chairman of the Joint Chiefs of Staff (JCS) and completed in March 1979, highlighted the criticality of our national C^3I network to force management, pointed out the fragile nature of the current system and recommended near term actions to ease some of our current deficiencies. The Study's recommendations, for the most part, centered on improving the present C^3 network with procedural changes and off-the-shelf equipment. All agencies, including the Unified and Specified commands, have now set about implementing the recommendations that were approved by the JCS.

One of the more significant early outcomes of the Study was the creation of a new Joint Chiefs of Staff Agency called the Joint Strategic Connectivity Staff. It is located at Offutt AFB, Nebraska, and is responsible for conducting continuing analysis of strategic connectivity systems and procedures and making recommendations to the Joint Chiefs concerning ways to improve both offensive and defensive C^3I capabilities.

Personnel from each of the Military Services have been assigned to the staff. The SAC Commander serves as the Director and RAdm. Paul Tomb, a veteran submarine officer, is the current Vice Director. I might add that throughout the entire Study and the creation of the Joint Strategic Connectivity Staff, the Navy was a key player and major contributor.

The second phase of SAC's recommended program to improve national connectivity will require a major change in how we currently operate: given the present state-of-the-art and that projected for the next five to ten years, survivable and enduring strategic connectivity in the 1980s requires increased reliance on airborne and space based command and control systems with minimum dependence on vulnerable terrestrial communications.

Many of the initiatives I am referring to in this phase are now programmed or have been accepted as technically feasible. An example is the installation of Automatic Data Processing (ADP) and appropriate warning receivers on board National Command Post and SAC aircraft. The program for single channel transponders on certain satellite systems is another example of a program which should be accelerated.

The expansion and acceleration of the E-4 Advanced Airborne Command Post program, with its related communications and airborne data processing equipment, also merit continuing consideration. SAC has proposed that a seventh E-4 aircraft be authorized and the entire program accelerated. By doing both, we fill the projected gap in the current procurement program and obtain the necessary communications and computer equipment sooner and at lower cost. In addition, SAC believes a seventh E-4 aircraft is needed to satisfy the command's "looking glass" Airborne Command Post requirement.

With regard to space systems, even with full deployment of AFSATCOM in 1983, we still will not have a C^3I network capable of enduring under projected wartime stress. To correct this and provide the NCA with the required connectivity, SAC has urged development of the Strategic Satellite System, or Triple S. We have been an ardent supporter of Triple S for several years and firmly believe it represents the best long-term solution for providing survivable, two-way connectivity that stretches from the National Command Authorities all the way down to the forces in the field. We also believe Triple S offers a means for direct control of the MX fields from the National or SAC Headquarters levels, thus providing important communications redundancy for the centerpiece of our nuclear triad.

The third phase of SAC's recommended program is an intensification of something we have been doing, with varying degrees of urgency, for 30 years—joint military-industry cooperation in the area of strategic communications. The goal would be to modernize and rebuild our terrestrial systems which have served us well for the last three decades but which have become highly vulnerable under the Soviets' massive strategic programs.

Joint Military-Industry Effort

Thus, this last recommendation centers on a proposed joint military-industry effort involving terrestrial based systems. To determine the feasibility and practicality of pursuing such a joint venture, SAC and the Joint Strategic Connectivity Staff have initiated a dialogue with the leaders of major communications companies across the nation. This is not to say we have not worked closely together in the past: we have, and quite successfully, I might add.

The type of joint project I have in mind has a parallel in the Civil Reserve Air Fleet (CRAF). As many of you know, a portion of the CRAF program involves the modification, at DOD expense, of a selected number of commercial aviation aircraft to allow them to augment military airlift in times of crisis.

Let me also point out that joint military-industry projects are not without precedent in the C^3 community. Based on DOD briefings to industry officials concerning threats to stateside communications, AT&T developed a transcontinental hardened cable network during the 1960s. Work began on the East Coast in 1961 and was completed at the shores of the Pacific in 1964. A second cable was laid along a more northern route which, conveniently, went through Nebraska and was finished in 1969.

Our present proposal deals almost exclusively with long-range planning and is designed to meet the needs of national defense as well as those of the communications industry. We will be investigating ways to blend future C^3I military requirements into those normal modernization efforts that would be undertaken by communication companies to upgrade their commercial circuits and networks. By so doing, a large amount of redundancy in key nodes and grids could be available for military use during periods of national emergency and at a greatly reduced cost.

As you can well imagine, bringing this concept to fruition will require detailed planning, extensive coordination and a great deal of cooperation on both sides of the military-industrial complex. But, by planning now, we think it may be possible to implement some of the following ideas:

1. Incorporate wartime survivability into the design of commercial satellites that would serve the commercial world in peacetime and the military in the event of war. For example, we should incorporate satellite protection against nuclear effects and add a secure command channel to every communications satellite placed in orbit.

2. Include electromagnetic pulse (EMP) protection in commercial switching centers, microwave relays and other transmission system elements. With the increased use of fiber optics by industry, EMP protection should also be considered in the design of future networks utilizing this important new technology.

3. Finally, as industry moves ahead with designs for new satellites and ground based facilities, we should plan and build an interconnect between these civilian systems and purely military systems to provide a mutual hedge against enemy attacks and natural disasters.

Conclusion

The three phase program I have just descibed is an ambitious one, but it is a challenge we must actively pursue if survivable, enduring, two-way strategic connectivity is to become a reality. Satisfying this national requirement will demand the absolute best the military-industrial team has to offer.

Finally, let me leave you with this thought. Modernization of our strategic connectivity systems requires the same funding priority as that enjoyed by our newest strategic weapons systems. Without it, the credibility of our entire deterrent policy is suspect. Without it, we will be unable to achieve the type of stable strategic relationship with the Soviet Union that provides an opportunity for responsible nuclear disarmament negotiations. And, most importantly, without this funding priority now, we have little to look forward to in the future in the way of relief from the continually growing defense burden on the Federal Budget.

In summary, our strategic connectivity priorities for the 1980s are evident: the technology is available and the need is absolute.

SAC Command, Control and Communications

Russell E. Dougherty

During this past year, Strategic Air Command (SAC) and the Air Force Communications Service (AFCS) have taken a major step in an attempt to "exploit C³ posture." As Commander in Chief, Strategic Air Command, I initiated the action to transfer over 5,700 SAC Communications personnel to the Air Force Communication Service, commanded by MGen. Rupert H. Burris, USAF.

I recognize that there will be some who may not accept fully my view that this action will help SAC to exploit command, control and communications (C³) technology or will improve our C³ posture; the merger was not without concern in some quarters. Those concerns were not ignored, but I considered this was a move that we should make and one that, on balance, promised greater opportunities to enhance both the technological competence and the ability of the communicators to serve Strategic Air Command's C³ programs and processes. Gen. Burris and I have "bound our wrists in blood" to that end; we are detemined to make this transfer effective for SAC, the Air Force and the entire Department of Defense (DOD).

The actual SAC command and control process, and the diverse and versatile array of communications systems we employ to support that process, were outlined in some detail in the article I wrote for *SIGNAL* in the March 1975 issue. So, I will not again develop these details; rather, I will focus on the managerial and technological innovations that are being employed by SAC and by Air Force Communications Service to improve the command and control process thoughout Strategic Air Command.

**This article originally appeared in the January 1981 issue of SIGNAL.*

Single Manager Concept

Obviously, consolidation of SAC communicators in AFCS will certainly further the overall single manager concept for Departmental communications-electronics. Gen. Burris, as the single manager, can achieve greater efficiencies and available economies while still providing SAC with it demanding, world-wide needs; and I am assured he will. Also, the single-managed structure can provide a streamlined way of exploiting fully the rapidly changing technologies of the demanding communications areas; and I am assured AFCS will do so.

The challenge, of course, is to keep the single technical management structure from looking progressively inward and keep it closely tuned to and responsive to the operational concepts and the practical requirements of the operational commands and their diverse weapons systems, and we will! Gen. Burris and I are confident that, by close contact, we can do this. We are determined to keep the intimate contact and understanding among the commanders, the planners, the operators, the producers of intelligence and the communicators that have existed among these elements throughout the 30 years that SAC communicators have been an integral part of this Command.

In July 1976, SAC communicators were formally transferred to AFCS; the AFCS Strategic Communications Area (SACCA) was created, with headquarters at Offutt Air Force Base. Col. Gerald Prather, the SACCA Commander, also served in a dual role as the Director, Communications-Electronics (C&E), DCS/Operations, Headquarters Strategic Air Command (SAC).

At the time of this transfer, I told those assembled for the change of command that:

> Communications have always been considered critical to the profession of arms. Chang Yu, about the year 1000, admonished officers and men to advance or retreat by "observing flags or banners" or "listening to the bells and drums" so that "the valiant may not advance alone nor shall the coward flee." And Admiral Mahan, a foremost naval strategist of past decades, flatly declared that "communications dominate war; broadly considered, they are the most important single element in strategy, political or military."

Certainly, the advances of weapons technology create the potential for an intercontinental nuclear attack to be but a decision of 30 minutes away. This awesome potential makes the communications capabilities, procedures and reliability of my Command even more critical than ever imagined by Chang Yu or even Admiral Mahan! It is my responsibility and the SAC Commanders' responsibility to our nation to insure that our weapons systems are used only in response to valid directives—and that they are used reliably and immediately, if properly directed.

I think that a more recent and updated view is that of Gen. Curtis LeMay (my predecessor as CINCSAC and the architect of much of the basic capability that has evolved into what is, today, SAC's command and control process), when he said that "The President can make you a general . . . but only communications can make you a commander."

I was very conscious of his admonition concerning the importance of SAC's communications at the time of the transfer of SAC communicators to AFCS, and I remain confident that we can make the new command relationships work well for the Command and the nation.

As a result of the Command realignment of the communicators serving SAC, an overall manpower reduction of 350 authorizations has already taken place and the opportunity for the communications-electronics professionals supporting SAC to increase the efficiency of the computer and telecommunications facilities serving our Command has never been better.

It is obvious to us that some of the most profound technological developments of this century have occurred in computer and telecommunications; both fields are developing at a rapid and accelerating pace. There is a strong symbiotic and synergistic relationship between computer capability and telecommunications. The single-managed Air Force communications structure offers the potential for widespread advantage in staying abreast (and hopefully ahead) of all of these changes and developments in a manner that will make them most useful for the operational forces of SAC.

In his book *Future Shock, SIGNAL* readers may remember Alvin Toffler quoting an 1865 newspaper editor as saying:

> Well-informed people know that it is impossible to transmit voices over wires and that, were it possible to do so, the thing would be of no practical value.

I wonder what the 1865 newspaper editor would think if he were given today's SAC C^3 briefing and exposed to the intense communication environment supporting our Command. I wonder what he would think if he saw high speed, digital transmission links operating at almost two billion bits per second—fast enough to transmit the contents of the Holy Bible in 1/100th of a second—with developments already underway to operate at eight times this rate.

It boggles the mind and challenges those who would master and utilize today's computer and telecommunications technology to contemplate computers that are capable of holding and transmitting more than a trillion bits; to plan efficient utilization of lasers, fiber optics and satellites that offer the potential for transmission beyond anything we thought feasible even as late as a decade ago. This is the way our technology is going and, unfortunately, this is the way the technology of "the other side" is also emerging. We must be capable of exploiting such developments and making the improvements that are feasible in our own weapons systems and in the C^3 systems that control them.

Credible Deterrent Force

We have always expected a great deal from our communications in SAC—from phone patch to remote missile control and reliable launch commands—

and SAC communications have always met our expectations. We are confident that our new communications realignment will enhance our ability to control the military power that has been entrusted to us by our nation and to apply it, if we must, with the speed and certainty that such weapons systems will demand fully to exploit their capabilities. In command and control of our weapons systems to execute, or not to execute, we are mindful of the requirements for positive control of our forces and weapons systems at all times, and our communications are absolutely key to that control, and key to my responsibilities as Commander in Chief of those forces.

As we see it, we must continue to meet the demands of a developing technology and must do it within budget constraints that, relatively, are becoming tighter and tighter. Since 1964, overall defense spending (including that devoted to military C^3) has dropped from 43 percent of the Federal Budget to a current level of less than 25 percent; and during the 13 years since 1964, the purchasing power of the Air Force decreased some 37 percent.

The expenditures for social and economic programs funded from the Federal Budget have increased to where they are now twice the defense outlays—and these relationships promise to continue. Notwithstanding this prior downward trend, however, our defense capabilities, and particularly our strategic capabilities, must remain relative to those of the Soviet Union or any other potentially inimical adversary. We can be encouraged, however, by the recognition of this potential asymmetry by the Administration in both the 1977 Defense budget, as submitted to Congress, and in the anticipated 1978 budget being formulated by the Administration at the time of this writing.

In both instances, real and needed growth in our strategic force programs has been included and supported. This turnaround, if implemented as programmed, will operate to reverse the downward relative trend that has become so apparent to those of us who are responsible for force management and force application.

We live in a very real and sometimes frightening world, a world in which, for the first time in the history of our nation, a potentially inimical and ideologically inspired nation possesses the operational power to attack this nation directly and possibly fatally. We are not going to be able to wish this away; nor will we be able to regain the levels of U.S. superiority that have been enjoyed by the United States in the past decades.

Ours is a world in which the Soviet Union has so applied weapons technology, resources and production that, in just a few short years since the early 1960s, the aggregate strength of its total forces has now equalled ours and in some areas is surpassing us. Ours is a world where America, which is at peace because it has remained strong, is thought by some to no longer need to be strong because it is at peace!

I think the need for us to remain strong has never been greater—and the need for us to remain tuned to the technological developments in telecommunications, data transmission and command and control processes is fundamental to the degree of strength we will maintain. Certainly it is fundamental to the stability we hope to create through the maintenance of a force that is relative in size, completely controlled and immediately responsive to the National Command Authorities.

As we collectively re-examine the value we place on deterrence and the means we choose to effect it, I think we will again come back to the basics of a credible deterrent force; i.e., a force that, in its totality: (1) is able to execute an assured, adequate and timely retaliatory attack against the military and economic targets of any agressive nation or group of nations, even after absorbing a massive counterforce attack; (2) is, at the very least, essentially equivalent with the forces of the Soviet Union or any other nation that achieves comparable strength; and (3) is able to execute response options to terminate hostilities at the lowest possible level consistent with our local interest—a force that will be sufficiently versatile, flexible and diversified that we do not become victims of our own inadequacy and our own inflexible planning.

The resources we allocate to command, control and communications systems associated with these forces must be commensurate with the important roles they play in the perception of deterrence and the indispensable capabilities that create these unequivocal perceptions. We in SAC operate under the premise that we must be responsible to authoritative direction no matter what, and that we must be able to do what we may be directed to do. We must reliably and instantly execute what we are told to do, and we must do nothing without such valid direction. Our capability must be real and must be controlled. The stability of force balance and the reflected deterrent strength of our force capability are dependent on the C^3 systems we acquire for the command and control of our weapons systems.

The former Secretary of the Air Force, Thomas C. Reed, has pointed out in the *SIGNAL,* October 1975 issue, that ". . . modernization of survivable C^3 systems could improve force effectiveness by a factor of two or more over a large range of threat levels." He estimated that "this capability could be obtained by an annual expenditure over the next several years of less than 4 percent of that portion of the DOD budget supporting strategic forces." On his authoritative bases of both payoff and cost, it is apparent that the modernization of survivable C^3 systems for SAC is both necessary and cost-effective.

New Programs

In SAC we continually evaluate our C^3 posture to determine needed, feasible improvements to insure that such systems, individually or collectively, do not represent a constraint in the flexible and responsive employment of SAC's forces. From these evaluations, several important programs are now underway which will enable us to exploit the new computer and telecommunications technology in a way that directly improves our C^3 posture.

The SAC Warning and Control System (SWCS) is a real time processing and display system which enables actions by the SAC Commander in Chief, upon receipt of missile attack information, to protect his forces pending decisions by the National Command Authorities.

Attack information is received over high speed communications links from several warning sensors operated by the North American Air Defense Command (NORAD) such as the Satellite Early Warning System, the Ballistic

Missle Early Warning System and the Sea-Launched Ballistic Missile Detection and Warning System. The hardware and software of the SAC warning and control system are continually upgraded to maintain state-of-the-art capabilities in conjunction with and compatible to sensor improvements.

Survivable, two-way communications in SAC have global dimensions, with satellite communications systems playing an important role in fully satisfying this continuing requirement. The first phase of the Air Force Satellite Communications (AFSATCOM) program will provide us two-way, record (teletype) communications between a family of airborne and ground terminals by special transponders which are installed "piggyback" on the DOD host satellites.

SAC will have four command post terminals at SAC headquarters, 8th and 15th Air Force headquarters, and 3 AD headquarters on Guam. SAC will also have command post terminals aboard the Airborne Command Post (ABNCP). Less complex terminals will be installed in SAC bombers, missile launch control centers, reconnaissance command posts, RC-135 aircraft, bomber and missile wing command posts and airborne launch control centers.

Together, the satellites and terminals which make up the AFSATCOM system will be used for the dissemination of Emergency Action Messages (EAMs) to SAC forces, for two-way communications with SAC forces to facilitate continuous force management and for the exchange of record copy information between the NCA and the other commands which manage Single Integrated Operational Plan (SIOP) forces.

SAC relies heavily on high frequency single side band (HF/SSB) radio, particularly in communicating with airborne elements of the force. Two former SAC HF/SSB radio systems, the ALPHA and BRAVO nets, have now been combined to form the GIANT TALK network. This network consists of four high powered CONUS transmitter stations and nine lower powered stations overseas; transceivers in all SAC aircraft, TITAN II launch control centers and MINUTE-MAN squadron command posts; and receivers in all other missile launch control centers.

SCOPE SIGNAL is a program aimed at improving HF/SSB capabilities. Under this program, our major HF/SSB transmitter stations will be consolidated with USAF aeronautical Stations—three in the Continental United States (CONUS) and seven overseas. We expect greater, more reliable coverage as a result. Moverover, in view of evidence that HF propagation windows quickly change in a nuclear environment, we expect an "adaptive HF" capability ultimately to be developed and employed.

In order to improve the performance of the Survivable Low Frequency Communication System (SLFCS), program 616A is now underway. Through this program, we expect to acquire improved modems and ancillary equipment that will result in better antijam capabilities, longer range, a more compatible interface with the Navy low frequency system and automated encryption and decryption capabilities.

The Data Transmission Subsystem (DTS) of the SAC Automated Command Control System is programmed to be replaced by the SAC Automated Total Information Network (SATIN IV). SATIN IV will be a record data communications system which is faster and more accurate, which has a larger message handling capacity, which has a higher degree of survivability and which offers greater reliability than the present DTS. In addition, it will extend two-way, automatically encrypted communications to all existing missile lanuch control centers for the first time. The system will be interoperable with the Automatic Digital Network (AUTODIN) and AFSATCOM, and it will offer multiple new ground entry points for communicating with the ABNCP battle staffs.

Conclusion

SWCS, AFSATCOM, SCOPE SIGNAL, 616A, and SATIN IV—these are technologically innovative programs that will indeed improve our C^3 posture. They are also programs which reflect the C&E professionalism which exists today in both government and industry.

In his book *Landmarks of Tomorrow,* Peter Drucker wrote: "Mercenaries can be satisfied with pay, professional people must see their work as a part of a grand design." To those of you who are the communications professionals involved in research, development, acquisition, operation and maintenance of the C^3 systems of SAC and the DOD, today and tomorrow, I can assure you that your work in these vital areas is indeed part of a "grand design." This grand design has as its ultimate purpose nothing less than to deter any strategic threat to or attack on this nation or our allies; and, in the event of any such attack, to terminate it quickly, at the lowest possible levels commensurate with our national interests, and on terms advantageous to the United States.

I wish you continued success, for our ability to deter conflict was never more important, and the consequences of failure never more catastrophic for our world and our way of life.

C^2 *in the Military Airlift Command*

Paul K. Carlton

The primary mission of the Military Airlift Command (MAC), global airlift of our military resources where and when needed in support of national objectives, faces demanding challenges. New realities in the international, economic and technical arenas dictate imaginative, realistic planning for the complex, demanding scenarios we envision for the future. MAC's command and control structure and its supporting automated management systems reflect an increased emphasis on realism, readiness and responsiveness.

Future operations are likely to demand the rapid intercontinental deployment of personnel, equipment and supplies. Airlift must provide this massive transport during the critical early stages of these deployments, as well as sustaining airlift throughout the operations.

MAC is realizing significant benefits from two recent changes in our organization and mission. All Department of Defense (DOD) strategic and tactical airlift has been consolidated under MAC, and MAC has become DOD's third specified command.

While consolidation offers significant economies during our normal peacetime airlift operations, the real payoff comes during wartime operations. MAC's specified command status, coupled with the benefits of consolidation, promise enormous gains in responsiveness and readiness. As a specified command, MAC's chain of command extends from the Secretary of Defense to Commander in Chief MAC (CINCMAC) through the Joint Chiefs of Staff (JCS). The JCS assigns priorities and MAC applies airlift resources to satisfy customer requirements.

This article originally appeared in the March 1977 issue of* ***SIGNAL.*

Peacetime and Wartime Control

MAC will apply airlift resources against these requirements based on precoordinated plans, and this is a key point. Wartime control under specified command status will provide for visibility of airlift forces at the JCS level, and resolution of the competing demands between unified and specified commanders as well as conflicting demands within and between theaters. Peacetime operations required minimal adaptations when tactical airlift was consolidated with MAC's existing strategic forces. But further modifications are required to greatly expand MAC's inherent surge capability during wartime operations.

Our peacetime role must certainly be performed efficiently to conserve resources while managing DOD airlift. But this peacetime role is actually a byproduct of our primary mission, combat readiness. During wartime conditions, MAC operations and responsibilities will expand by a factor of three.

An inherent surge capability must be incorporated and maintained in our peacetime operations. Already, MAC's planning, Management Effectiveness Inspections (MEIs) and exercise scenarios all reflect a new look and focus on the realistic evaluation of greatly expanded wartime operations.

The MAC Command Center at Scott AFB, Illinois, is the nucleus of our worldwide command and control system. This center directs all MAC resources, including 348 C-5 and C-141 aircraft based at seven the Continental United States (CONUS) locations and more than 260 C-130s at five CONUS and six overseas locations. The Command Center also monitors other MAC resources: Air National Guard and Air Force Reserve aircraft flying MAC missions and exercises; civil aircraft under DOD contract; and nearly 200 aircraft assigned worldwide to MAC's Aerospace Rescue and Recovery Service (ARRS).

Not to be overlooked is MAC's responsibility for airlift of the President and other senior government officials. We also operate a worldwide aeromedical airlift system using C-9 Nightingale aircraft in the United States and within overseas theaters, and reconfigured C-141s intercontinentally.

This mixed force of aircraft can respond quickly to several missions, which include strategic airlift of troops and equipment to any suitable airfield in the world, tactical airlift and logistics support to combat troops in the field, or aerial delivery—at times through cloud cover during adverse conditions—when runways are not available. Our C-5s are capable of intercontinental delivery of outsized cargo, non-stop if required, through inflight refueling from SAC's KC-135 tankers. On a normal day, over 170 MAC aircraft are away from home station on missions to airfields in 20 countries with up to 800 arrivals and departures in a 24 hour period.

The MAC Command Center has real time, voice communications with the enroute airlift fleet. The USAF high frequency/single side band system (HF/SSB) aeronautical sites, manned and operated by the Air Force Communications Service (AFCS), provide MAC with 24 hour-a-day command and control voice communications on a global basis. MAC's validation of this concept resulted in expansion of the USAF HF Aeronautical System to include a command and control capability for Air Force use.

Tactical Mission

Consolidation of all airlift in 1974 resulted in a new mission for MAC, when in addition to the strategic, the tactical mission was gained. MAC's command and control structure is composed of two systems: one covers the application of strategic and tactical resources to CONUS/intertheater missions; the other intratheater airlift missions.

The CONUS/intertheater system includes centralized command and control from the MAC Command Center located at Scott AFB with decentralized operational control exercised by Airlift Control Squadrons (ALCSs) collocated with the two MAC Numbered Air Forces (MAC AFs).

Twenty-Second Air Force (22d AF) at Travis AFB, California, is responsible for airlift in the western half of the United States and westward to the Indian Ocean. There are three strategic airlift wings and three tactical airlift wings assigned to the 22d AF. Twenty-First Air Force (21st AF) at McGuire AFB, New Jersey, is responsible for airlift eastward to the Indian Ocean, where it meets the 22d AF area, as well as the eastern half of the United States. Three strategic airlift wings and two tactical airlift wings are assigned to 21st AF.

The intratheater system provides for the management of those selected airlift resources designated by the JCS as theater-assigned airlift forces. MAC's Theater Airlift Manager (TAM) provides the means through which the Air Force Component Commander (AFCC) exercises operational control of these theater-assigned airlift forces. The AFCC receives the supported CINC's operational requirements, and the TAM, as the on-scene MAC airlift specialist, can best employ the available airlift resources.

The TAM, having complete knowledge and visibility of other MAC-assigned resources working in the theater, arranges airlift with efficiency and responsiveness through coordination with the appropriate CONUS-based MAC AF's Airlift Control Squadron (ALCS). Under certain contingency conditions, and during exercises, the AFCC will receive extensive augmentation forces.

In addition to airlift resources in the theater, fighter, reconnaissance, tanker, helicopter and rescue units may also be operating. These diverse air activities require the integrated control provided by the Tactical Air Control System (TACS). The heart of the TACS is the Tactical Air Control Center (TACC), and MAC provides appropriate airlift control elements to interface within this system.

Although the ALCSs, as part of the CONUS/intertheater system, work for the MAC AF Commanders, they are self-sufficient and this provides the MAC AF Commanders with maximum forward mobility. This has led to a concept known as MAC AF Forward, which is designed to rapidly expand and augment the overseas TAM function during certain contingencies. Under this concept, the MAC AF commander and selected staff elements will be deployed to the overseas theater, or to the exercise area, to become the Commander Airlift Forces (COMALF) and the TAM.

Within the Air Force Reserve (AFR) there are two Reserve numbered air forces, the 4th and the 14th Air Forces (Reserve), which are associated with the MAC AFs. When mobilized, the appropriate Reserve numbered air force will assume the mission previously performed by the deployed MAC AF, thereby providing continuity in the management of airlift in the CONUS. The Air Force Reserve will assume the MAC AF's CONUS mission, using existing command and control facilities and some MAC AF elements, including the ALCS, which do not deploy.

The MAC AFs, ALCSs, AFR units and the TAMs already work together during normal peacetime operations and thus provide an integrated, trained and reliable surge capability for wartime operations.

As the scope of MAC's worldwide responsibilities broadens, the need for a truly reliable and responsive automated command and control system becomes pressing. During the mid-1960s, MAC realized that the jet fleet of the 1970s would require a quantum jump in the volume and types of managerial information. This requirement marked the beginning of our subsequent development of the MAC Integrated Management System (MACIMS).

MACIMS

In 1972 and 1973, MAC acquired three of the 35 Honeywell 6000 computers which were purchased for the Worldwide Military Command and Control System (WWMCCS). We have since developed computer programs to support nearly every aspect of the MAC mission.

Beginning with operational planning, individual MACIMS subsystems have been designed to monitor the requirements and qualifications of air crews; produce the flight schedules for standard MAC routes; provide passenger reservations for CONUS outbound missions; follow cargo shipments through the six MAC stateside aerial ports; maintain current mission and enroute aircraft status; monitor engine and supply items in the MAC system; and, to complete the cycle, bill the customers for the services rendered.

Two of the 16 MACIMS subsystems focus more directly on the worldwide movement of the airlift force. First, the primary automated system for command and control is the Military Air Integrated Reporting System (MAIRS).

As MAC aircraft operate around the globe, coded MAIRS arrival, departure and advisory air movement messages are transmitted to the force control computer at Scott AFB. These Automatic Digital Network (AUTODIN) messages, in a secure format if desired, are routed directly to the H6080 where they are processed and formatted into the appropriate mission records. Headquarters MAC, the Numbered Air Force Headquarters and CONUS MAC wings all have access to the MAIRS data through over 150 remote terminals.

Last year, an AUTODIN Query/Response terminal was installed at Ramstein AFB, Germany, in support of the Military Airlift Center in Europe (MACE). Answers to predetermined queries are received in approximately two minutes from the time of initial query input. Similar support is available to MAC managers in the Pacific, Panama and Alaska.

Effective command and control communications extend to wherever MAC must deploy. The TAM can deploy combat control teams in drop zones and Airlift Control Elements (ALCEs) to airfields where MAC facilities do not exist or which require augmentation. Both the ALCEs and combat control teams can input MAIRS data through the use of UYA-7 digital communications groups over HF radios and HF transceivers. The UYA-7s transmit secure MAIRS data to the flexible digital terminal which is connected directly to the force control computer at MAC headquarters.

The MAIRS system is complemented by the Logistics Readiness Center Support System (LRCSS). While MAIRS follows our aircraft from departure to arrival, logistics is concerned with the aircraft's status between arrival and departure. LRCSS monitors MAC airlift aircraft delayed for logistical reasons. On request, LRCSS data is displayed on cathode ray tubes in the Logistics Readiness Center at Scott AFB. This data is also available to our subordinate units tasked with operational control. If necessary, sufficient information is available to deploy maintenance personnel and parts to support the delayed aircraft.

MACIMS is being developed incrementally to keep pace with the growing MAC mission. Efforts are currently underway to link the planning, scheduling and flight (monitoring) subsystems of MACIMS into an Airlift Implementation and Monitoring System (AIMS). AIMS will provide MAC with a closed-loop management system designed to measure planning information against actual execution data. Required adjustments can then be made to future plans.

A second capability planned for MACIMS is the Consolidated Aerial Port Subsystem (CAPS) which will provide MAC with a standard worldwide automated transportation system. CAPS is work center oriented; as such, it includes the use of remote management terminals located in strategic points within the aerial port complex as well as hand-held digital terminals to be used for capturing data directly off the cargo being shipped.

Some MACIMS data (which essentially monitor daily activities) also provide valuable inputs to our MAC Resource Optimization (MACRO) models. The models are versatile management tools which significantly enhance our planning for future operational scenarios.

The recently developed MACRO programs are large-scale, dynamic computer simulation models which duplicate the essence of MAC's global airlift system. These realistic, cost-effective computer simulations are used to evaluate MAC's various options during peacetime conditions and wartime contingencies. Used to simulate a planned contingency deployment, MACRO might well indicate potential problem areas. Further simulation, focusing specifically on the problem area, can then be used to evaluate the effectiveness of suggested solutions.

Future Systems

These models can also be used to evaluate the impact of future systems proposed for MAC. For example, we can visualize the expanded options offered by the proposed Advanced Cargo Tanker Aircraft (ATCA) which is being evaluated as a means of increasing MAC's airlift force capability through air refueling.

Yet another MACRO program could be used to simulate MAC's wartime surge capability. This simulation would help MAC maintain peacetime operations sufficient to support our inherent capability for greatly expanded, short notice wartime requirements.

MAC's airlift force is a vital resource, and future operations will take place in demanding scenarios. To meet these challenges, MAC operates a realistic and versatile command and control structure. Automated data support systems are key elements allowing MAC to effect reliable, versatile performance. MAC plans to take advantage of benefits offered by new technology, improved command and control data systems and enhanced operational realism. Events demand no less.

Tactical Air Force Command and Control

Robert J. Dixon

Tactical Air Force Command and Control is a complex, demanding and an increasingly vital process. It involves a spectrum of air missions and a diversity of weapon systems planned for contingencies varying from small to catastrophic in scope, intensity and significance. Additionally, multiservice and multinational interfaces are common, and global responsiveness is required. The increasing significance of General Purpose Forces in the context of national/North Atlantic Treaty Organization (NATO) survival affects directly the criticality of assured command and control of tactical air power.

The Challenge

This increasing dimension gives us an imperative, a do-it-or-else challenge to solve the basic problems inherent in controlling tactical air forces. The people of Tactical Air Command and the tactical air forces in Europe and in the Pacific are taking apart and examining the command and control process from the applications of the basic principle of centralization to the hardware which supports the applications.

The tactical air forces are not working alone. Equivalent Army elements are interested and participating, as they must be if joint requirements are to be satisfied.

All parts of the rethinking are important, but the greatest need and downstream payoff for the Air Force comes from developing the attitude, the

This article originally appeared in the March 1977 issue of **SIGNAL.*

thought processes, the awareness to deal with theater air force management in the aggregate in real time as opposed to piece by piece, after the fact. Tactical air command and control and tactical air employment are inseparable processes.

The effective application of tactical airpower is of vital concern to the Army as well as to the Air Force. The air-land battle—too narrowly defined as close air support—is a mutual Army-Air Force problem that is best solved jointly. Joint perceptions are needed to determine how to apply air and ground forces against enemy ground forces for the greatest total effect. Joint operational concepts and procedures have to be challenged, tested and exercised repeatedly to produce the most effective control and integration of combat assets.

These are tough problems, but Tactical Air Command and the Army's Training and Doctrine Command (TRADOC) are working on them now. Solutions will be measured against only one criterion—joint combat effectiveness.

Principles and Requirements

History and logic provide us the principles of employment of theater air forces. The primary principle is that of centralized control and decentralized execution. This principle survives as a guideline for Air Force command and control, since it is derived from the nature of the forces controlled. Because of the inherent characteristics of airpower—flexibility, mobility and firepower—centralized control allows an air commander to apply force theaterwide at the right place, at the right time and in the right numbers. To this end, the command and control process must obtain the information to make decisions and the control mechanisms to maneuver the force.

If we go to war today (or tomorrow), we need to be able to make rapid, accurate decisions in coordination with or on behalf of joint forces. Current speeds of maneuver combined with the lethality of modern weapons leave little time to consider the response, and error exacts heavy, perhaps irreversible, penalties.

Because of time compression and the strength of potential adversaries, the results of acting indecisively could be disastrous for air forces and others as well. Therefore, the air commander needs real time information. Today, real or near real time data are available in quantities which present both a problem and a solution.

The Big Picture

As a case in point, consider the revolutionary impact the E-3A Airborne Warning and Control System (AWACS) has on international and national—as well as theater and Air Force—command and control capability. By extending surveillance horizons far beyond the limits of ground based systems, the E-3A can provide civilian and military leaders an aggregated, real time, organized view of air, ground and maritime operations, completely and accurately, in a way previously not achievable.

The E-3A has proven its capability to be the most significant single improvement in command and control since the advent of radar itself.

Last November, in the Southwestern United States, using the Nellis and Edwards Air Force Test Ranges, an environment for the E-3A was provided which permitted the time and space compression of opposing air operations of sufficient density to represent the threat environment potential in Central Europe. The purpose was to stress the system—man and machine—so that the operational capability could be assessed. Over 400 Air Force, Navy, Air Force Reserve and Air National Guard aircraft were employed from 21 bases in nine states, with the Red/Blue ratio established at slightly more than 2:1.

The Red Force was given ground control elements and extensive electronic countermeasures (ECM) capabilities to employ against the E-3A which provided the Blue Force's only source of surveillance data. With the E-3A providing real time prehostility situation assessments, the Blue Force commander was able to concentrate his forces to meet the main threat. Despite intense ECM and over 150 Red Force flights to counter, the E-3A enabled the air commander to achieve effective battle management of a massive defensive/offensive air battle.

A new and powerful dimension now exists in command and control. No air commander should be surprised or outmaneuvered tactically again, if he is educated to use E-3A capabilities and has the means to exercise control of adequate forces.

Current Applications

The Tactical Air Control System (TACS) provides the organizational structure to work the theater air management problem. Centralized control of tasking is retained at the Tactical Air Control Center (TACC), which is the air commander's operations center.

Decentralized execution is achieved by the Control Reporting Center (CRC) controlling the airspace management and air defense function, the Direct Air Support Center (DASC) operating the ground force support functions and the Airlift Control Center (ALCC) managing the theater airlift task.

Strong mission control is exercised by the TACC through the planning and tasking cycle and, to a lesser degree, the execution cycle. This centralization meets the requirement to integrate all of the categories of tactical air employment into a single air campaign plan and operation. For example, employment airlift sorties may require fighter cover, defense suppression and electronic countermeasures support. TACC planning must assure this integration.

The Tactical Air Control System has worked successfully for several wars and continues to do so in joint exercises. It works in a structured environment that permits a daily plan to be devised, target by target, mission by mission, where a relatively smooth schedule of air activities is possible. The commander is involved in the development of a concept of operations, force mix and beddown, the apportionment process and approval of the daily allocation and application plan. In the relatively structured environments of past wars and current exercises, the execution of the plan typically has not required minute by minute participation of the commander.

We are concerned whether or not the system will work as well in the intense, relatively unstructured environments of potential current and future conflicts. Joint operations — where major minute by minute decisions must be made, where the formal daily plan is subject to considerable instant and critical variation during execution and where the commander's direct participation is essential—stress the system.

Predictably, data management will become so difficult that system efficiency will decline, and the ability to manage mission by mission will erode with it. The commander could be presented with too much information, much of it contradictory, and his ability to make decisions could be impaired. Current joint exercises are designed to work within system capabilities.

We need to question old concepts and test new solutions at the saturation level—to analyze, fix and repeat the process over and over again. Tactical Air Command has just begun to do this in BLUE FLAG at Eglin Air Force Base.

Remove the Barriers

The first step was to get the right people working together. The Combat Intelligence Center (CIC) accumulates all-source intelligence data, fuses it into a meaningful product and presents the results. However, combat information comes from separate operations and intelligence sources which are interdependent. The physical separation of the systems that produce and present what is really a single type of data slows the decisions making process and limits the quality of data analysis.

As a step to solve this, Tactical Air Command now integrates the CIC and the TACC into a single facility. Experts still do their own work in gathering data, but the analysis of those data occur with operations and intelligence personnel working together minute by minute to decide what the data mean to their mutual problem—tactical air force employment. The traditional differences in disciplines must be blended into individuals and groups of people well schooled in operations and intelligence processes as they relate to combat information and tactical air force employment. This must be done if we are to produce the practical thought process required in the short term.

After getting the right people to work together, the next step is to get those people to take apart, analyze, examine the command and control process and find ways to improve it. One part is now being worked—the study of the information requirements needed to support the commander's decision making. Inherent in that study is the need to categorize the operational decisions the commander needs to make. Current perceptions indicate the following categories:

1. Decisions that produce a concept of operation to achieve joint force objectives;
2. Decisions that determine force mix and bed-down;

3. Decisions needed to participate in the apportionment and allocation processes;

4. Decisions on force posture and dispersal; and

5. Decisions to adjust weight of effort by task and locale during employment.

This preliminary list is being refined, and specific subsets are being developed. The timeliness standards and the sources of data are being examined. What we want from this are criteria against which information gathering systems and methods of presentation can be measured.

Test and Evaluate

The third step in the process is to give the people working the problem a mechanism for testing and evaluating their concepts. So we created one, put it in the Eglin AFB complex and called it BLUE FLAG.

BLUE FLAG's emphasis is directly on the air commander, his staff and the elements which support the decision making and execution process. Individuals are brought together; given a situation, resources, information gathering systems and a command and control system; and are asked to solve a problem.

BLUE FLAG exercises educate individuals through hands-on participation and, in addition, provide the vehicle for saturating the command and control process with decision requirements and combat information to locate the deficiencies in the system, the information and the process. Current concepts such as the mission by mission management at the TACC will be tested, challenged and improved.

BLUE FLAG situations will not be failure-proof—mistakes are expected—but lessons learned from the failures will enhance individual learning and give direction to the system repair process. Once the fixes are made, the cycle will be repeated again and again. Eventually, TAC will be able to modify, prove and field a command and control process to operate at peak efficiency in the most intense scenarios.

BLUE FLAG is currently working on Air Force problems, but joint problems are not being neglected. The concepts of operations, procedures and the organizations needed to better plan and direct tactical operations (the air-land battle) at Army Corps level are being examined by TAC and TRADOC personnel. Enhanced joint combat effectiveness is the goal—the right people have been brought together, They are thinking through the problem and their results will be tested and evaluated eventually in joint exercises.

Automation

The proper presentation of well organized data requires automation; it is the key to the management of mountains of data. The activity levels required to do the joint command and control job now and in the future cannot be satisfied with manual speeds and capacities. Programs in varying stages of development designed specifically for tactical air command and control include the automated TACC, the Display Control/Storage and Retrieval (DC/SR), a segment of TAC's intelligence system and the proposed automated fusion division of the CIC.

These systems collectively will provide the required capabilities, but they must be employed and integrated with joint service systems to support the information needs of the commander and the rest of the command and control elements. If care is not taken, automation will provide more unstructured data. BLUE FLAG will be used to help us determine what information must be automated—automation for its own sake is not the answer. To see that it is done right is one of the primary goals of the Tactical Air Command.

Clear Concept of Goals

The challenge is clear. Confronted with an ever increasing, dynamic, offensively oriented threat, the tactical air force must maximize its combat capability by every means possible.

Tactical command and control—tactical air employment—properly employed is a key force multiplier. Command and control on a tactical air force scale must be understood and practiced in the joint force context.

Because of the threat, the education must be accomplished in a hurry. Present and future commanders must be educated to be capable of extracting the last drop of effectiveness from available forces. Present and future systems must be designed—or redesigned—to support the commander's needs.

Is the Past a Prologue?

Gordon T. Gould, Jr.

During and after each major military conflict, a popular expression reappears which accuses the military of fighting today's war with yesterday's doctrine and weapons.

There is an admitted risk in preoccupation with history. But I am not sure there is not equal hazard in any arbitrary rejection of experiences and lessons learned from past conflicts.

My own career spans three major military conflicts. As I periodically review the situation in Southeast Asia (SEA) with my staff, I cannot help but note a certain similarity with events encountered a decade past in Korea and 25 years ago in what was then called the China-Burma-India (CBI) Theatre. The political map of that area is vastly different today, but the geographic and the social-economic picture is little changed.

Many of the problems confronting communicators in SEA today are the same as those which faced their counterparts a quarter of a century ago. To admit these similarities invites an inevitable question—why do the same problems recur again and again? Certainly our military conflicts do not suffer from lack of historical documentation. Have we perhaps failed to adequately record our Communications and Electronics (C&E) experiences in such a way as to be useful to those who follow?

Or does history have, in fact, a useful message for us? Are the resemblances between SEA and CBI more than superficial? I believe that there does exist a meaningful and useful basis for comparing lessons learned from these two earlier conflicts. But, we should be careful in the choice of the questions we ask of history.

**This article originally appeared in the March 1968 issue of SIGNAL.*

Solutions to certain problems we have encountered in the past remain valid today. In some instances we appear to have profited from experience. Conversely, there are some cases in which we apparently have forgotten. On the other hand, we must not overlook some fundamental differences between SEA today and the conditions encountered during World War II; differences so basic that they tend to outweigh other considerations. Such a review may be useful to those of you who are engaged in some aspect of planning or designing communications equipment.

I will cover some of our management philosophy after World War II, then touch briefly on the manner in which the Korean and post-Korean experiences caused us to alter our concepts. Next, I will describe the effect that these alterations had on our ability to cope with SEA. Finally, I will discuss some of the technical incompatibilities which are a product of past doctrine and which have caused us much difficulty in SEA.

Management Concepts

There was a concept, fundamental to our operations in the Pacific in World War II, which required a unit to carry everything it needed to do the job—or be able to beg, borrow or steal—or *be prepared to do without*! Frequently the most valuable man in the outfit was the sergeant who would appear at the critical moment with the item you so desperately needed. Also, communications engineering was not always the complicated task then that it is today. For instance, a communications unit usually was capable of constructing a suitable antenna field, or installing a telephone outside plant. Of course, there were factors which simplified tasks in those days. The demands for telephone service were modest by today's standards. Many communications units were authorized quonset huts or other prefabricated buildings in their table of allowance and most had the organic capability to erect these structures. Usually they carried most of their spare parts and other supplies. In short, they generally had, on hand, within the unit itself, the people, the skills and the tools to do the job. The result was a very high degree of self-sufficiency and quick responsiveness.

The same basic philosophy still existed 15 years later in Korea. We then had an Air Force with its own communications elements and the same principle of self-sufficiency generally applied and was embodied in what was called the SCARWAF program—special category Army with Air Force. Some of these units were Signal Corps heavy construction companies assigned to the Air Force to give our communications elements a significant outside plant capability.

However, shortly after Korea, subtle changes began to appear in our philosophy. These were influenced by at least two factors. First was the emphasis on strategic nuclear weapons, based in the United States and operating from relatively few large, fixed installations and employing extremely complex and sophisticated electronic equipment. This caused us gradually to consolidate certain communications functions and to strive for a high degree of centralized control over certain critically short engineering and installation resources. It was no longer feasible or even necessary for each communications unit to be entirely self-sufficient in terms of engineering, installation and construction capability.

The second of the two influencing factors was a change in strategy and tactics. Warfare based on strategic, fixed based concepts is largely a static situation. It is also a situation which emphasizes science and technology and weapons sophistication. Under these conditions, it was believed that future needs could be forecast with reasonable accuracy, and that the planning and acquisition cycles could be safely lengthened to accommodate the emphasis on new technology.

Factors of mobility and self-sufficiency were largely neglected, except for those few organizations specifically tasked for contingency operations. In very general terms, the new planning and implementation cycle was suitable for this new concept of warfare. Certainly no unit commander was ever completely satisfied with a system which often required from one to four years to translate a statement of requirements into a completed, operational facility. On the other hand, these time limitations did not prove intolerable.

Within the Air Force, these two factors I have described led to consolidation of much of our ground C&E business into two organizations — Air Force Communications Service (AFCS), to do the planning, programming, operating and maintaining of certain communications and flight facilities; and the Ground Electronics Engineering and Installation Agency (GEEIA), to perform the engineering, the installation and some of the heavy maintenance.

By and large this concept proved to be suitable for the post-Korean environment I have just described. However, it had a subtle, unintentional and adverse effect on our ability to react to situations which fell outside the static model. Not for a moment did we assume that the normal organizations of AFCS and GEEIA would be completely suitable for tactical operations. Indeed, a significant portion of the AFCS capability will be found in its five Mobile Communications Groups, deployed in Europe, the Pacific and the United States. But, to a great extent, the capabilities of these mobile units; the organizational structure; the equipment, manning and *modus operandi*, had been influenced by our experience in Korea. We had learned the essentiality of mobility — *AIR* mobility. The limited size of our transport fleet had placed stringent limitations on weight and cubage.

As a result, much of our planning for new equipment and the concepts for their deployment were heavily influenced by the need to hold weight and cube to an absolute minimum. We supported many technical approaches toward this objective, frequently at the expense of ruggedness and staying ability. These concepts of deployment were validated and even reinforced by our experiences in the Middle East, Africa and the Caribbean in the late 50s and early 60s. Such crises developed with little warning, expanded rapidly and were of short duration. The U.S. response was modest in terms of commitment of military forces, and the deployment periods were short.

Crises Communications

Just what did these factors mean to the military communications planner? Short alert notice generally precludes anything but air transportation for deployment. The small size of the military force to be supported meant we did not have to contend with large demands for communications service. For instance, we thought in terms of fully mobile 200 line telephone exchanges and communications trunking systems of about four to 24 voice channels. The period of deployment was short, so we were not overly concerned about equipment having "staying" power. Ninety days was the duration assumed for planning purposes. Finally, because the equipment was extremely mobile and self-contained, little if any external engineering and installation assistance was required by the operating units. These organizations were self-sufficient. This concept proved quite satisfactory in supporting the crisis management roles that the United States has played in such places as the Dominican Republic.

The AN/MRN-12 is a good example of equipment designed specifically for rapid and short term deployment. It is a mobile airfield control tower which can accommodate just two operators. Visualize, if you will, a mobile tower coping with air traffic at Kennedy or O'Hare International airports. Add to the 15,000 to 30,000 monthly operations, tactical fighters returning from combat strikes, low on fuel, with pilots fatigued from the strain of air refuelings and penetrations of some of the most dense and accurately directed ground fire ever encountered. I think the difficulties in such an operation are apparent.

I think it may be useful to examine the circumstances which led to the use of mobile equipment for this sort of an operation. I mentioned earlier that we had gradually adopted a communications-electronics programming and implementation concept in the 1950s, which was geared to a multiyear design, acquisition and implementation cycle. Implicit in such a system is the ability to accurately forecast C&E requirements well in advance; and, the ability to sustain an effective operation with equipment in hand, until the programming-acquisition cycle is completed.

Beyond the limited, emergency mission capability of the AFCS Mobile Groups, we had little with which to respond to short fuse requirements. The 1st Mobile Communications Group has been filling each new mission requirement in SEA since 1962—and this is exactly what it was constituted to do. But its equipment and deployment concept was based on the short-term 90 day mission. Mobile deployments became a one-way street. Everything was going out of the warehouse and nothing was coming back. Its mobile control towers—for one thing—soon became very fixed and very permanent installations.

It is obvious that we had identified a gap in our equipment inventory. Even under the most favorable conditions, it takes at least 16 months to construct and equip an adequate, fixed control tower, suitable for the demands of this type of operation. We needed a tower facility with sufficient equipment and space for operators to control a high volume of traffic — which could be deployed and erected in less than 30 days and which could sustain operations indefinitely.

Let me give you one more example—base telephone exchanges. Following the same pattern I have already explained, we planned for relatively small, lightweight mobile telephone exchanges, employing field wire for an outside plant. The MTC-2 is a good example. It is nominally a 200 subscriber system, which under extreme conditions can accommodate up to 300 lines. Cam Ranh Bay was an extreme case. There were soon almost 8,000 people on that base and MTC-2 was completely saturated within two weeks of its installation. The 90 day deployment, for which it was designed, soon became a year and finally 14 months before a fixed, dial central office could be constructed. The Cam Ranh exchange soon required 3,000 lines.

Once again we have encountered an equipment gap. What we need is a modular type of telephone exchange—including its building—which can be erected and commissioned in about four months, with a minimum of 600 subscribers. Such a facility must be capable of rapid growth, as the base mission increases. The exchange must be expandable, in easily installed modules. We must be able to easily and rapidly expand the size of the central office building, without recourse to expensive and time consuming construction procedures. Finally, we must be able to alter and expand the outside telephone plant on relatively short notice. These were all lessons we learned in SEA—learned the hard way.

Our communications planning fell short of the mark for at least two reasons. First, as I have already explained, our planning time frame was based on a 90 day deployment. We had little in the inventory to plug the gap when the 90 days stretched into months and years in some instances. But there is a second and more fundamental reason. We apparently did not fully grasp the real significance of what is popularly called the "Information Explosion." Reduced to simple terms, this means that our complex society—its individuals and its organizations — have come to be dependent upon electrical communications to an extent that few yet appreciate.

Let me be more specific. In the Air Force, we have structured many of our maintenance and supply concepts around exploitation of high data rate record communications systems and private telephone and radio systems. Among other things, this has permitted a high degree of centralized control over maintenance and logistic operations. Rapid and reliable communications, in concert with air transportation and electronic data processing, have permitted elimination of conventional overseas supply depots and reduction of stockpiles elsewhere. Many of our Air Force maintenance and supply organizations deployed in Vietnam with this organization and operational doctrine intact and found base and long haul communications which simply could not begin to satisfy their needs in terms of quantity or of quality. These organizations suffered a temporary loss in efficiency while we undertook a number of expensive crash programs in order to provide the type of service needed.

New Demands

Let me try one other way of illustrating the new demands which are being made on communications systems in a combat theater. Command and control over military forces extends, quite literally, from the White House to the foxhole. This concept poses some interesting and challenging technical problems for the communicator. The field telephone has become an element of the worldwide Defense Communications System.

It probably never occurred to its designer that it would ever be connected, over 12,000 miles of microwave, tropo, submarine cable and satellite circuits, to the Pentagon or White House. Consequently, no thought was given to such an eventuality in its design. What I have just said about this particular telephone instrument applies to practically every piece of communications equipment available for deployment to SEA today.

The problem is electronic compatibility—or rather the lack thereof. I have saved this particular "lesson learned" until last for several reasons. I consider it to be one of the more basic and important lessons. Also, I think this particular problem needs to be widely discussed and thoroughly understood by every individual, civil and military, who plays any kind of a role in the design and development of military communications equipment.

Integration Problems

In SEA today, we are experiencing difficulty in achieving an acceptable degree of integration of the many types of tactical communications equipment employed by the services. In order for you to more fully understand this problem, it may be helpful if I briefly discuss some of the circumstances which lead to the development of so many technically incompatible equipments.

Up until very recently, the military C&E community tended to divide its facilities into one of two categories: those which were integral to and in direct support of military field units—which we called *tactical*; and those which were a part of the worldwide, common-user, long haul communication facilities of the Defense Communications System (DCS).

It was commonly understood that the worldwide DCS could not evolve without the prescription of and rigid adherence to technical standards, methods and procedures. Unfortunately, what was not so well understood was that current concepts of command and control demanded toll-quality communications right down to the lowest operating echelon of command. These end-locations, more often than not, were served exclusively by tactical systems. The design of our tactical systems had never been predicated upon a total system network consisting of tactical and long-lines equipment.

This is one example where reference to history would not have proven very useful. I do not mean to imply that the problem of interfacing tactical and non-tactical systems is peculiar to SEA. But, in previous wars, the interface usually took place at relatively few geographic locations, frequently collocated with major headquarters units having recourse to material and skilled personnel. This is certainly not the case in SEA today. The DCS does not enter SEA and stop at some convenient point—such as Saigon—for conversion into tactical systems. It interlaces the whole peninsula. It is, in fact, the backbone system. The original system was installed under an Air Force contract in 1962 and was regarded essentially as a tactical system. Nevertheless, because it was a high, wideband system, its technical standards closely approximated those of the DCS.

Closed-Loop Operations

Our difficulties began with the rapid build-up of forces at locations other than major terminals of the backbone system. As each of these units was activated, it tied into the existing backbone system, utilizing whatever tactical equipment was readily available. Bear in mind that very little effort had been made in the past toward technical standardization of purely tactical equipment. Each had been designed on the basis of a closed-loop operation and exploited those electronic techniques best suited for its primary purpose.

Let me give you one example of the manner in which such a system evolved and why certain techniques were selected—techniques which sometimes cause problems today. Our experiences with tactical air control in the Korean conflict — a highly fluid and dynamic environment — demonstrated the need for very mobile tropospheric scatter radio terminals.

The type of cargo aircraft expected to be in our transport inventory during the life of the projected system demanded rigid adherence to weight and cubage limitations. If you are trying to conserve weight in electronics equipment, a good place to start is with the power supply. This is the principal reason we adopted the 400 hertz power supply for certain tactical equipment. For prime electric power generation, we procured turbine units, mainly because they offered promise of something like a six-to-one reduction in weight and considerable savings in size.

Within the context of a closed-loop tactical air control communications system—located on or close to airfields employing jet fighter aircraft—it is not startling to visualize turbine powered generators burning JP-4 jet fuel. In a dynamic, tactical situation, you can afford the higher fuel consumption and relatively shorter operating life of turbine generated power.

However, the logic of this decision becomes obscure when you encounter turbine power units installed on a mountain top in Vietnam, as part of a fixed, static long haul communications system and far from a source of jet fuel. Non-standard power is just one of the problems you begin to encounter in adapting purely tactical equipment into a fixed system. DCS stations may be operated and maintained by Army, Navy or Marine Corps, and the impact of peculiar equipment on their maintenance and supply system is considerable. Training problems become formidable also.

But training and logistics are only a part of the total problem. Another facet is technical interface complexities arising from non-standard techniques. The fact that different electronic techniques are employed in various tactical equipments comes as no particular surprise. Indeed, no real effort had ever been made to achieve any particular degree of standardization in tactical equipment. There had never been any demonstrated need for extensive compatibility between systems. The enormity of the problem slowly came to the surface as the Services and DCA attempted to integrate these tactical systems into the backbone and still retain the original toll-quality of the basic system.

Automated Organism

Let me expand on the term "toll-quality." I am just talking about *voice* quality circuits. I mentioned earlier that the information demands of the modern military organization—on the command and control side as well as the logistics side—differ enormously from those we encountered in any previous situation. There are many electronic computers in operation in SEA today and more on the way. The military machine is a highly automated organism, both at its home base and in a combat theater. Data terminals require high quality communications circuits—2,400 baud being about the minimum today. It is proving difficult to obtain that quality of circuit in a lash-up of different tactical systems.

We have encountered technical incompatibilities at practically every point where they could possibly occur. The systems manager must resolve differences in signaling and pilot frequencies, input impedences, signal levels, channel spacing, frequency shifts and many, many others. Most of these interface problems have now been indentified, and corrective actions have been completed or are underway. The price we have paid has been a degraded communications system in terms of quality; high cost, due to field modifications of equipment; and, added complexity in circuit and system management.

However, in the process we have learned some valuable lessons. We have discovered some basic conflicts in technical standards which are now in existence. We have indentified some areas where necessary standards are virtually non-existent, and we have found standards which need strengthening.

Basic Philosophy

However, the mere establishment and promulgation of technical standards will not, by itself, completely solve these problems. I think it might also be helpful to examine our basic philosophy concerning the nature of communications in the command and control process. The last decade has brought about a fundamental change in the concept of command and control of military forces, and these changes have had and will continue to have a profound effect on the communicator.

It seems to me that, from a purely technical viewpoint, we must cease attempting to categorize communications as either tactical or non-tactical. We must surely recognize that practically all communications equipment, without regard to its intended end-use, must be designed so that it either is, or can readily be made, technically compatible with the DCS. If the President or others in the National Command Authority wish to speak directly with the pilot of a reconaissance aircraft, the skipper of a ship or a platoon commander in the field, it should be our task to insure that no technical impediment exists to such a request.

I do not suggest that these changes are going to be made overnight. Certainly the variety of technical approaches to a single communications problem are, at least in part, a reflection of procurement policies and even our patent regulations. These are considerations which must and will be carefully studied. However, as is true with any problem, the solution must necessarily begin with the recognition that a problem exists.

We have learned—and in some instances re-learned—some valuable lessons from our experiences in SEA. We have found that the principle of self-sufficiency is every bit as valid today as it was in Korea and the CBI. We have again confirmed that equipment and tools needed to cope with a variety of unexpected contingencies must be organic to a unit—not held in emergency reserve in the United States. We have found that our mobile communications units need a limited engineering and construction capability to construct and expand radio and telephone facilities.

Time Frames

We have found that a "tactical" situation can cover at least three time frames, and that different types of equipment and capabilities are needed for each. We have found that, in general, existing concepts of mobile equipment are suitable for periods of up to 90 days. We have identified the need for a family of equipment, to replace mobile facilities, which can be erected on several months' notice and can fill the gap until fixed equipment can be programmed and constructed.

We have discovered that it may no longer be possible to design a piece of communications equipment for a purely "tactical" use and in total disregard to the technical standards for the DCS. We know that the total system performance will be determined by the lowest quality component and that today's concept of command and control of military forces may require that any set of equipment be incorporated into the worldwide system.

These have been a few examples of some of the lessons learned. In some cases history was a good teacher, but we failed to heed. In other instances, we have found that the nature of the use of communications has changed radically over the years and past experience was of little use. I have concentrated on our problems and difficulties because I think it absolutely vital that the military-industrial electronics team use these experiences in a profitable manner so that we will not repeat these mistakes.

There is an admitted difficulty in maintaining proper perspective in any presentation which concentrates on difficulties and problems. Aerospace power is playing an effective and vital role in the conflict in SEA today; however, the layman probably does not fully appreciate the manner in which communications and electronics systems permit airpower to be exploited so successfully.

I have been fortunate enough to have visited SEA several times in the past two years and I am consistently impressed with the ingenuity with which our military and civilian technicians have adapted and modified the tools we have provided. These technicians and their equipment are performing splendidly. The challenge to us, as I see it, is to continue to strive to provide them with the proper tools for the job to be done.

Future Tactical Communications Requirements for the USAF

Robert L. Edge

Tactical communications is an area upon which the Air Force is focusing an ever increasing amount of attention. The Air Force doctrine of centralized control and decentralized execution for tactical airpower is hinged upon the capabilities of our tactical communications system. The demand for speed and flexibility in application of air weapons systems requires corollary attributes in the tactical communications that support them. In simple terms, we must have enough reliability, capacity and mobility in our tactical communications systems to insure that they are a contributing factor in the effective application of tactical airpower.

I believe that the current and immediate requirements for the USAF Tactical Communications are well known to industry. The 407L program and 485L program have been acquiring equipments—and are still actively acquiring. So, I will emphasize the word "future" in my discussion. But before proceeding directly to the future requirements, *per se,* let me examine how we like to approach the specifics of our requirements—particularly how the methodology is viewed from a resource manager's chair.

The first step—everybody would agree—is to describe the task that has to be performed. Next—and again everyone would agree—we ask, "what are the

This article originally appeared in the September 1972 issue of **SIGNAL.*

tools available to perform the task?'' But in answering this question we sometimes begin to get into difficulty.

There is a natural and legitimate desire to exploit technology. But when this desire takes the form of ''we *can* do it this more sophisticated or more modern or more exotic way, therefore, we *must''*—this is when the difficulty starts.

Instead, we should insist on settling for the best cost-effective mix which *just barely* does the job effectively. Why ''just barely?'' The simplest answer is that we should want to not accept the penalties attached to overdesign. These penalties can take many forms—increased weight, increased cube, decreased reliability with its increase in maintenance demand and increased numbers of specialist personnel, etc., all of these result in diverting needed transportation to the tactical area; and, of course, the increased cost and delayed use which generally results from increased sophistication.

So a very general statement of our tactical communications requirements is: The simplest possible means of voice or record communications—and at the same time reliable and maintainable, rapid enough for the purpose, secure enough for the purpose, with suitable compatibility—in the information sense, in the electronic sense and in the electromagnetic sense.

We need to look at all of these things from a system viewpoint—not just a black-box or even a subsystem viewpoint.

A Systems View

Another way of describing this system view of things follows the weak link analogy—if you want to strengthen a chain, you do so by beefing up the weakest link—not by minor improvements to the strongest link.

For instance, let us look at manpack or hand-held radios. I have seen the electronics packaged in a small match box, including the transmitter power amplifier. But the prime power to support the radio is at least as large as a small brick. So why does it make sense to continue working on radio miniaturization when the weak link in the cube and weight problem is the prime power source? Maybe we should be instead trying to improve the antenna gain so the power requirement can be reduced, at the same time trying to get more energy from smaller batteries. Since antenna gain will most likely come from greater directivity, we should at the same time be able to reduce susceptibility to intercept and radio location and improve spectrum utilization.

The same thesis applies to high power radio equipment. The mainstay of our tactical systems is tropospheric scatter. Our need here is to achieve the same kinds of quality, reliability and channel capacity that we have in our fixed station sets. But jacking up a fixed station and putting wheels on it will not satisfy the problem. We need the quality and reliability without paying a penalty in loss of mobility.

The weak links in mobility are not only weight and cube—but also set up time. To satisfy our requirements, we must attack the weakest links first. Changes to the radios may be worthwhile and necessary to decrease weight and size. But miniaturizing the radios will not solve our problem if the antenna erection takes hours—and reducing the size of prime power generators does not help if we must pay a penalty in increased fuel consumption. Larger sources of prime power—generators—are still a major weak link. We still need a lightweight, high duty cycle, quiet, reliable generator which uses fuel more efficiently.

But I, for one, would be totally unwilling to pay any premium to replace any of our equipment already in the inventory without a significant—a really significant—improvement in one or more of the important parameters. Again, this statement holds in the general case—across the board—not just for power generators.

The Economic Law

What we face is the economic law of decreasing marginal utility which Professor Paul Samuelson at MIT so well describes. This law says that when operating at that point on the curve where obtaining a much diminished increment of improvement costs us much more than it is truly worth, then it is time to redirect our attention. In other words, we know it will be very costly to obtain 100 per cent perfection, so we must ask ourselves how close we want to get to overall perfection, given the cost of trying.

We have already said that we are generally not willing to buy perfection in one attribute at the expense of ignoring other attributes which are known to be relatively low. The great challenge to all of us here, in the satisfaction of U.S. Air Force (USAF) tactical communications requirements, is achieving the optimum trade-offs in useful technological innovation—trade-offs among attributes as well as trade-offs versus cost.

We must select those innovations which will allow us to move forward on a broad front, with each added improvement or change contributing significantly to shoring up the weakest link—and that weak link will continue to shift as individual links are strengthened. Weight, cube, set up time, transmission quality and capacity, range and many other factors may surface as the point where the maximum gain may be achieved for each dollar.

It is in balancing these trade-offs that we need the support and assistance of industry and the military acquistion agencies. The more realistic the choices we have to select from, the better chances we will have to achieve our goals. But they must be honest choices. If we set a particular goal that is too difficult or for which technology is not yet ripe, we must know that, too. For as we move further into the era of automated command and control systems, our communications requirements will continue to become more complex and difficult to satisfy.

The increasing demands for data and the quality needs of that data traffic will drive us to new balances in tactical communications between capability and mobility. Achieving the best balance at any point in time is the task we must all share.

Employment of Tactical Air Control Radars

Robert E. Gardner

The employment of tactical airpower in response to the wide range of threat scenarios, from global nuclear through theater conventional to limited contingency operations, calls for a responsive and flexible command and control system. The Tactical Air Control System (TACS) fills that need today and is evolving to meet the changing environment of the modern battlefield.

The TACS is a network of facilities, equipment, people and procedures that permits the Air Force Component Commander to plan, direct and control tactical air operations. In short, it is a battle management system. As a theater battle management system, it must have certain characteristics. Tactical air forces are mobile, responsive and may be employed at all levels of conflict. The command and control system associated with those forces must be equally as dynamic. It must provide the Air Force Component Commander with the ability to deploy, shift, concentrate and redeploy forces to meet changing requirements. A deployed TACS has neither a definite size nor a definite number of elements. It is tailored to meet the particular command and control requirements of either large or small scale operations.

The Tactical Air Control Center (TACC) is the senior TACS element or the operations center for the Air Force Component Commander. It acts as the focal point for all tactical air operations, prepares air tasking orders and directs the execution of tactical air missions. The TACC monitors and supervises day-to-day operations to ensure that missions are executed as planned or, when needed, are redirected against high priority requirements dictated by the

This article originally appeared in the November 1982 issue of* ***SIGNAL.*

emerging, fluid battlefield situation. From the TACC, the various TACS elements are managed:

1. The Wing Operations Centers (WOC);
2. The Airborne Battlefield Command and Control Center (ABCCC) aircraft;
3. The Airlift Control Center (ALCC);
4. The Air Support Operations Center (ASOC); and
5. The ground based and airborne air surveillance and control radars.

It must be re-emphasized that while all of the elements or subsystems listed above can be used in a tactical air control system, there is no TACS which is "standard" or "classical." The number and type of elements deployed will be tailored for each operation. TACS is structured to meet the needs of the Air Force Component Commander, and other criteria such as the force to be employed, the expected threat and the geographical area of operation.

This article deals with the surveillance and control radar subsystem, a suite of specialized equipment dedicated to that function. The purpose and employment of each type of radar unit will be discussed as well as future enhancements.

Radar Units

The radar equipped units that form the TACS air surveillance and control subsystem are the Ground Based Warning and Control System (GWACS). The ground based radar units are designated as the Control and Reporting Center (CRC), the Control and Reporting Post (CRP) and the Forward Air Control Post (FACP), depending on assigned tasking.

The CRC is the focal point for decentralized execution of air defense and airspace control functions. The CRC directs air defense operations, provides aircraft guidance or monitoring for offensive and defensive missions, relays mission changes to airborne aircraft and supervises the integrated activities of other radar elements.

The Control and Reporting Post (CRP) performs surveillance and control within an assigned subsector. It does not perform the CRC's identification or weapons commit functions; however, it can assume any or all of the CRC functions in an emergency. One or more CRPs may be employed depending on area size, terrain features and the anticipated level of air operations.

The Forward Air Control Post (FACP) is a small, highly mobile radar element which is normally deployed into forward areas to extend radar coverage for control of air operations early warning. It is the most mobile of the ground radar elements and can be moved rapidly to support changing tactical situations. All three ground elements, the CRC, CRP and FACP, are equipped with the AN/TPS-43E radar.

The E-3A Airborne Warning and Control System (AWACS) provides detection and threat warning of targets below and beyond the coverage of ground based radars. The AWACS extends radar and radio ranges beyond the ground elements and permits air defense warning, aircraft control and changes to tactical missions at distances well beyond the Forward Edge of the Battle Area

(FEBA). The AWACS is also capable of operations where deployment of ground TACS elements is not possible or practical. Because of its rapid mobility, it is often the first TACS element to deploy into an area of operation and has become the diplomatic battleship of the 1980s.

GWAS

The inherent flexibility of the TACS is particularly evident in the mobile radar units of the GWACS. The CRC, CRP and FACP can all be tailored in size (equipment and personnel) to meet specific scenario requirements. The TPS-43E radar is common to all three unit types regardless of unit sizing. The TPS-43E utilizes a stacked beam antenna configuration to provide range, azimuth and height information. The radar set is equipped with two control scopes and air-to-ground and ground-to-ground communications. The complete radar set is housed in two packages—electronics shelter and antenna pallet. It is deployable by helicopter, mobilizer, M-35 trucks or C-130 aircraft.

The CRC and CRP are usually identical in size and equipment configuration. Again, the distinction between the two units is the aircraft identification and weapons commit function reserved for the CRC as the senior radar element. A CRC and/or a CRP is comprised of over 200 personnel, approximately 100 M-series vehicles, plus integral air-to-ground and ground-to-ground communications equipment. The CRC/CRP Operations Central, when integrated with the AN/TPS-43E radar, functions as the major control agency of the TACS by performing all functions of surveillance and weapons control in an assigned area of responsibility. Two Operations Central modules are joined together by an inflatable shelter to form a single operations room or cell which houses personnel, manual display boards, radar consoles and communications equipment. To this basic complex is added ancillary equipment, data processing and air conditioning modules to form a minimum six scope operations center.

A total of three cells are joined together to form the normal CRC/CRP 14 scope configuration. Two modes of operation are possible:

1. Computer assisted, in which the CRC/CRP HM-4118 computer processes surveillance tracks, computes weapons control data, generates console displays and allows data linking with other command and control (C^2) facilities; and

2. Manual, in which only raw radar data and the selective identification feature (SIF) are displayed. In the computer assisted mode, exchange of air situation data and C^2 orders/information are available via the Tactical Digital Information Link (TADIL) system.

The Message Processing Center (MPC) is another modular facility normally interfaced with an operations center to provide management of the TADIL system through expanded secure digital data exchange capability. TADIL B is used to interface a CRC or CRP with another CRC/CRP, the Army TSQ-73 Missile Minder and the USMC Marine Air Command and Control System (MACCS).

TADIL A is used to provide secure digital connectivity between the AWACS and MPC, U.S. Navy (USN) ships and E-2C aircraft. The MPC is the facility which forwards information from TADIL A units to TADIL B units and vice versa. Additionally, the MPC is equipped with the North Atlantic Treaty Organization (NATO) Link 1 to provide the interface between the NATO air defense system and U.S. C^2 units.

FACP

The FACP is considerably smaller and more mobile than either the CRC or CRP. It has about 65 personnel assigned with a fleet of approximately 22 M-series vehicles. Using an ancillary, two control scope Operations Central, the FACP provides an optimal four scope manual-only capability. It can be fielded in a "mini" two scope role with just the TPS-43E radar set and supporting power equipment. The distinguishing feature of the FACP is that it functions only in a manual mode, having no automated aids.

The E-3A Airborne Warning and Control System has combined a narrow beam F-band radar using pulse doppler technology, a crew to man nine multipurpose control scopes and a Boeing 707 airframe to provide a highly mobile extension of the overall TACS radar system. Operating at an altitude of 29,000 feet puts the radar horizon at approximately 200 miles, well past that of ground radar units. With the ability to "look down," the AWACS seals gaps in the ground based radar coverage caused by terrain masking and unit placement.

Fully automated, the AWACS performs the standard air surveillance, identification and tracking functions found in the GWACS. High frequency (HF), ultra high frequency (UHF) and very high frequency (VHF) organic communications equipment gives AWACS data and voice link capability. Using TADIL A, the AWACS and GWACS and exchange air situation and C^2 information automatically. The TADIL A net is also accessed by appropriately equipped U.S. Navy ships to provide further extensions and enhancements to the tactical air situation/picture. The aircraft crew complement consists of 13 E-3A systems specialists and four flight crew members. Crew assignments vary depending on the mission scenario.

The control of a defensive counter air (DCA) mission illustrates the flexibility and cohesiveness of the TACS. Based on the threat the TACC allocates a number of fighter sorties for DCA. The sorties are serviced as airborne combat air patrol (CAP) and ground alert missions. Using the AWACS deeplook capability, the FACPs are deployed near the FEBA with the CRCs/CRPs further back. The AWACS, detecting an aircraft flight over enemy territory, sends track information via TADIL A to all TADIL A units. The MPC receives the data and relays it to the CRC where the track is classified friendly, unknown or hostile based on the rules of engagement.

If the track is a hostile, the CRC either commits fighters from the CAP, selectively scrambles alert aircraft or commits Army Surface-to-Air Missiles (SAMs). CAP fighters, in all probability, will already be under the control of either the AWACS or an FACP, dependent upon assigned areas of control responsibility. If the intercepting aircraft are scrambled, they pass through the control areas of the CRCs and CRPs and are provided with continuous intercept vectors. The scrambled aircraft may be turned over to the AWACS or an FACP for the final phase of the intercept or they may remain with a CRC or CRP depending on the designated control agency for air space in which the final phase of the intercept is conducted.

A Real World Application

While the TACS is routinely configurated and deployed in support of exercises, both in the Continental United States (CONUS) and overseas, the ongoing ELF-ONE Command operation in Saudi Arabia is an excellent example of its inherent flexibility to support real world applications.

In the fall of 1980, the Saudi Arabian government requested E-3A AWACS enhancement of its air defense system. To link the AWACS to the fixed Saudi radar system, selected elements of the GWACS were also deployed. A single cell CRP was deployed to interface with a Saudi Sector Operations Center (SOC). Air situation and C^2 information from the AWACS and other C^2 systems operating in the area were transmitted to a remoted MPC facility via TADIL A. This information was then forwarded via TADIL B to the CRP operations center. An FACP was also deployed to enhance the overall radar picture and manually interfaced with the CRP via a tactical radio relay system.

Several initiatives are underway to increase the mobility, survivability and flexibility of the GWACS. Improvements in communications capacity and speed to be realized by the TRI-TAC and the Joint Tactical Information Distribution System (JTIDS) programs will be complemented by two significant actions: the replacement of the Operations Central nodes in the CRC/CRP/FACP and the TPS-43E antenna. The Operations Centrals of the CRC and CRP, the FACP operations van, the MPC facility and several radio vans will be replaced under a program known as Modular Control Equipment (MCE).

This program will use an AN/TYQ-23 Operations Module being designed to replace equipment at the U.S. Marine Corps (USMC) Tactical Air Operations Center (TAOC) under a program known as TAOC-85. Air Force versions of the USMC module will contain computer systems to provide the four control scopes with automated track data: HF, UHF and VHF radios; a TADIL A/B capability; ancillary equipment; and environmental control unit, fiber optic cables; and a War Readiness Spares Kit (WRSK). Up to five MCE modules may be interconnected by fiber optic cables to increase the capabilities of a unit. The modules can be transported by rail, air (fixed or rotary wing), mobilizer or truck.

The TPS-43E antenna will be replaced by an ultra-low sidelobe antenna (ULSA) designed to reduce jamming susceptibility. This antenna is a development of the slotted waveguide technology utilized in the AWACS radar system. The flat-faced antenna is an 11 x 18 foot traveling wave array protected by a honeycomb radome. It uses a single shaped transmit beam and six simultaneous receive beams for elevation and height finding. Folded during transport into a package approximately 11 x 8 x 18 feet, the ULSA retains the mobility qualities of the present antenna.

While this article has discussed the operational use of the TACS surveillance and control radar subsystem, the critical nature of its communications is apparent. This subsystem, like the other TACS elements, is inherently reliant on communications, from the present voice, teletype and TADIL systems to the new higher speed data systems soon to be provided. Without communications, the Air Force Component Commander cannot control, let alone command, the employment of tactical air power. If the Tactical Air Control System is the heart of the effective application of tactical air power, integral communications is its life's blood.

Communications in Transition

Paul R. Stoney

For at least a decade, we communicators have woven a common theme. That theme has been called, varyingly, the Information Explosion, or the Data Deluge. Whatever name was selected, the message was the same. That was the growing demands being placed on our services.

In the late 50s and the early 60s, we talked with some justifiable pride of the contribution communications was making to the effectiveness of our national policy of strategic deterrence. The demands of that era were, in retrospect, not too difficult to meet. Certainly this was true if one views the problem from the standpoint of technology. All that was really needed was a single, one-way voice circuit with guaranteed availability and reliability. We used a lot of redundancy to reduce the uncertainties. Only at the very low and the extremely high ends of the radio frequency (RF) spectrum did we even begin to push state-of-the-art.

Technical Problems

Still, we communicators continue to talk about the information explosion, albeit with an occasional glance of apprehension over our shoulders. Because now, for really the first time, some signs of real technical problems begin to crop up. We are starting to talk less about volume and more about standards and compatibility, words that had little relevancy in the days of 100 percent analog voice. Still, however, the relatively low data rates of the early data systems placed no real demands on technology.

All we really did was tighten slightly our technical management of our analog systems over which we were pushing digital tones. and we paid a bit

**This article originally appeared in the May 1973 issue of SIGNAL.*

more attention to error rates. Even one or two of what we called high speed data circuits five years ago could be easily accommodated in groups in our Frequency Division Multiplexing (FDM) radio systems. Data loading limitations were essentially theoretical problems, which we encountered only in textbooks. Power budget management and intermodulation distortion also fell into the same category of theoretical problems.

Communications Explosion

But now we are well into the seventies and I am here talking to you once more about the communications explosion. But, where once I was almost boasting because our business was booming, now there is a note of real apprehension. Our once hypothetical and theoretical problems are upon us in a most real and direct sense.

Communications, both military and commercial, are at a crossroads, and some very tough decisions must soon be made. We have hit a technological barrier. We just cannot simply "up" the keying speed of the modem any longer. The circuits will not take it. To be technically honest, we must recognize that we have not wrung all the capability we can from our present generation of FDM radio systems, and our procedures, techniques and instrumentation must improve significantly to get the rest of the performance.

For reasons of national security policy, we are being asked to find ways to encrypt all military radio communications. The best, in fact, the only practical means of doing that today is digital; and given present technology, that means truly high speed data. The computer is also pushing us to even higher data speeds.

The Money Crunch

The initial reaction to all this is to tell the computer people what we are sorry but they must slow down and that they are disturbing our procedures and costing money. A little analysis, however, clearly discloses that the computer people are not the ones really pushing for higher speeds. The money crunch and the complexities of life have forced the Air Force, as they have many other agencies and businesses, to centralize most management functions.

Centralization of pay, logistics and personnel are already far advanced, and many more such actions are underway, or pending. Such centralization saves people, and for most routine actions, saves time, all of which mean money. The money the Air Force is saving the taxpayer by its Advanced Logistics System is very significant. Not only are there less people required, but dead inventories can be greatly reduced. We all know what a problem it is trying to discuss a special issue with a computer. Nevertheless, the money to be saved drives the world to this centralized approach.

This, together with the Federal Communication Commission's (FCC) Foreign Attachments decision, has brought all manner and sorts of people and equipment into the telecommunications business. Let us be charitable about this phase of our growing up period and say that it has been characterized by something less than a system approach. We are seeing a lot of those textbook problems come to life, in a hurry.

Couple this pressure to move lots of data around with not only a rising personnel cost per man year, but also a declining budget, and you immediately see why everyone, including the military communicator, is trying so hard to get 9600 bits out of a single analog voice circuit, four times what we traditionally do now.

Tough Barriers

We are suddenly finding that the management of our systems in the era of highspeed digital communications is an order of magnitude more complicated than life just five years ago. Hum and clicks, that is impulse and idle channel noise, that went virtually undetected a few years ago suddenly are a major problem today. Noise is throttling some of our most critical systems. In many base telephone cable plants, noise is cancelling out all the hard-fought improvements we have made in cleaning up the backbone structure. In other words, we break our backs getting a tone pack halfway around the world and cannot get it from the base gate to the computer center because 25 years ago somebody put the telephone and power cables in the same trench.

I do not think that our problem is unique. We are not the first technology to encounter a tough barrier. In my mind, our current situation is somewhat like that encountered by aeronautical engineers with the sound barrier. Their answer to that boundary was rather straightforward. It was solved by a whole new technology of airframes, jet and rocket propulsion. The aeronautical people do not build strictly upon the past hardware. They used the technology of earlier aircraft, but they did not take a B-29 or B-36 and merely add new parts or more power to make a B-52.

Aerospace people are to be envied. A communications designer is bound, by both tradition and cost constraints, to use everything already in place until it can no longer be maintained. We have much in-service equipment that would be replaced if pure economic life considerations were paramount, but total aerospace-traditional starting over from scratch is an unrealistic dream in the communications arena.

What disturbs me very much are the schemes of some of today's developers to solve communications problems as though we were not faced with these real world facts. By that, I mean the adoption of not only a whole new technology, but a proposal to implement worldwide and immediately. The commercial and the military communicator stand together at this barrier and we each must make a choice. Of course, the barrier I am talking about is analog to digital conversion.

Commercial Communications

Actually, the commercial world has partially made its decision. A commercial carrier has announced it plans to incorporate a relatively low 1.5MB rate on some voice paths. They are far advanced on their plan to install a new parallel structure for all digital transmission and a new independent route underground. The other commercial carriers seem to be moving in a similar direction. For them, there seems to be no other alternative. In some respects, they have selected the aerospace method for making the transition through the barrier.

The commercial communications world will continue to employ the present voice FDM structure, and for several years will in fact still be adding to it. It will mix in some data riding on the lower end of some FDM and will also construct totally new Time Division Multiplex/Pulse-Code Modulation (TDM/PCM) systems.

The new digital transcontinental routes will be underground, where rain, space, frequencies, fog, interference and the like are fairly well evaded. But the "build on the old structure" principle still holds. When the new digital streams get to the destination, they still must come up for air where they flow through the city distribution switches. They will also interface with all connected subscribers and therefore must not antiquate any existing portion of the present structure.

Today's Budget

However, the military communicator cannot follow this exact same technological path. There are two reasons for this. One is technical and the other is economic. Let us take economics first since it is the most easily understood. We have a billion plus investment in FDM radio systems, new ones now being installed and more still coming down the pike in a few years. Military analog transmission is going to be around for a long time if for no other reason than we simply cannot afford to replace it. However, and this is extremely important, I would strongly oppose a wholesale changeout to TDM/PCM even if we could afford it.

The military communciator has been under considerable pressure for the past few years to abandon analog transmission for the alleged greater economies and efficiencies of the all digital world. Of course, we are technicians too and we are just itching to do just this. But we have studied, in some depth, the course being followed by the commercial world and find little correlation between military needs and the problems being solved by the use of TDM/PCM in the commercial world. Generally speaking, they are putting TDM/PCM in high density traffic urban areas and, more importantly, they are putting it underground. More significant still, they are not trying to inject segments into a FDM backbone structure.

Second, it is highly unlikely technically, and, in the political environment of today, impossible, to parallel our structure with digital elements. Can you imagine the uproar if we asked our European allies for permission to dig ditches all over Europe and put in manholes every several miles or so? Obviously, what is good for industry is not necessarily good for the Air Force and Department of Defense (DOD).

Hybrid World

Hopefully, I have by now convinced you of what we are not going to do in military communications, at least for the next decade, or even two. What we are going to do is move, but with reason, into a hybrid world. What then is the message for the commercial carrier or the equipment manufacturer who must help the military through this transition? First let me address the carrier. You must help us achieve and sustain the high quality circuits your tariffs promise, and for which we pay.

Our experiences to date on our attempts to pass 9600 bit on a 4KH voice channel are not encouraging at all. It is not because the transmission path cannot be made to carry the load, now and again. The problem is that it will not do it consistently. If those circuits are going to carry 9600 bits-per-second of vital command and control data, you can bet that we are going to insist that the quality of service ordered is the quality delivered. The margins are narrow and it is not easy, but it can be done. However, it will require a drastic change in performance assessment philosophy from that in use now.

When susceptibility to noise, phase jitter and the like of our terminal device is very close to the optimum performance of a particular transmission path, maintenance takes on an entirely different specter. A higher degree of instrumentation is obvious, as is constant measurement of critical circuit performance parameters. But even more important is constant analyses of the trends of these measured parameters, because a condition moving from the green, or all clear condition, toward the red, or danger point, must be corrected in the in between state, or amber region.

By the amber region, I mean we have not yet reached the unacceptable error rate, jump phase lock point or whatever that draws the failure, or red line. We in Air Force Communications Service (AFCS) have faced up to the need for this type of tight preventive performance assessment maintenance. I am afraid our commercial providers still have not.

Transition Challenge

In the military, we have learned another lesson which should have general applicability. The old day of keying system performance maintenance to customer complaints is gone. Customers are too tolerant and by then, it is way too late anyway. One great advantage of FDM systems over TDM is that the former degrade gracefully in quality. This will be a great help in meeting the transition challenge.

If you can characterize peak performance in the system and then constantly monitor a few circuit parameters on unused channels, you can pretty much detect these declines in quality before you are in real trouble. You can detect and fix them before the customer notices degradation, that is, while they are amber. This was not too important in the old days of all analog service. It is of the utmost importance in the current hybrid world.

I am not going to get involved in the legal or financial aspects of this challenge I have thrown down. That has been and will continue to be debated in the proper arena. I am simply saying that constant performance assessment must be done by the transmission media provider. Commercial and military systems are linked together throughout the world. We must find ways to work together in a systematic, harmonious and cooperative way to continually assess the true performance of our systems. Less than optimum circuit quality is the most serious problem in my command today. Phase jitter is a major difficulty. I have already mentioned impulse noise. It is completely disruptive of digital traffic. Envelope delay is another.

The military tech controller and his commercial counterpart must work together to fight these enemies, to continuously maintain circuit quality so essential to high speed data traffic. The military is now embarked on a very expensive program to equip our tech controls with the kinds of sophisticated measuring devices needed for this quality assessment job. The commercial carrier must do likewise. It is the only answer to the analog to digital transition phase.

So much for the carriers. What can the equipment manufacturer do to help us through this painful transition period? By now, you should be aware that you can forget about any proposal to swap out our entire system of FDM equipment. You can and you must help me find ways to double and even triple the bit rate of transmission through the present FDM systems, with conversion and conditioning equipment that can be kept operational.

We must look at the present hardware and see where component changes can help improve the subsystem performance on a cost-effective basis. Replacing receiver front ends is a good example. The parametric amplifier was promised to solve all of our receiver sensitivity problems not too many years ago and we bought them. Well, it did not. Its contribution to performance in the laboratory turned out to be an absolute detriment in the field.

What else can we do to our RF and multiplex equipment, our modems, to make them less sensitive to noise, phase jitter and envelope delay? How much automatic circuit conditioning should we buy? We need to identify these areas and to develop an investment strategy which proves the necessity and the economy of selected, phased improvements.

Performance Assessment

We badly need the capability for continuous performance assessment in our systems in service, without shut down. Every jet aircraft has a thrust indicator which tells the pilot if he can get his bird into the air. Why can I not have built-in instrumentation to tell me how my equipment is really performing? Why do not some of you come to visit our test facilities and let me share with you the results of our growing data base on actual, measured performance of your equipment? I think you would be amazed. Some new equipment has noise figures worse than 20 year old stuff. Understand my problems before you come up with the solution.

What have I tried to say? I have said it seems very clear that we will be working our way into a digital world. But not in one day. For a long time to come, we must find ways to straddle the hybrid voice/data world. We must convert the increasing digital traffic load to analog form and find ways to stuff more of it through voice circuits or through an FDM group. Or if we are to use the commercial offering at 1.5 bits, we must be ready to use exactly what we need. We must achieve an orderly transition and continue to provide 99.99 percent reliability throughout the entire time. It cannot be done on old maintenance philosophies.

My main theme to anyone who will listen is technical performance assessment management. There is no other way. It is clear then that we all must work together if there is to be an orderly and well managed transition.

Base Communications Mission Analysis

William Bettis

The Air Force Systems Command (AFSC) mission analysis of "Air Force Base Level Communications and Information Transfer Systems Through 1985" has been completed. The Mission Analysis Study Group was formed on August 1, 1972, at the AFSC East Coast Study Facility, L.G. Hanscom Field, Bedford, Massachusetts.

The purpose of the study was to take an in-depth look at all communications systems and at those functions and operations that require communications support at Air Force bases and stations to define needs and objectives for Air Force Base level communications and information transfer systems through the year 1985.

Most Air Force intrabase information transfer takes place within the following three base services—base communications, base administration and base data automation. To determine the current and future user needs required of these sevices, it was helpful to consider the historical base upon which today's operations are founded.

Base Communications

When the USAF became a separate service in 1947, it "inherited" from the Army Signal Corps the various elements of the existing base communications facilities. Although these systems were then inadequate for most on-base user

*This article originally appeared in the August 1974 issue of **SIGNAL**.*

requirements, an even more serious problem was the slow response time of the long haul record communications system.

As weapons systems became more sophisticated, both direct command and control and support functions experienced unacceptable delays in the record communications area. The installation of "Plan 51" in 1951 replaced some 27 major teletype relay stations in the continental United States (CONUS) with five semiautomated stations. In 1955, "Plan 55" extended that concept to include overseas areas. With the establishment of the Defense Communications Agency in the early 1960s, further improvements in long haul message traffic handling times were noted.

The implementation of the Defense Communication System (DCS) worldwide AUTODIN System in 1967-68 has improved the long haul situation to the point where routine messages are transmitted halfway around the world in a matter of a few minutes; that is, from communications center to communications center. But the process (and the time required) to get the message from the releaser's signature to the AUTODIN entry point (local communications center) at the originating base is very much the same as it was in 1947 (or 1937). The same is true for the process (and time) at the receiving end. The net result is that while we have greatly improved the long haul communications system, we have done little or nothing to improve the terminal facilities and procedures.

The story of base telephone systems is similar. These systems usually consist of a central office or switching exchange and a distribution plant of multipair twisted wire cables. A large majority of base telephone exchanges (whether government owned or commercial leased) are of the so-called "step-by-step" electromechanical type. The basic design for this type of exchange was completed in 1922 and equipment first went into service in 1928.

Although the "step-by-step" exchange is relatively expensive to operate and maintain, it still provides the basic user service for which it was installed—on-base dial telephone (voice) service. The same can be said about the outside plant or distribution system — it is old, requires extensive maintenance and is inflexible to changes in user needs—yet it still provides the distribution for on-base voice telephone service.

In the late 1950s and early 1960s, the increased use of digital data and ever increasing data rates soon proved that neither the switching exchange nor the distribution plant could be used for high speed data transmission. As a result, specific needs have been met by designing to them or through "engineering by addition"—hence the proliferation of individual solutions.

Base Administration

The Base Director of Administration has responsibility for many areas that are vital to intrabase communications. Of these, none is more important than the base distribution system, which is a key element in the writer-to-reader flow pattern for all forms of written intrabase communications. This system includes all means of transporting written correspondence (intrabase couriers, pneumatic tubes, etc.) from one office to another. In fact, the base distribution system handles nearly every sort of on-base communications at one time or another, including mail, correspondence, messages, reports, forms, maps, charts and parcels, and most of these pass through the Mobile Distribution Center (base "admin run," "base courier," "bread truck").

But the mechanics of the base distribution system have changed no more than those of the base communications system over the past 30 years. In fact, the most significant changes over the last three decades have been the introduction of the electric typewriter and the exchange of bicycles for the now familiar "bread trucks."

Base Data Automation

Data Automation as a means of information exchange has a much shorter history than either base communications or administrative services. Although the Air Force was the first major military user of the large scale computers—the Semi-Automatic Ground Environment (SAGE)—the widespread use of their computational ability for a broad spectrum of activities came much later. Most early installations (and the necessary communications links for them) were specific in their applications of computer technology. Only recently has there been a trend to more generalized applications. For example, the 1971 Inventory of Automatic Data Processing Equipment in the U.S. Government showed the Air Force having 1,271 of the 5,961 computers in the inventory.

As recently as 1959, the total inventory in the Federal Government was only 403. As hardware and systems proliferated, massive interfacing problems began to surface. Only now are we beginning to consider the use of netted computer systems to eliminate unnecessary duplication of effort and to facilitate the exchange of information among various data bases.

Base Information Transfer Needs

As previously stated, the agencies primarily concerned with providing intrabase information transfer on most Air Force bases are:

- Base communications;
- Base administration; and
- Base data automation.

All of the functional elements on an air base—mission, support and tenant—are users of intrabase information transfer services.

The Representative Base

Given a description of current services and facilities and a valid statement of user needs, one would think that difference between the two would represent current needs. But the problem was to detemine valid user needs for all Air Force bases of all commands. The study group applied statistical sampling techniques to a list of 161 Air Force bases worldwide to determine the profile for a typical base. Testing the results of this exercise revealed that a typical base did not exist.

User Group Survey

The study group decided to categorize support and mission functional areas such as personnel, civil engineering, fighter unit, comptroller or tanker unit, as user groups. These functional user groups and their requirements for information transfer then became the building blocks for construction of a "representative" base.

Twenty user groups were selected as most representative of those major mission and support areas normally found on a base. Eight air bases were surveyed to provide detailed data on current information transfer methods and requirements for those 20 user groups. (See Figure 1.)

	SUPPORT									MISSION										
USER GROUPS / BASES	COMMAND & ADMIN	BASE OPS & WEATHER	PERSONNEL	COMPTROLLER	DATA AUTOMATION	LOGISTICS	PROCUREMENT	CIVIL ENGINEERING	HOSPITAL	BOMBER / TANKER	TAC FIGHTER	FIGHTER-INTERCEPTOR	AIRLIFT	DEFENSE CONTROL	FLYING TRAINING	DEPOT	MAJ. CMD. HQS.	TECH. TRAINING	DEVELOPMENT	SEP. OP. AGENCY
EGLIN					X						X								X	
KEESLER									X									X		
McGUIRE							X	X					X							
LANGLEY	X			X													X			
ROBINS		X														X				X
MOODY			X			X									X					
HANCOCK														X						
GRIFFISS										X		X								

Figure 1. Selected Bases and Surveyed User Groups

The size, structure, population distribution and prevalence of the various user groups were determined and the locations of the heaviest information transfer requirements were identified. The data also provided for a fine grain representation of information transfer modes, traffic volumes, and their relative percentages for each user group and its subgroups. Thus, mission and support elements from the 20 user groups were selected to depict the representative base and its information transfer needs.

However, two significant factors cast some doubt on the validity of user needs as determined by this method:

- A detailed listing of user identified deficiencies is not sufficient to evaluate the adequacy of today's base information transfer systems. The way the user conducts the day-to-day business of the Air Force is heavily biased by the communications facilities provided to him.
- Most information does not flow through the base communications system; i.e., not through wired or radio systems. Instead, it is transmitted through letters, forms, face-to-face conversations or publications being shuttled around in the base distribution system.

Information Flow Study

The study group developed a different approach to eliminate bias and arrive at true communications requirements. This approach entailed studying the causes and mechanisms of the information flow process. The ultimate reason for information flow is to enable personnel to accomplish their various missions. Twelve basic actions, whether initiated by man or machine, cause information flow on air bases.

DE	DATA ENTRY	CO	COMPARE
A	ADDRESS	R/S	REFERRAL / SEARCH
SE	SIGNATURE ENTRY	D	DISCARD
ST	STORE	S	SELECT
SO	SORT	VQ	VERIFY QUERY
CY	COPY	VR	VERIFY REPLY

Figure 2. The Actions

Using the coding for the actions shown in Figure 2, and a simple 3 by 3 matrix, 41 flow studies in 10 different on-base areas were made. Figure 3 shows the actions and information flow associated with changing an Air Force member's Air Force Specialty Code.

Figure 4 illustrates the same activity (AFSC change) assuming a fully responsive information transfer system and no changes in internal procedures or personnel function management. Using this dual approach (a user group survey and an information flow study), the study group constructed a representative air base against which concepts could be compared, costed and tested. In addition, the representative base provided a detailed set of current information transfer needs that did not assume hardware or media solutions and that could be used as a backdrop for the projection of realistic mid-term and future needs.

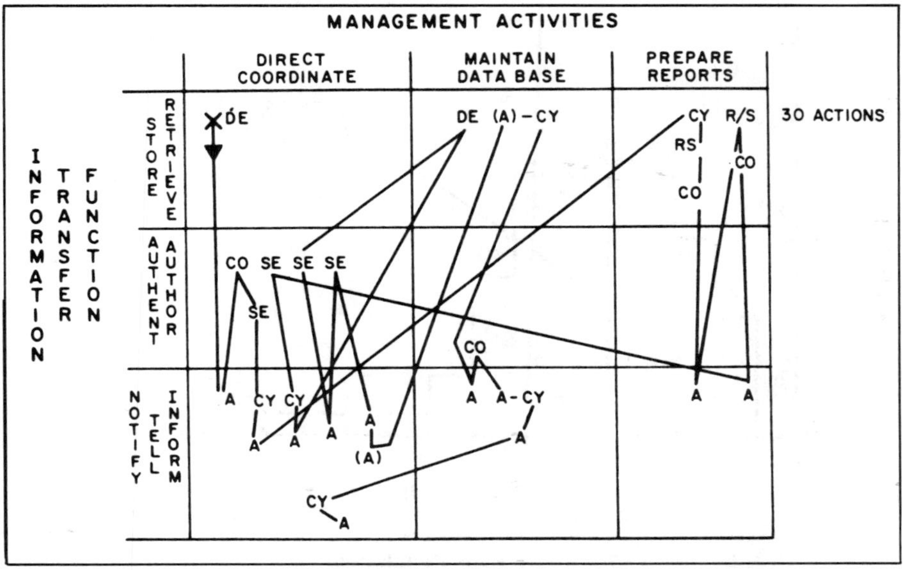

Figure 3. Current Information Flow—AFSC Change

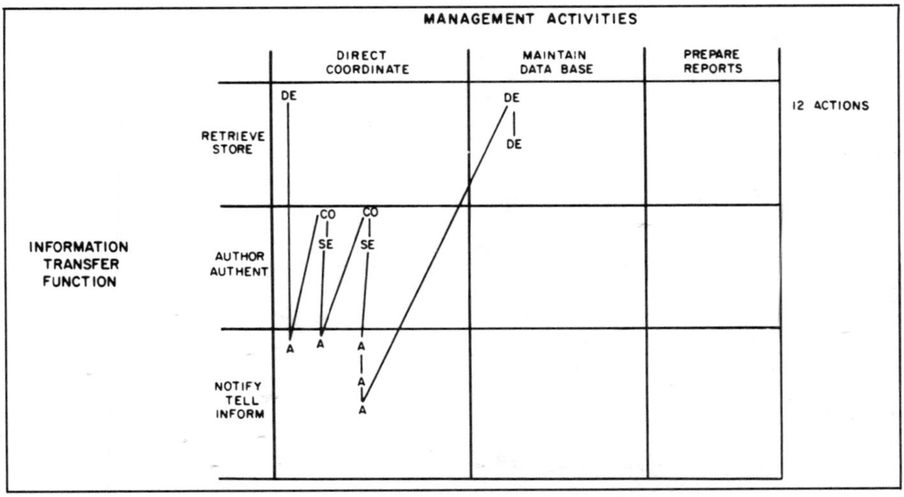

Figure 4. Minimal Flow—AFSC Change

In summary, the representative base is a single wing, medium-sized base with a population of 4,640. Figure 5 shows the representive base on a plot of population distribution for 97 CONUS bases.

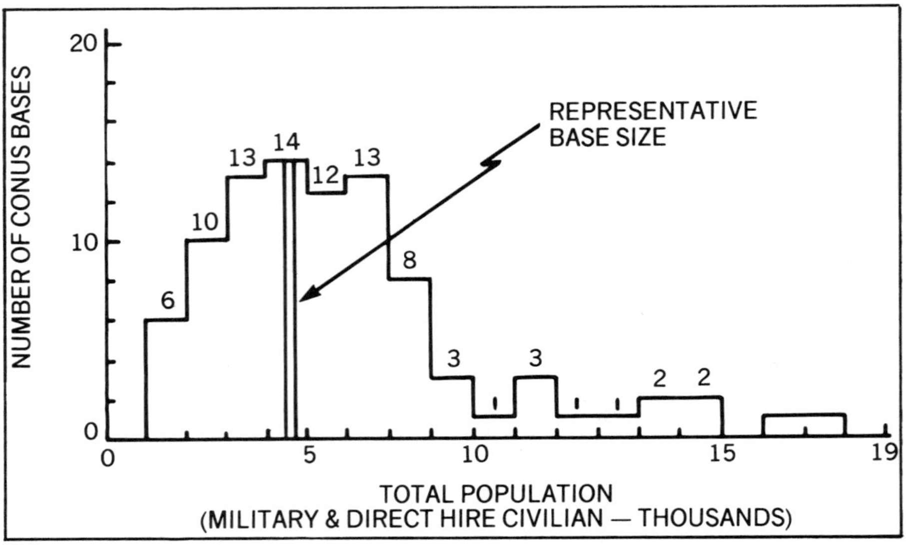

Figure 5. Distribution of 97 CONUS Base Populations

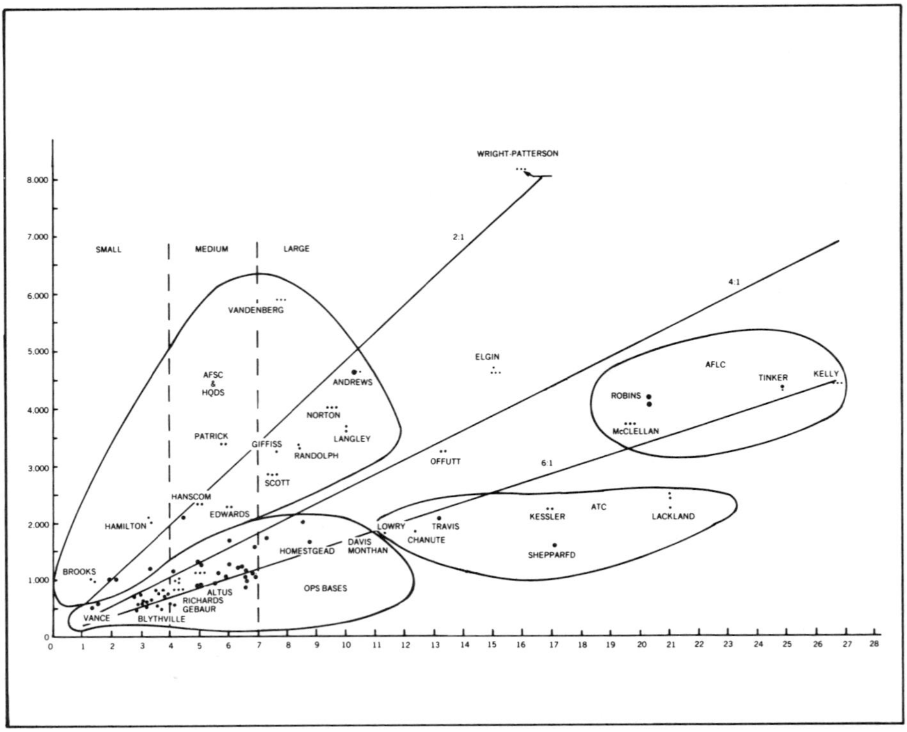

Figure 6. Active Telephone Lines Versus Total Population—97 CONUS AF Bases

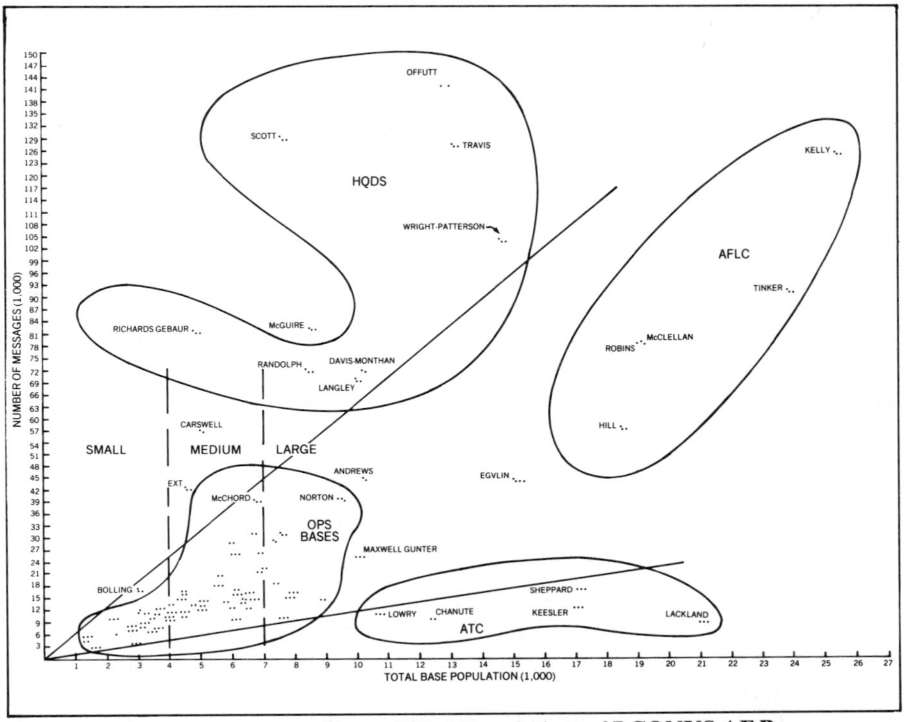

Figure 7. Number of Messages Versus Total Population—97 CONUS AF Bases

Although base population and communications density are not statistically related, they do correlate strongly as a function of base mission. For example, Figure 6 is a plot of CONUS base populations and active telephone lines. The bases group themselves into classes according to the same or similar missions. A plot of one month's AUTODIN message traffic (Figure 7) shows the same grouping tendency. (The representative base falls well within the medium size operation base category on both plots.) To verify the representative base concept, user group models were mailed to some 85 addresses including interviewed user groups and various other user groups and headquarters. The study group made some minor changes as a result of the replies received and conducted some additional surveys before verifying the representative base.

Present Needs

Developing a statement of current information transfer needs was a critical step in the work of the Base Communications Mission Analysis. Such a statement is essential to the determination of mid-term and far-term needs and is the baseline for concept development for future information transfer systems.

An ideal statement of current communications needs would be in terms of needed capabilities and would be independent of hardware, media, traffic volumes, etc. Extensive interpretation of the collected needs data was necessary to approach this ideal statement.

The first step was to assemble a composite list of needs categories (e.g., telephone-type service, record-type service, etc.) from the user group surveys and the information flow studies. (Although those two studies were different in method and did not attempt to gather the same sort of needs, they were in fact compatible and complimentary.) A matrix was constructed that showed the needs of the user group under each of the indentified information transfer categories.

During the course of the user group surveys, the needs panel developed a unit to normalize information transferred by different modes. This unit, the Base Communications Mission Unit (BCMU), normalized traffic to a representative three minute telephone call. (One BCMU equals 1 letter page, 1 message page, 12 data cards, etc.) The BCMU was useful in weighting user needs to eliminate the bias introduced by large differences between high and low volume users. A weighting factor, the number of BCMUs per assigned person for each functional user group, was applied to the unweighted user needs matrix.

One might question the value of developing a list of qualitative user needs; although they can be used directly for system concept development, they cannot be used to apply those concepts to the representative base. What is required for the development of information transfer system concepts on the representative base is a set of quantitative user needs for the 1973-1985 time period.

Future Needs (1973-1985)

The first step in determining future information transfer needs was to examine oncoming programs and projects for their impact on current needs. Since there is no central listing of all such Air Force programs, it was necessary to compile a list through contact with project/program officers at Headquarters the U.S. Air Force (USAF), the major Air Commands, the Data System Design Center and various research organizations. A candidate list of 42 programs was reduced to 25 programs through the elimination of those obviously having no impact. Trial criteria (i.e., approved vs. unapproved, funded vs. unfunded and single base vs. multibase) were applied to further reduce the number of programs to be examined.

The needs panel abandoned this approach when the number of permutations proved it unworkable. Instead, the panel assumed that all programs would be implemented and performed an analysis with the same matrix, BCMU and weighting factors previously used in developing the current needs. Although each program was evaluated for its individual impact, the matrix results reflected the combined impact of all programs. In general, the impact of the programs increases the need for record services and data base availability relative to improvements in telephone, mail and radio services.

The impact of these programs on future needs indicates the trends in user requirements; although each of the 25 programs examined came into being to satisfy a particular user need, oncoming programs do not address all future needs. The first analysis of these oncoming programs deliberately assumed that all of them would be implemented, even though this was obviously incorrect. The second analysis considered each of the oncoming programs in terms of its impact on each of the functional user groups on the representative air base, using a set of subjective questions:

1. Does this program apply to or affect this user group?
2. Does it satisfy all or part of known future needs?
3. What needs remain, assuming implementation?
4. What is the probability of implementation in 1973-1985?

An argument can be made that this determination of future information transfer needs for the representative air base is predicated on "best guess" answers to the questions. Our defense is that the "guessers" were a large cross-section of day-to-day users at all levels, and that the "interpreters" were technical experts from each of the information transfer areas.

Table 1 shows the total information transfer summary for the representative air base. The categories listed across the top of Table 1 are intended to mean "types" of service, i.e., dial telephone means a service that provides the equivalent of dial service.

USER GROUP SIZE (POP)	TOTAL BASE COMM MISSION UNITS *	DIAL PHONE	HOT LINES & INTERCOM	RADIO	MAIL & COURIER	ELECTRICAL MESSAGE	DATA	GRAPHICS	FACE TO FACE & CCTV
SUPPORT (1820)	23199	8687	1297	715	3757	75	7066	73	1529
HOSPITAL (220)	2894	686	1243	21	441	2	87	29	385
MAJOR MISSION (2100)	14565	5226	1570	831	2277	21	120	4	4512
FLYING TENANT MISSION (500)	2392	1100	330	230	110	20	34	7	561
REPRESENTATIVE BASE TOTALS (4640)	43,050	15,699	4440	1797	6585	122	7307	113	6987

* NORMALIZED UNIT OF MEASUREMENT FOR ALL TYPE INFORMATION TRANSFER ACTIONS

Table 1. Total Representative Base Information Transfer Summary

Concepts and Technology

The principal tasks of the Mission Analysis in the concepts and technology areas were to develop and describe information transfer system concept options to satisfy Air Force Base level needs through 1985—and to prepare a technology baseline and forecast. In the early stages, it was neccessary that these two efforts proceed in parallel. The question was often asked, "will technology support this particular concept or portion of a concept?"

Later, as the technology baseline and forecast progressed, it became apparent that today's technology is fully capable of satisfying current — and most future — user needs. System concept options were developed using qualitative user needs furnished by the needs analysis. Quantitative needs were not considered until a set of concept options was selected for analysis and application to the representative base. The selected concept options were defined only to the level of detail sufficient to permit life cycle cost estimates and comparison when configured to the representative base. It is important to note that the system concepts are precisely that — concepts — and that no design or engineering was attempted or accomplished.

The Concept Options

Four system concept options were selected for detailed study. These options range from essentially continuing the present "system" to providing a totally new integrated broadband system to handle all the voice, data and pictorial services required for intrabase information transfers. A one-line description of the selected options is provided below:

- Concept 1 — Respond only to individual program developments.
- Concept 2 — Maximize utilization of the circuit switching system.
- Concept 3 — Add on a digital data subsystem to produce a hybrid system.
- Concept 4 — Develop an all new totally integrated system.

Concept 1

Concept 1 (Figure 8), was the last of the four concepts to be developed. While it is intuitively apparent that the present communications "system" is inefficient, obsolete and not economically effective, that contention cannot be supported without analysis and comparison. In the absence of such analysis and comparison, the possibility exists that to do nothing different is actually the best approach. Concept 1, then, continues to operate and maintain the various

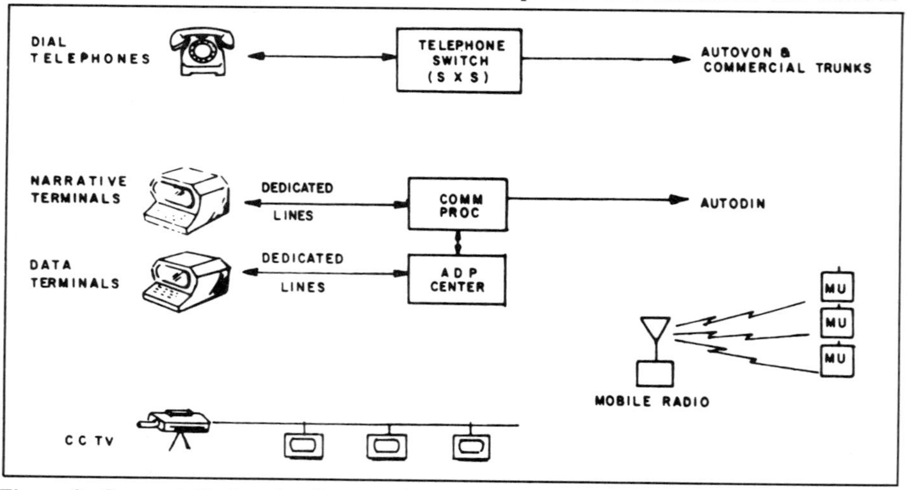

Figure 8. Concept Option 1—Present Approach—General Configuration

parallel "systems" that presently comprise the base information transfer system.

The primary common user subsystem is the base telephone plant. Services the telephone system cannot accommodate are provided by special dedicated circuits. These circuits are either part of an on-base facility, i.e., intercom, hotline, radio keying lines, computer network, etc., or the tails of DCS network, such as AUTOVON or AUTODIN. On the average, special circuits account for approximately 45 percent of the usage of the present base wire distribution plant. Concept 1 provides for incremental upgrading of the present wire system to handle increased communications requirements as they arise.

To allow a reasonable comparison of Concept 1 to the other concepts, the same set of user requirements was used. In this context, Concept 1 provides the same user services as other concepts in kind, but not necessarily in degree. For instance, new data services are provided by direct connection of a data terminal to its associated processor via dedicated circuits. Because each terminal accesses only one computer, a large number of terminals are required. (This is partially offset by the absence of any data switching requirement).

Secure message and data traffic to selected terminals continues to be provided by existing inventory type cryptographic equipment on a link-by-link basis. Pictorial services, except for stand-alone and closed television systems, are provided through physical transport of hard copy as is now the practice. Non-tactical radio service is upgraded by providing a single, multifrequency radio system. This system provides station-to-station telephone type service through dynamic assignment of dual frequencies.

Concept 2

The principal reason for the increase in special or dedicated circuits on base is the inability of the base telephone switch to handle the variety of user services. Future needs for greater data transmission—both in volume and in data rates—will further increase the number of special circuits. Concept 2

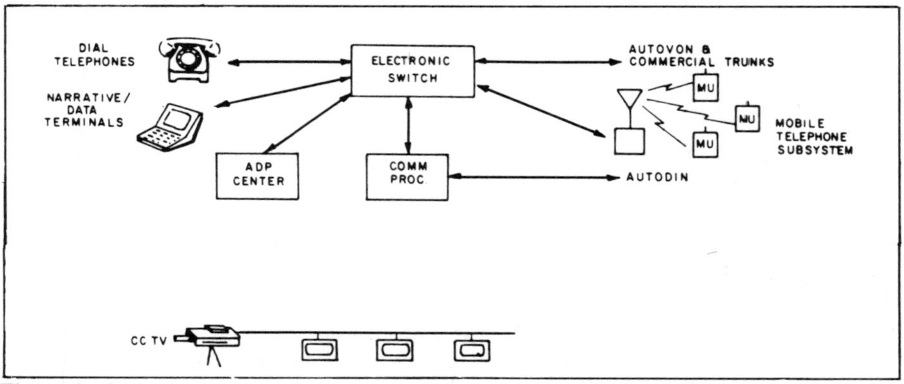

Figure 9. Concept Option 2—Circuit Switched System—General Configuration

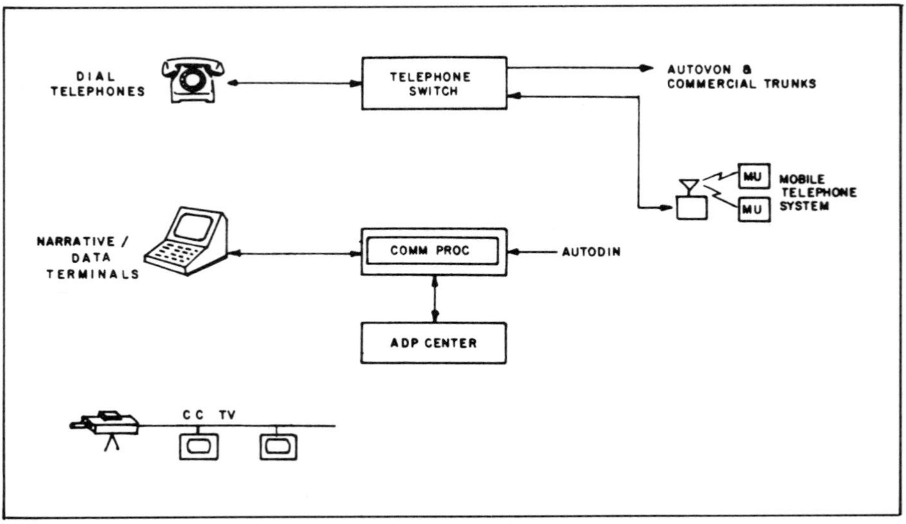

Figure 10. Concept Option 3A—Comm Proc Digital Add-On—General Configuration

(Figure 9), provides a new electronic, 100 k Hz bandwidth, four wire, analog circuit switch as the common user telecommunications service.

This new switch permits integration of most present special circuits into a common system (those few isolated video circuits remain dedicated). Analog and digital voice, data terminals, computers and message centers, regardless of data rates and code structure, are handled in a single system. Centralization of most telecommunications services into a single powerful circuit switch, on a planned systematic basis, meets the current and future user requirements.

The present cable plant, with minor upgrading, can support such a wideband circuit switch and high speed data rates for typical on-base distances. At some selected bases, the installed electromechanical switch is reasonably new or has been maintained in such condition that it can be used for switching local medium speed data circuits. Concept 2, then, provides the option of implementation on a case-by-case basis at certain locations using the existing equipment.

Concept 3

Two methods of approaching the digital analog hybrid system were investigated. For convenience, these are referred to as Option 3A and 3B. Under Option 3A (Figure 10), a conventional store-and-forward message communications processor handles all messages, narrative and data traffic. The present twisted pair cable plant continues to provide the distribution for this traffic. Analog (voice) switching continues as in Concept 2, but the switch is purged of all digital-type traffic; pictorial services are supplied by separate facilities as in Concepts 1 and 2. In this context, Option 3A is not a true hybrid system. While 3A offers a relatively straightforward implementation approach using well

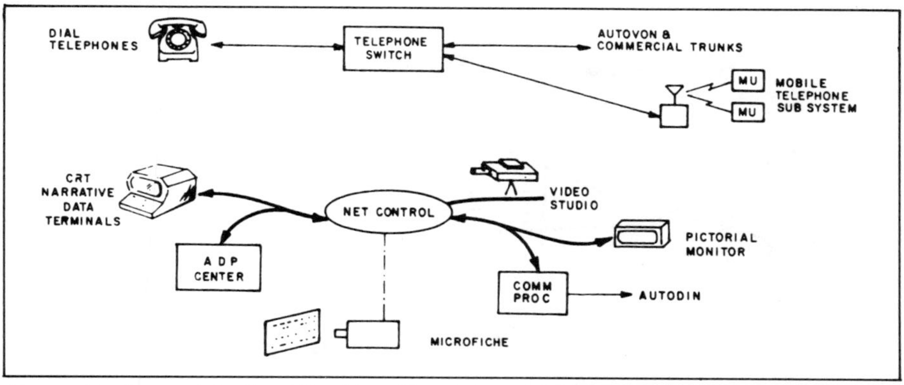

Figure 11. Concept Option 3B—Broadband Tree Digital Add-On—General Configuration

developed techniques and appears to be a reasonable extension of Concept 2, it is a "dead-end" as far as achieving a totally integrated system.

Option 3B (Figure 11) provides a broadband transmission medium—dual coaxial cable in a tree configuation—capable of handling all digital and pictorial services in a single distribution plant. Again, the analog (voice) services continue to use the Concept 2 switch and the twisted pair outside plant. With the digital/pictorial capability provided by the add-on wideband transmission/distribution plant, cathode-ray tube (CRT) or equivalent type terminals are required for display of video as well as alphanumeric data.

The digital add-on system also permits utilization of remote access to automatic microfiche retrieval systems that can be employed for data base storage of static documentation, pictorial and graphics material. Significantly, this option permits consideration in the wideband distribution plant of future integration of all electrical information transfer methods into the single base distribution plant. Thus, it represents a logical step in an evolutionary approach to a totally integrated system.

Concept 4

This concept provides a common integrated telecommunications system to service all electrical information transfer on a base including voice, data and pictorial. Basic to such a system is the wideband transmission/distribution outside plant (like the coaxial cable mentioned above), but several design and implementation alternatives are available. Again, as in Concept 3, two options were investigated—but as opposed to the options under Concept 3, the options here are not really options but rather alternatives—that is, there are two implementation methods, both of which reach the end objective of a totally integrated system. For convenience of discussion and for later reference, these two approaches are referred to as alternatives 4A and 4B. Alternative 4A

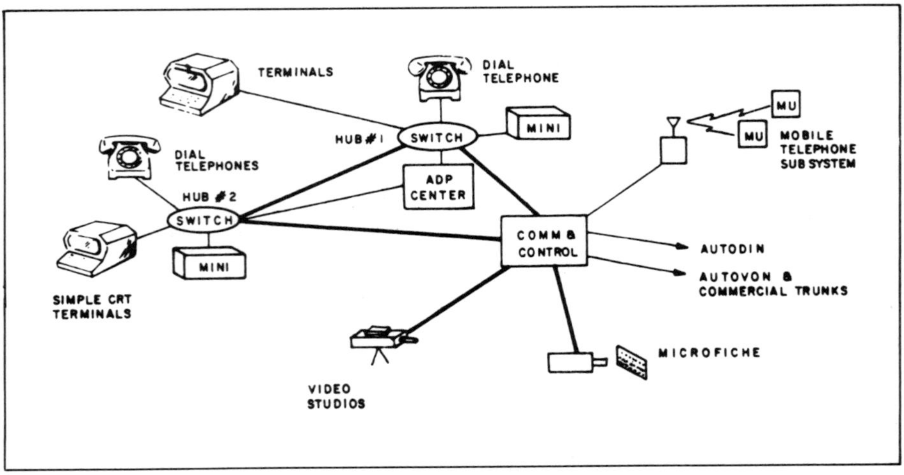

Figure 12. Concept Alternative 4A—Broadband Hub Integrated System—General Configuration

(Figure 12) provides a set of distributed hub switches under the control of a common center. Subscribers are connected to the nearest hub via coaxial subscriber loops, and the hubs are interconnected by broadband coaxial cable trunks.

The two switching hubs postulated for the representative base are sufficient to limit subscriber loop length and eliminate the need for line amplifiers. Voice and data are digital, with voice using 64 kilobit Pulse-Code Modulation (PCM) (switching is by means of a Time Division Multiplex matrix at each hub). An analog/digital modem (similar to the AT&T Codecs) is required at each telephone mainstation and at analog truck termination points. Pictorial services are provided by a postulated simple high resolution display driven by shared refresh and character generators located at each hub. A space division matrix performs video switching as an integral part of the hub switch.

Alternative 4B (Figure 13) is an extension of the digital add-on developed in Concept 3B. One or more of the frequency slots in the Frequency Division Multiplexing (FDM) on the broadband cable are further divided into narrowband frequency slots to provide for the required number of analog telephone channels. Each telephone is equipped with a multifrequency synthesizer and attached directly to the cable with selection of the proper narrowband channel in the FDM slot providing connectivity. A central switch is not required for telephone, data or pictorial service. Data and pictorial services are the same as described in Concept 3B, hence a fully integrated FDM system is provided.

Resources Analysis

The basic objective of the Resources Panel was to obtain, analyze and structure all cost data necessary to the economic analysis of each system

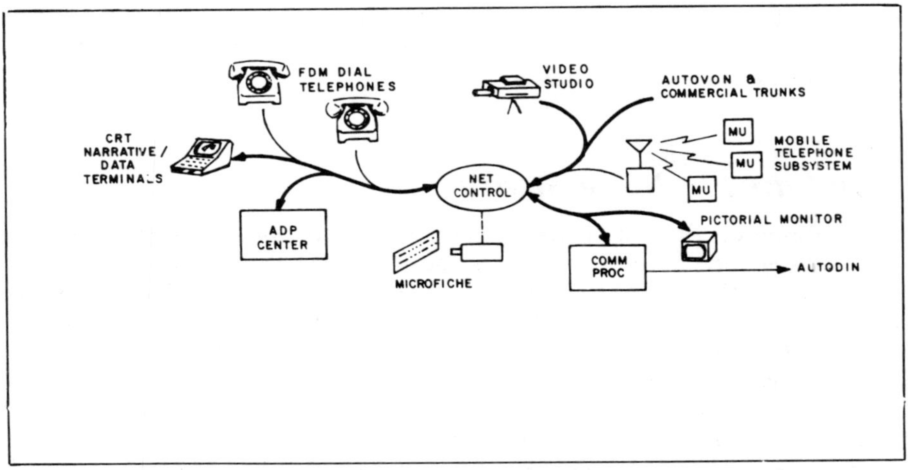

Figure 13. Concept Alternative 4B—Decentralized FDM Integrated System—General Configuration

concept option. The first step was to develop an economic baseline from which to proceed. A pair of parallel studies were conducted—one a detailed study of selected bases, the other a "top down" study independent of bases. Figure 14 shows the major stages in those two studies. The second step in developing the

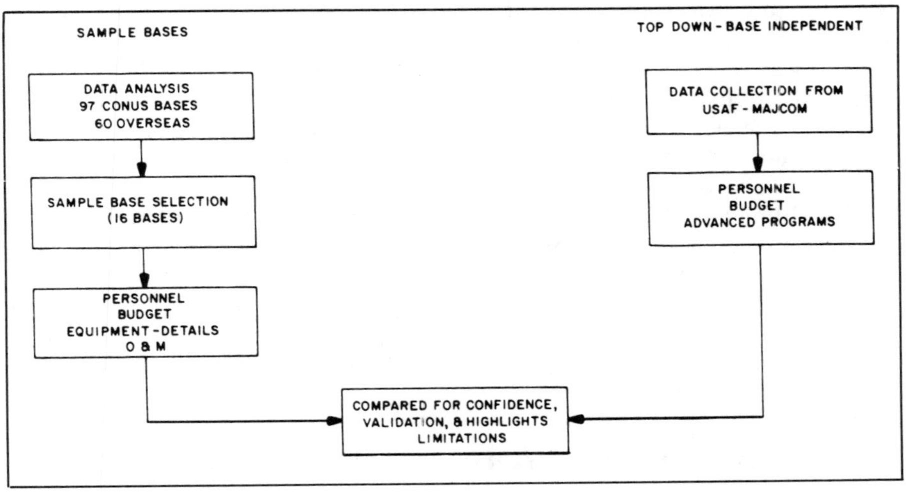

Figure 14. Base Communications Mission Analysis Baseline Development

economic baseline was its testing, simulation and verification. The final step was the application of the economic baseline to the representative base in order to compare various concept options.

The Sample Base Study

The Resources Panel developed a sample base program to estimate information transfer costs at individual bases and to generalize the results Air Force-wide. Statistical sampling techniques reduced the initial list of 97 CONUS and 60 overseas bases to a set of 12 CONUS and four overseas bases to be studied in detail. Cost data associated with information transfer services (Base Communications, Base Administration and Base Data Automation) was collected by total for the base and by total for each service area.

A detailed analysis of equipment, personnel training, spares, age and technical data costs determined investment costs for each base. A similar analysis of personnel, pay and allowances, equipment lease and rental, spares and supplies, equipment maintenance and facilities (building, utilities, transportation, etc.) determined recurring costs. The panel then aggregated and analyzed investment and recurring costs to develop the total life cycle cost for information transfer services at each of the bases. Since the bases were scientifically chosen using criteria including size, population, command, mission and communications traffic volume, the total life cycle costs resulting from this study was a powerful tool in establishing the economic baseline.

The "Top Down" Study

For purposes of comparison and verification, the "top down" or base independent study was directed at the same life cycle costs determination as the sample base study. All cost data identified with base information transfer services was collected from the Air Staff and the Major Air Commands, analyzed and compared with that developed in the sample base study. The Resources Panel identified several gaps in cost data through this correlation process; they examined each gap carefully, obtaining and verifying new source data where appropriate.

A study of nine Department of Defense (DOD) Program Element Codes associated with communications operating costs at the Air Staff failed to substantiate previously collected data. Further investigation revealed that many communications costs are submerged in program elements that are not primarily communications oriented. Administration was the most difficult area in which to obtain cost data because the services provided are spread over many functional units at all levels.

The Economic Baseline

Briefly, the economic baseline is a set of computer programs accompanied by the most comprehensive, detailed, organized and useful body of cost data this baseline, it is possible to determine life cycle costs for any individual base, for the entire Air Force and for a particular base, given only the command, base population and average monthly message traffic. The baseline program is also structured to provide for a number of sensitivity tests. The study group developed the economic baseline specifically for system concept analysis; however, communications planners and programmers may find the data base most useful for other applications.

Economic Analysis

Some interesting sidelights (based on annual life cycle costs for information transfer services) that arose during the baseline development are:

1. Average cost for CONUS bases is $7.6 million per year.
2. Average cost for overseas bases is $2.2 million per year.
3. Average cost for total Air Force is $800 million per year.
4. Average cost Air Force-wide is $778 per person per year.

To establish the bench mark for the economic analysis of proposed concepts, the Resource Panel applied the baseline in the representative base, using a consistent set of rules that could also be applied to each concept. For example, some of the rules were:

1. Life cycle costs are determined using a 16 year base (comparisons may be made using either total or average annual).
2. For the baseline, current (1972) dollars are used for all costs.
3. A government owned inside and outside telephone plant exists on the representative base.
4. All concepts result in a government owned information transfer system.

One significant but unsurprising fact shown by the analysis is that recurring costs are higher than investment costs both for the baseline and for all evaluated concepts; personnel costs represent more than 50 percent of these recurring costs.

The Baseline

The baseline is the life cycle costs of the present information exchange system on the representative base. Whether these costs are expressed as average annual, total 16 year period, discounted or current dollar, the baseline represents the "now" situation and is the basis against which concepts are compared. An important point in the development of the baseline is that it does not include investment costs; the sunk costs associated with continued use of the existing plant were included in the maintenance costs. Recurring costs, then, are the basis for the life cycle costs of the baseline. The average annual life cycle cost of the baseline is $4.65 million.

The Concepts

Each concept was analyzed as installed on the representative base. Figure 15 is a summary chart showing average annual life cycle costs for each option. (The shaded area of each bar represents the investment portion of the costs; the clear area represents the recurring portion. The figure near the top of each bar is the life cycle cost total; the figure near the shaded area is the investment cost.) The difference between total and investment costs is the recurring costs for each option. The average annual life cycle costs range from $2.93 million for Concept 4a to $4.45 million for Concept 1.

The most significant fact is that all concepts have a lower life cycle cost than the baseline. A similar analysis was done using 10 percent discounted costs — with the result that discounting does not change the relative ranking of the six concepts. Figure 16 shows each concept (annual life cycle costs undiscounted) analysis with investment and the principal elements of recurring costs shown separately.

Findings and Conclusions

The findings and conclusions listed in this section are not all inclusive, nor should the reader expect to find the supporting rationales for them here. This list contains only those items considered significant for inclusion in this summary.

1. Present systems in base communications, data automation and administration do not satisfy today's user needs.
2. The system concept options identified and defined in this Mission Analysis will meet current and future user needs; all are more economical than the current "system."
3. The 1985 user needs (as opposed to requirements) will be essentially the same as those identified for 1973.

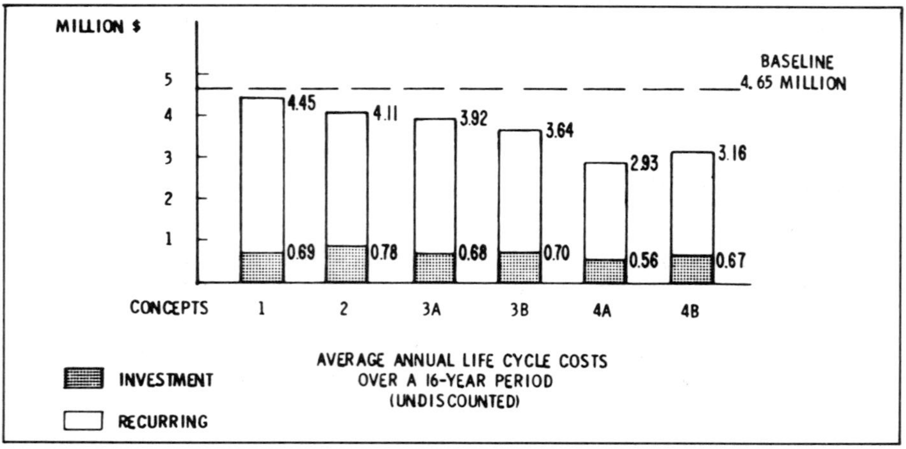

Figure 15. Concept Evaluation Life Cycle Cost Summary

4. Voice communication will continue to be one of the most important modes of on-base communications.

5. The requirement to move more information in a shorter time span will be reflected in efforts to automate source data entry at the terminal.

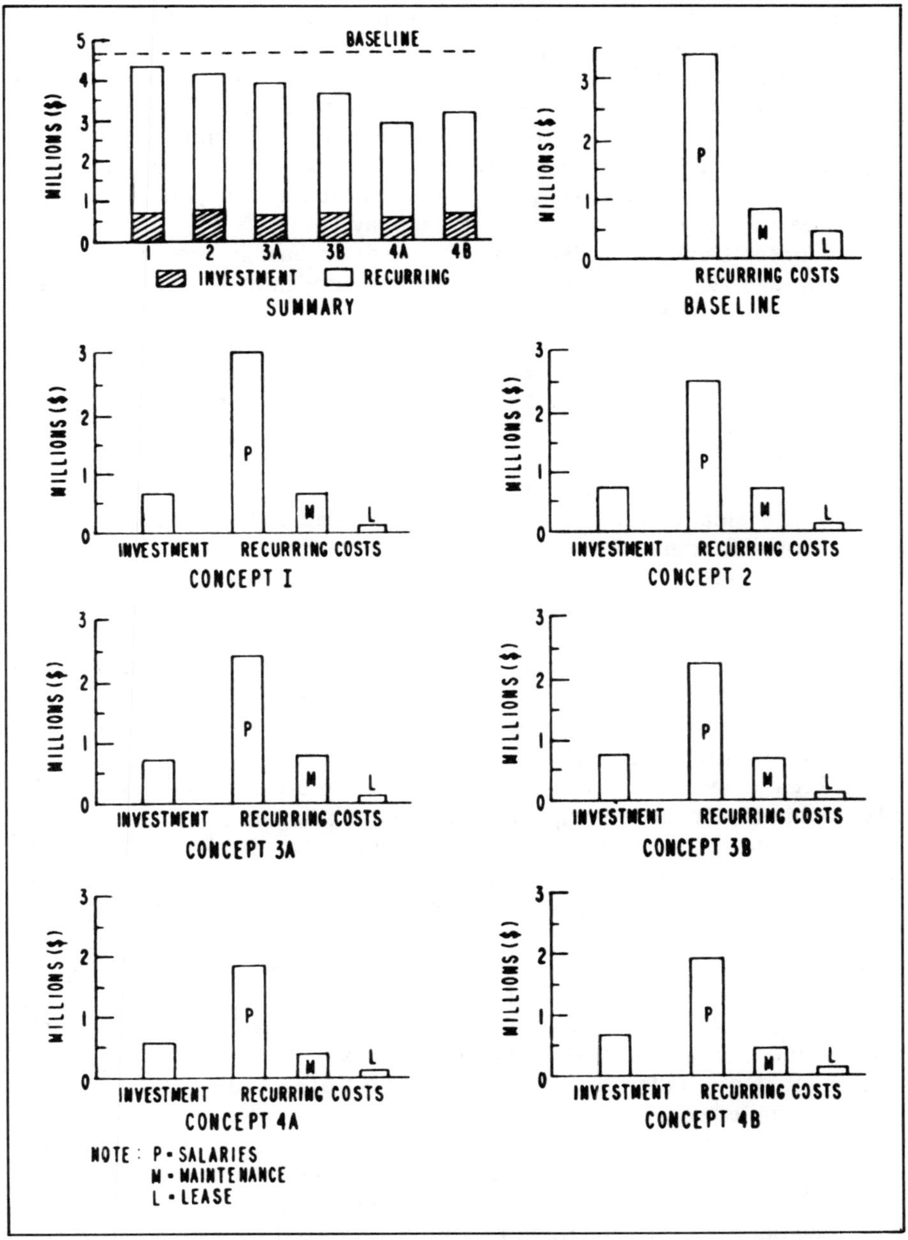

Figure 16. Average Annual Life Cycle Costs — 16-Year Period — Inventment and Recurring Costs Elements

6. The capability to automate various means of source data entry exists today, but no single terminal has been found that meets today's needs, or those of the future, at a reasonable cost.

7. System acquistions and programs have responded to identified deficiencies in specific user areas rather than addressing the total information creation, transfer, storage and retrieval problem.

8. The need for critical examination of new programs in light of existing and planned systems is becoming more crucial (and difficult).

9. By 1985, the separate functional areas of base communications, data automation and adminstration will probably merge into one area that treats information creation, processing and transfer. (The functions within these separate area are being merged by technology today.)

10. Concepts have been developed to satisfy on-base information transfer needs through 1985. Each concept is a separate application of technology and hardware and has been costed separately.

11. Based on economic considerations only, and for those "real" bases that approximate the representative base, the concept choice narrows to either 4A or 4B. The choice between these two will depend on other factors: i.e., risk assessment, engineering trade-offs, implementation ease, available level of funding and achievement of potential personnel savings.

12. The representative air base was developed only for the purpose of testing and evaluating the system concepts developed by the Mission Analysis.

13. Providing users with a totally responsive integrated information transfer system will result in major savings in man-hours expended in the functional areas. Some manpower savings will accrue in the communications and administrative functional areas.

14. True user-to-user information transfer is dependent on both communications facilities and management practices. It serves no purpose to provide a sophisticated data terminal to a staff member who has neither the authority nor the responsibility to initiate message traffic.

15. Technology is available today to provide the transmission media for 1985 requirements. The principal obstacle to providing user-to-user data services (entry, store, access, retrieve and transfer) is the availability of a reasonably priced keyboard/display terminal. A standard terminal with several configuration options is required. (Standardization at the major component level.)

16. Developments in large scale integration and plasma display technologies offer potential for the development of a cost-effective data terminal.

17. A significant gap between current and forecasted technology is the man-machine interface problem. Different approaches to direct voice input/output to computers are under investigation, but most are speaker dependent and have very limited vocabularies and low reliability.

18. Technology in discrete address, multiple access mobile and non-tactical radio systems to solve frequency congestion problems is currently available.

19. The national policy of a totally secure environment is a desired goal and should be pursued; the economics of the required technology will not permit full achievement by 1985.

20. In FY 72 there were 70,182 Air Force personnel authorizations involved in intrabase information transfer — this represents approximately seven percent of the total USAF authorizations for the same period.

21. On all the surveyed bases, the greatest investment costs applicable to base communication systems are in the base telephone system. The most expensive single component of the telephone system is the central office switch.

22. The bulk of annual life cycle costs for base communications are recurring costs; personnel costs account for well over 50 percent of these recurring costs.

Air Force C^3 Technologies

Robert T. Marsh

In providing improved surveillance and warning, a major concern is improved ability to monitor bomber approach corridors and provide adequate warning. An overall update program, the Joint Surveillance System (JSS), is scheduled to be operational by the end of 1982. JSS will replace the current North American air defense surveillance and control system, which is costly to maintain and uses radar systems that were built in the 1950s.

To achieve longer-range coverage of ocean approaches, we have an Over-The-Horizon Backscatter Radar in development. If our experimental system in Maine works out, and it looks promising, it will enable us to look far beyond the earth's curvature, out to 1,800 miles, by using the ionosphere to reflect radar waves beyond the horizon.

For missile surveillance and warning, we rely on our satellite early warning system and land based radars, such as the Ballistic Missile Early Warning System and PAVE PAWS (Submarine Launched Ballistic Missile Phased Array Radar System), to provide corroborative data on intercontinental and submarine launched ballistic missiles.

Another critical area is rapid, reliable detection and monitoring of satellites. Our space surveillance network now consists primarily of land based sensors. The latest is the Ground Based Electrooptical Deep Space Surveillance System which will watch satellites out to geosynchronous altitudes of 23,000 miles.

To pass on the data received through these sensors and for command and control of our forces, we must have highly reliable, survivable communications. One program designed to do this is the Air Force Satellite Communications System (AFSATCOM), which provides two-way, jam resistant teletype communications using miniaturized keyboards and printers. This is a big step

**This article originally appeared in the September 1980 issue of SIGNAL.*

forward, particularly for our bombers, since we have always had difficulties communicating with them in the polar regions.

In the force management area, there are three national command centers, one in the Pentagon, another at a remote underground location and the E-4 Airborne Command Post. Due to its mobility, the E-4 has increased survivability since fixed command posts, even if hardened, are vulnerable to a concentrated nuclear attack. Its full range of very low to super high frequency communications and data computation equipment enable the decision maker aboard the airplane to receive much of the same information that flows to ground command posts and to take the same actions.

Future Strategic C³

With that brief review of current programs, let us examine the promising potential of new capabilities and improvements for future strategic C^3. In the surveillance and warning areas, we must improve our ability to characterize the objectives of an enemy attack. In doing this, we need to be able to answer, literally in real time, questions such as what are the weapons, their number and their intended targets. While our current sensors, primarily ground based, provide some ability to assess the attack, we see the need for a warning system in space to provide both early warning and assessment information.

Toward the end of the 1980s, we should have space based electrooptical surveillance sensors capable of this function. Basically, these are telescopes with new mosaic array sensors which will provide detection and assessment. Progress is being made in integrating programmable signal processing capabilities into the arrays.

These sensors offer important new capabilities, but we must look further ahead to a radar system based on space for an all-weather surveillance capability. The Rome Air Development Center (RADC) at Griffiss Air Force Base (AFB), New York, working with the Defense Advanced Research Projects Agency (DARPA), is moving toward a ground level validation and demonstration of this capability. Emphasis is on space deployable phased array antennas and high reliability solid state transmitter and receiver modules.

By the mid-1990s, we expect to have the ability to deploy a space based radar system which will provide multiple target search and tracking capability with minimal electro-magnetic impact on the environment.

As for satellite communications beyond AFSATCOM, the Air Force has requested funds for a follow-on Strategic Satellite System (SSS) which include a dedicated satellite segment and related terminal improvements for command centers and force elements.

The future of satellite communications is keyed to using the higher frequencies of the super high frequency (SHF) and extra high frequency (EHF) regimes. This capability in the SSS will offer major advantages over AFSATCOM, which operates at ultra high frequency (UHF) and is limited primarily to low speed teletype. These higher frequencies offer a greater probability of getting through the nuclear disturbed ionosphere, and their much wider bandwidths provide for greatly increased data rates and greater jamming protection.

Impressive Capabilities

By the early 1990s, we expect to see a spaceborne laser communications system with impressive capabilities, a very high data transmission rate and an extremely narrow beam. The narrow beam provides covert communications since it is extremely difficult to intercept and dramatically increases the problems in trying to jam it. We expect the most important application will be for satellite cross-linking, lessening dependence upon ground relay terminals located on foreign soil. Combining space-to-space laser communications with space-to-ground EHF and SHF systems will permit global communications with a high assurance that they will be secure, unjammed and capable of penetrating the disturbances of nuclear detonations.

In the vital area of strategic force management, we are concerned about the survivability and endurance of strategic command centers. The E-4 is a survivable command center, but this and other airborne command posts are dependent on their ability to get into the air to escape attack and remain airborne. As we look to new weapon systems, such as the MX missile, we must be able to extend the endurance of our command centers by complementing this airborne capability with other mobile alternatives.

The Electronic Systems Division is closely working with the Strategic Air Command to develop mobile force management teams. These teams will be highly dependent upon automatic data processing capability to process large amounts of data. These processors must be small, limited in their power requirements and possess high capacity memory for storing the needed data bases. In meeting these needs, the ongoing electronics explosion offers exciting prospects.

By 1985, we may see, on one chip, the processing capability of an IBM 370/158 computer. Bubble memories and charge-coupled devices (CCDs) will displace electromechanical bulk storage devices. Bubble memories and CCDs are now on the market and are maturing rapidly. Still more dynamic potential exists in gallium arsenide and Josephson Junction Devices, which can take us well beyond the physical limits of conventional silicon technology.

Current Tactical C^3 Overview

Turning now to the tactical area, let us look first at surveillance. We have made great improvements in this area. For example, the long-range, look-down radar of the E-3A Airborne Warning and Control System provides a composite picture of the air battle such as we have never had before. Additionally, we have other ongoing programs to improve our ground target acquisition capability including such things as radar to penetrate foliage, infrared sensors to aid in reconnaissance day or night and when visibility is low, as well as moving target detectors and emiter locators.

Scanning the status of present communications highlights a key area where we are making good progress in achieving new capabilities. We presently do not have a jam resistant secure voice communications system for fighter and attack aircraft. We currently have in production a quick fix, jam resistant, secure voice communications system called HAVE QUICK, for fighter aircraft. This system will enhance jam resistance of UHF radios by adding an existing radio function which provides electronic counter countermeasures capability. A longer term program called SEEK TALK will replace current radios with new ones that employ spread spectrum technology and adaptive antenna nulling arrays to significantly improve jam resistance.

Having good surveillance and communications, though, will not by itself guarantee the commander a picture of the battlefield. We must provide automated aids to insure commanders receive information from sensors in a timely and useable fashion. Information from reconnaissance photography, voice reports, radar, electronic signals and other sources, can seldom be manually synthesized and interpreted quickly enough to be effectively utilized.

One of our programs to provide automated assistance to decision makers is the Operational Applications of Special Intelligence Systems (OASIS), which we are developing in stages at the North Atlantic Treaty Organization (NATO) War Headquarters in Europe. OASIS involves a series of improvements to the data processing system, as well as to the inputs, interfaces and environment for that system, enabling it to better combine, analyze and handle intelligence information in near real time.

We have already installed some new hardware under the OASIS program. The first software deliveries are scheduled for this fall, and we are working on ways to accommodate new inputs as well as on other major enhancement efforts.

Future Tactical C³

Looking ahead in the tactical area, we see opportunities for impressive advances. Within five years, we will have the capability to internet all the Army and Air Force ground radars operating in a battlefield environment. This will enable us to link the radars to command centers and provide commanders continuous tracking of all aircraft in a theater. Combining this with the E-3A's inputs will give us increased ability to detect low level penetrators and track them during their flight.

In the 1990s, we will have the capability to tie fighter aircraft such as the F-16 and A-10 and spaceborne sensors into the system. With that addition, the radar network will give us the capability to provide a theaterwide view, allowing commanders to plug into the system and receive data on the situation in their sectors.

The future of tactical communications could not be discussed without mentioning the Joint Tactical Information Distribution System (JTIDS). With it, the Air Force and the other services would be able to provide better communications for our tactical forces. It would give us a communications system that is jam resistant, secure and survivable, and one that allows up to several thousand users to share a common communications network. The JTIDS program exploits Time Division Multiple Access which splits every second of time into fractions with each communicator pulsing his message into the system as a digital burst when his fraction of time comes up.

Three terminals could be developed. A large "Class I" terminal has already been installed on the E-3A and will also be used for ground sites. For the E-3A and the ground sites to communicate with tactical force elements, a smaller "Class II" terminal for fighter aircraft, small ships and mobile ground platforms is needed. The "Class III" terminal calls for an even smaller, lower cost, limited capability JTIDS unit that could be used in remotely piloted vehicles, guided weapons or have potential applications for use as a manpack communications set.

Basically, we do not have the technology in hand at the moment to build this small Class III terminal until we can come up with a greater reduction in size and weight. The technology we need is Very High Speed Integrated Circuits (VHSIC). Applying VHSIC to a piece of JTIDS equipment will let us reduce the weight from 80 pounds to about 10 and reduce cost by a factor of 15 or more.

Man and Automation

With the great flood of data coming from all these systems, commanders will be heavily reliant on automated decision aids. This leads me to fusion, the capability to rapidly and accurately interpret and distribute information collected by surveillance systems. The Battlefield Exploitation and Target Acquisition (BETA) program is one of these efforts. BETA is a test bed using commercial off-the-shelf equipment to demonstrate and evaluate the ability of automated centers to combine information on ground targets from many varied sensor systems.

These aids will not replace the human in the decision making process. Decision making is and always will be a human process, but humans are limited in their ability to process and cope with the vast amounts of information generated in a modern conflict. The tactical force management system of the future will undoubtedly look to machines to log, compile and correlate sensor information; subtract bombs when they are dropped and fuel when it is consumed; maintain the target list; keep track of sorties available and provide other automated assistance.

By the year 2000, commanders will have the ability to communicate with computers in conversational language. This will permit commanders to gain a comprehensive portrayal of the battle situation by "talking" to the computer and to obtain, in real time, an estimate of what the consequences will be for specific responses against changing threats.

Summary

As you have seen in this brief look at C^3, we are making good progress in strengthening strategic and tactical C^3 capabilities and the forecasted technologies and planning underway make us optimistic for continued progress. New spaceborne sensors, radar internetting, adaptive high frequency (HF), laser communications and the other efforts I have discussed will greatly expand our surveillance and communications capabilities.

As for force management, the automated aids under development will substantially increase the decision maker's ability to know and control the battle situation, in real time.

Today, a dominant fact is that two superpowers, the United States and the Soviet Union, face each other in a worldwide competition for influence in global affairs. In this competition, the balance of military capability plays a crucial role. And in that military balance, no element is more important than the capability to effectively employ aerospace power.

In the 1980s and beyond, our primary objective must be to support our national strategy of deterrence at all levels of conflict. To do this will increasingly require that our combat forces have improved and survivable surveillance, communications and force management capabilities.

Applications for Airborne Digital Communications

Nicholas P. Janus, John J. McConnell, Lynn S. Nored and Paul H. Phillips

For military communications applications, aircraft provide a reasonable alternative to vulnerable fixed-ground communications facilities. However, airborne communications present additional constraints which are not inherent in most ground facilities. For example, aircraft are essentially required to operate in an analog "broadcast" mode, thus exposing themselves to alien monitoring and tracking. Also, aircraft are limited in payload and power, rendering them vulnerable to a variety of disruptive mechanisms such as jammers. Further, because of flight geometry, they are vulnerable to hostile activities up to a 300 mile range (air-air). Digital communications present an attractive method for circumventing many of these constraints.

While there exists a large array of digital communications systems being adapted for airborne military use, from high power very low frequency (VLF) transmitters to advanced satellite terminals, this article only addresses Line-of-Sight (LOS), Air-to-Air (A/A) and Air-to-Ground (A/G) communications requirements.

When compared to analog systems, there are a number of advantages to be realized with airborne digital transmission systems. Fundamentally, since communications must be conducted in a noisy medium, the atmosphere, digital techniques offer methods to engineers against errors or noise introduced by the variable properties of the medium. Security is also more easily realized by digital processes. No less important is the capability to process larger informa-

This article originally appeared in the September 1978 issue of* ***SIGNAL.*

tion quantities, necessitated in part by the requirement for secure voice communications.

The basic technology of digital processes and devices is at hand and offers much in the application to military airborne systems.

Statement of the Requirement

Survivable airborne command and control platforms require the effective interchange of critical information in both benign and hostile environments. This requirement depends on reliable communications which do not emit a detectable radio frequency (RF) signature and which do not allow the possible monitoring of traffic. This requirement also depends on countermeasures against threats posed by hostiles attempting to prevent communications. The requirement can be supported by a Low Probability of Signal Intercept (LPI) and by an anti jamming (A/J) system. Unfortunately, both solutions are mutually exclusive and introduce their own complexities.

Threat

Hostile detection and monitoring platforms may include other aircraft, satellites, shipboard or ground based sites within LOS of the aircraft. Ordinarily, the hostile does not have sufficient knowledge of the signal structure to use correlation or matched filtering techniques. Therefore, to detect a digital (LPI) signal, the detection tool is a radiometer. For signal interception, the emitted signal-to-noise ratio must be sufficient for detection within the allowable integration time. The maximum integration time is determined by, among other parameters, the maximum length message you are transmitting. To minimize detection, the emitted signal-to-noise ratio must be excessively small, especially when operational requirements dictate a range advantage to the hostile interceptor.

Once intercepted, more sophisticated techniques can be used to analyze, identify and monitor the signal. Detection allows the aircraft to be tracked and its position located via a number of Direction Finding (DF) techniques, including antenna array systems. These systems use both amplitude and phase information to determine the direction of arrival of the incoming signals. Monitoring is essentially accomplished by demultiplexing and decoding the individual information channels of the RF link.

A number of options are available to the hostile once basic signal detection and spectrum identification have been accomplished. The most simple is broadband Gaussian noise jamming. This threat would be used if only the operational frequency and the spectral bandwidths were known. If the benign margin is low, a limited amount of noise jamming could be used to force the emitted signal to a more decipherable level. The jamming threat can be raised by using repeat jamming, or Initiative Communication Deception (ICD). Signals which are transmitted long enough may be detected and repeated by the jammer, confusing or blocking altogether the intended receiver.

Countermeasures

Four effective types of communication countermeasures against detection and jamming can be employed:

1. Reducing the signal strength, preferably below the noise threshold;
2. Increasing the transmitted power to try to overcome the jammer power;
3. Using cancellation or antenna null steering techniques to reduce the jammer's effectiveness; and
4. Using non-linear codes to protect message integrity.

The first countermeasure, reducing the transmitted signal strength, can be accomplished by spread spectrum signal processing which broadens the bandwidth of the processed signal for a given transmitted power and thus supports LPI. Figure 1 illustrates the LPI characteristics of spread spectrum

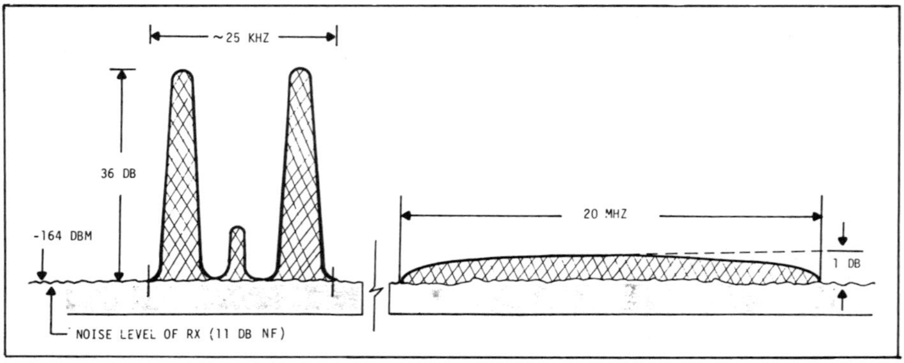

Figure 1. Comparison of Spread Spectrum and Conventional UHF/Frequency Shift Keying

signals. A typical spectrum of 3 kbps Frequency Shift Keying (FSK) data for a 200 nmi (LOS) link is compared to a 10 MHz spread spectrum signature. The goal is to embed the desired signal deeply in the noise, thus preventing detection. Correlation techniques recover the desired signal at the receiver output. (See Figure 2.)

There are four basic types of spread spectrum modulation techniques, with hybrid combinations of these types commonly used in order to enhance their effectiveness:

1. *"PN (pseudo-noise) sequence" or "direct" sequence spread spectrum:* This technique consists of modulating the carrier in discrete steps under the control of a PN sequence whose clock rate is much higher than the information signal rate.

2. *"Frequency hopping" spread spectrum:* This technique includes shifting the basic carrier frequency in discrete steps under control of a PN sequence. The total available bandwidth for carrier frequency selection is much greater than the bandwidth of the information signal.

3. *"Chirp" spread spectrum:* This technique uses a pulsed FM carrier where the carrier is swept over a wide frequency band during a pulse interval.

4. *"Time hopping" spread specturm:* This technique involves shifting the time of transmission randomly under the control of a PN sequence.

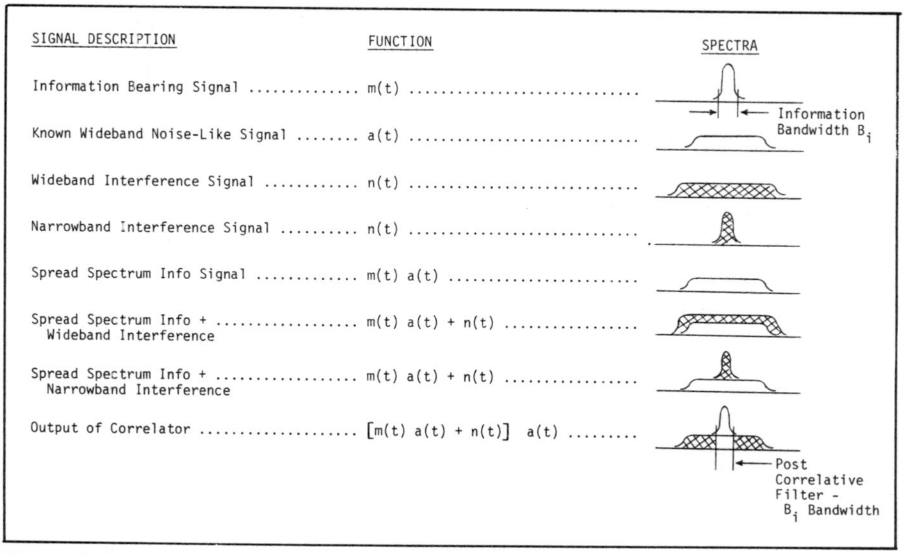

Figure 2. Results of Spread Spectrum Processing

PN sequence and frequency hopping are the two most commonly used systems which support LPI. Unfortunately, PN systems have several limitations:

1. The RF bandwidth required is equal to twice the code bit rate. Practical code bit rates are limited to about 100 Mbps due to the accuracy required of the digital circuitry.
2. The RF bandwidth obtainable at other than microwave frequencies is considerably less than 200 MHz, e.g., 20 MHz at ultra high frequency (UHF).

Because practical frequency synthesizers have setting times from one to several milliseconds, frequency hopping techniques are less desirable than a PN sequence spread spectrum signal. For example, an instantaneous frequency measurement receiver or a comb filter receiver can detect signal presence during the synthesizer dwell time. To improve LPI, hybrid combinations of the PN sequence spread spectrum and frequency hopping systems are possible. The broadband frequency hopping forces the hostile to search across more spectrum and the PN sequence reduces the probability that any single dwell frequency is detected.

The second countermeasure, increasing the transmitted energy per bit (E_b), provides maximum jamming protection by increasing the effective transmitter power output in the direction of the intended receiver while at the same time rejecting part or all of the jammer's transmitted power. However, increasing the transmitted maximum power output negates the LPI requirement low signal strength. Maximum jamming protection is, however, consistent with maximum bandwidth spreading. The jamming margin is related to bandwidth by:

$M_J = G_P - L_S - (E_b/N_o)$ (in dB) where

L_S = system implementation losses

G_P = processing gain $= \frac{BW_{RF}}{R_{info}}$

BW_{RF} = occupied bandwidth

R_{info} = information data rate

E_b/N_o = energy per bit/noise spectral density required at the demodulator.

Wide hopping bandwidths are available with frequency hopping systems. Again, the trade-off introduces the possibility of a jammer monitoring and repeating a signal during the frequency synthesizer dwell time. Also, the information error rate is easily damaged by a jammer hitting a single frequency unless more than one frequency hop is set for each bit of information. This information redundancy, however, decreases error rate at the expense of information throughput.

The third countermeasure, cancellation or antenna null steering techniques, is used to reduce jammer effectiveness by presenting an effective antenna pattern in the direction of the receiver while presenting nulls in the director of the jammer. A simple method is to use a pencil beam antenna. Unfortunately, this implies operation at microwave frequencies and requires tracking antennae. For multiple simultaneous user applications, the aircraft constraints preclude the use of these tracking antennas. A more effective approach is to use correlation; techniques to insert nulls electronically in the direction of the jammer.

The fourth countermeasure, using non-linear codes, is used purely from the point of view of security and message integrity. Superficially, this should be easy as a simple "20-stage circuit can generate 1,048,575 bits." Linear codes, even those which are very long, have the fatal flaw that the number of intercepted bits necessary to break an encoded text requires only two times the number of shift registers minus one, plus a similar portion of that text in the clear. Thus to counter intelligent jammers, the code must be non-linear, i.e., non-repetitive and therefore difficult for the hostile to detect and particularly to capture. In order to obviate signal monitoring, such non-linear codes could support bulk message level encryption.

System Capability

Communications from aircraft offer significant command and control capability for military authorities because of the mobility and flexibility offered by operating in an airborne environment. Yet this same platform imposed constraints on the design of communication systems. In planning improved communications for aircraft, the limitations of space, prime power, endurance and maintenance capability, as well as jamming susceptibility, must be recognized and addressed concurrently with the desired improvement features. These imply a realistic analysis of system traffic sources, link capacity, connectivity, response times, security and interoperability in the system planning phases.

Digital data interface requirements for an improved airborne LOS communication system include secure voice operation, secure message and data transmission and an interactive Automated Data Processing (ADP) access capability. Together, these requirements indicate a nominal data rate of approximately 250 kbps with projected growth to 500 kpps on each communication link from the aircraft.

Physical constraints have a significant effect on the complexity of communication equipment that can be implemented in an airborne system. These include available space for mounting of equipment, the removal of internally dissipated head and system weight.

Another major consideration is that of antenna type, quantity and location. Very limited space exists for antennae. Furthermore, the external mounting of high-gain antennae can significantly impact aircarft flight endurance. The application of larger aircraft (wide body) as communication platforms, may reduce but not eliminate some of these limitations.

Other physical constraints are equipment reliability and maintainability. The command and control of military forces dictates a high system availability implying the use of redundant patchable equipments. High system availability also requires fault indicators and isolation methods be designed into the equipment.

Electrical constraints include power consumption, Electromagnetic Compatibility (EMC) with other on-board systems and the ability to survive nuclear effects, inclulding protection from Electromagnetic Pulse (EMP) effects as well as those of nuclear radiation.

Frequency assignments for the airborne communications must conform with both the flux density levels at the earth's surface, as specified in United States and International regulations, and avoid specifically excluded frequency bands. Frequency allocations and assignments within other countries are also primary items of consideration. Finally, determinations of the RF link parameters, including path loss, rain attenuation and nuclear disturbance must be made in the evaluation of link availability.

From the threat discussion, it is apparent that LIP dictates spreading bandwidths wide enough to place radiated energy below the noise level but retain the capability for power levels sufficient to communicate at maximum data rate and maximum range.

For maximum jamming protection, power output should be adjustable to the maximum attainable, spreading bandwidths must be large, data rates must be adaptive and codes should be non-linear to counter the intelligent jammer. Typical noise density thresholds for UHF and super high frequency (SHF) are −196.2 dBW/Hz and −201.46 dBW/Hz, respectively. Thus, for a 75 bps data rate and 20 MHz or spreading, transmit power cannot exceed six watts or this noise threshold would be exceeded. Yet, for maximum distance, UHF LOS A/A links, only a milliwatt of power is required for communications at 75 bps.

For airborne links, maximum power must be emitted in order to counter even minimal jammer threats. Figure 3 illustrates this situation for a UHF jammer having a moderate randge advantage and no antenna pattern null processing gain. For example, assuming a data rate of 75 bps (equivalent to 18.75 when measured as dBs), a bandspread of 10 MHz and a jammer power of 50 dBW, the communication system power requirements can be scaled directly above and are equal to 30 dBW Effective Radiated Power (ERP).

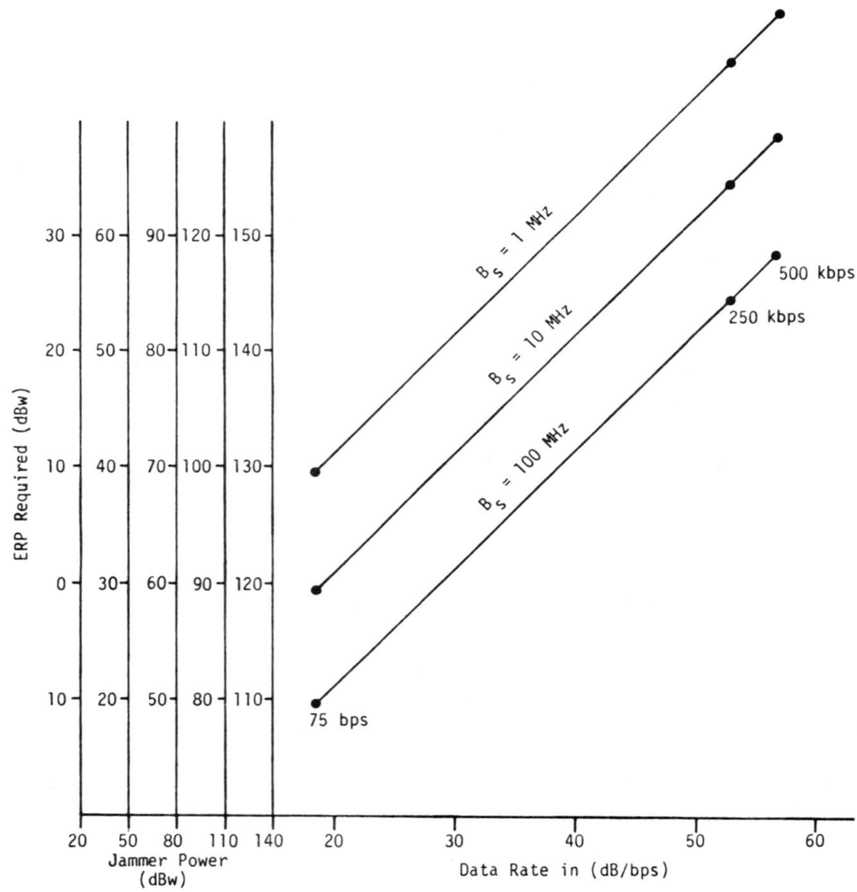

Figure 3. ERP Required at UHF vs. Threat Level, Spreading Rate, at Data Rate

For high power jamming, a combination of null processing, reduced data rates and spread spectrum techniques is necessary to counter the jammer threat, particularly at UHF.

System Considerations

Conversion from analog to full digital transmission offers many advantages to the system designer; however, it also introduces difficulties. These include the stability of frequency standards, synchronization of appropriate terminal devices, doppler shift introduced by aircraft relative motion and link recovery after interruption.

Before commenting on these factors, the components of a typical system (single communications link) should be noted and are shown in Figure 4. User terminating devices are connected to the ports of a digital switch where they may be connected to each other or routed to the radio equipment.

Security devices may be inserted between the user device and the switch as necessary. The digital switch automatically routes the data to the time division multiplex (TDM) unit. After combining with other user information, a single TDM bitstream is provided to the modem where it is converted to a modulated Intermediate Frequency (IF) signal.

Modern modulation techniques, discussed earlier, may be of many varieties including spread spectrum. However, the availability of an accurate reference frequency is an important factor in high data rate digital systems possessing long or non-linear codes. Atomic clock, such as the cesium beam standard, provide accuracies better than one part in 10^{11} and are adequate for this application.

Initial establishment of each communication link requires total synchronization between the modem, assuming that coding is applied to the modulation, the time division multiplexer and if included, the security device. Since these devices are in series, link synchronization time is no less than the sum of the time required for all devices. Since devices such as the modem and TDM have the ability to "flywheel" (retain initial timing) during momentary loss of signal, significantly less resynchronization time is required after a link interruption.

Momentary interruption can be expected during link transfer situations, as in a transition of A/G communications from one ground facility to another. Retention of data integrity during link interruptions would require a fully adaptive system capable of transferring data in blocks, and while such a system is feasible, it becomes very complex for airborne applications.

Correction for Doppler is most easily provided in the modem detection and combining networks. Adjustment for Doppler is most easily provided in the modem detection and combing networks. Adjustment for Doppler error created by the relative motion of aircraft is well within the state-of-the-art and exists in currently avaiable hardware.

Implementation of a coordination channel or "order wire" in an all digital system can be provided easily using one of the multiplexed channels on each communication link.

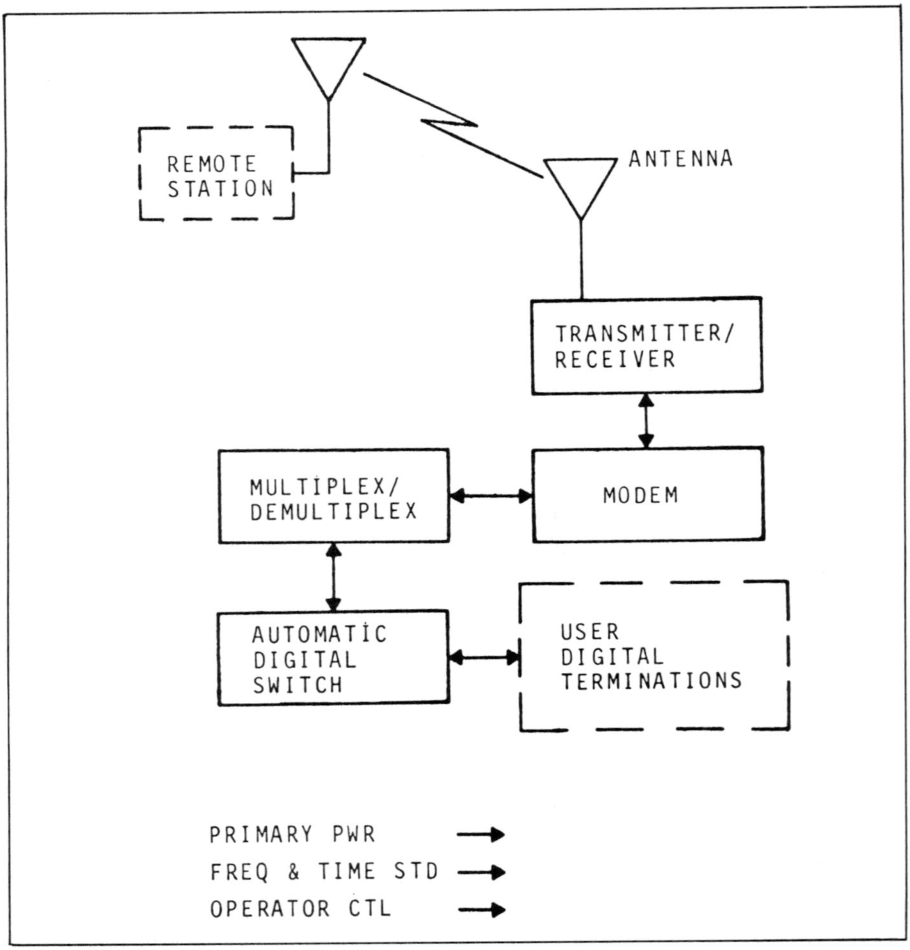

Figure 4. Typical Aircraft Digital System

A recent state-of-the-art survey indicates that techniques and hardware are presently available to permit a sufficiently high power, wide bandwidth, communications system to be readily designed. Broadband, Low Noise Amplifiers (LNA) are obtainable to support almost any degree of spreading desired with the spreading limit basically imposed by the digital control circuitry employed in the modem. The digital area requires a trade-off between switching speed and power consumption, e.g., the use of microprocessor control versus Transistor-Transistor Logic (TTL) many sacrifice speed of operation for lower power consumption.

In the UHF frequency band, one kilowatt solid-state power amplifiers are currently practical and available. Junction dissipation of each is the basic limiting factor along with the number of semiconductor devices, operating in parallel, that can be incorporated into a single unit.

Although in the UHF band many high power vacuum tube amplifiers are in service, they are generally discounted for any new application because of their low reliability and large power dissipation. At the higher frequencies, (through approximately 10 GHz) Traveling Wave Tubes (TWT) are generally available for any desired power output up to 10 kilowatts. Beyone 10 GHz, the power output capability of the TWT begins to decrease.

Antenna

Antenna availability is dependent on the system requirements imposed, i.e., RF power, mounting capability and most importantly, whether it is to be directive or omni-directional. Omni-directional blade antennae are commonplace in the UHF band and are easy to install. High gain antennae, while readily available at almost any frequency, impose installation and endurance problems on aircraft as well as difficulties in acquistion and tracking for A/A communication. Techniques involving the use of electronically steered phased arrays for communications purposes required further reseach and development. Null steering or correlation techniques used to improve A/J performance are currently in the developmental stages.

Radio

Radio equipment which has potential application for airborne wideband digital communications is available at UHF frequencies. Existing capabilities include 10 MHz bandwidth, 25 kHz tuning increments, one kw transmit power and fast tune synthesizers (approxiately one ms); such units are small and lightweight.

At approximately 1000 Mhz, a 20 MHz bandwidth is available, and it appears feasible to increase the transmit power from the present 100w level to one kw. In the 7-8.5 GHz band, current capabilities provide a 100 MHz bandwidth and 10 kw transmit power capability.

The advantages to be gained by moving to the higher frequencies are greater bandwidth and a less crowded spectrum. However, RF power requirements also generally increase as do size, weight, cooling and primary power and become significant considerations for aircraft implementation.

Modem

Many sophisticated digital modulation techniques have been developed in the last 10 to 20 years. These techniques make efficient use of transmitted power and, given enough bandwidth, can easily sustain hundreds of kilobits/second of traffic.

Typical techniques include Spread Spectrum of Code Division Modulation (SSM/CDM) and Multilevel Frequency or Phase Shift Keying (MFSK/

MPSK). Today's codes are long enough to be considered non-repetitive and thus provide a certain level of encryption at no extra cost. Time Division Modulation is not acceptable because it is a burst-type of modulation technique and cannot satisfy the requirement to use low powered signal for minimum detectability.

Recent technological advances in Large Scale Integration (LSI) logic and the inherent increases in switching speeds have finally made these modulation techniques a reality outside of the laboratory. Today, digital modulators provide frequency hopping, pseudorandom sequence and hybrid combinations which provide spreading rates limited primarily by current radio constraints.

Frequency Band Evaluation

In the application of digital communications to airborne use, the operational frequency must be evaluated in light of national and international frequency spectrum policy.

A frequency survey from the UHF Military Band (225-400 MHz) to approximately 40 GHz, and evaluating communication performance at selected intervals, established the feasibility of conducting the desired airborne digital communications. The equipment parameters sected for evaluating performance were typical, in that they exist in currently available hardware or are well within the state-of-the-art.

State-of-the-art gain and power limitations preclude operation above 8.5 GHz. While the higher frequencies provide greater jamming resistance, they would also face higher technology jamming threats than would the UHF frequencies. Therefore, the UHF band offers the best possibility of resistance to hostile jamming. However, the UHF band is extremely crowded, and coordination is difficult both within existing aircraft and on external communication links.

The higher frequencies are easier to coordinate and are not as restrictive with respect to flux density. Further, internal and external aircraft EMC problems force the frequency selection to higher than UHF. The L-band region is an acceptable compromise, but currently, policy restrictions prevent its use. However, more developed equipment is available at UHF. If cost, physical constraints and reliability are weighted more than performance, then UHF is "optimum." If more performance is desired, higher frequencies must be sought.

Conclusions and Recommended Approach

The survey of techniques and capabilities for implementing high speed digital communications aboard aircraft reveals that the necessary methods and hardware for countermanding threats are currently available to the system designer. Considering the application for military aircraft, certain recommendations can be made for a typical system.

When equal weighting is given to communications performance compared to cost, reliability and physical parameters, the results dictate selection of the lower part of the UHF band (225-400 MHz). As discussed previously, radios with the adequate bandwidth, power amplifiers and low noise amplifiers are available in this band to meet the communication link requirements.

In order to obtain maximum A/J protection and maintain a low probability of intercept capability, a hybrid spread spectrum sequence modem which utilizes non-linear codes is recommended.

In the use of the UHF band, performance can vary 40 dB or more depending upon specifc system design. To reduce costs and complexity, to improve aircraft performance and to allow simultaneous A/A, A/G operation, omnidirectional antennae are recommended. Upper limit jamming performance must employ:

1. State-of-the-art transmitters;
2. A means of directing antenna beams;
3. Maximum spreading techniques; and
4. Null processing techniques.

At the same time, jammer detection and power management must be used.

Military Applications of Laser Communications

Charles E. Whited

As the strategic command, control and communication systems of the United States become increasingly complex, the Department of Defense faces an expanding need for new and improved communication capabilities. Among those capabilities must be secure, jam resistant, survivable, mobile systems that can send tremendous amounts of data literally in real time through hostile environments, such as a nuclear disturbed ionosphere.

These communication systems must carry enhanced surveillance and warning data, secure voice channels and computerized interpretations of reconnaissance photography, radar, electronic signals and other source data. As LTGen. Robert T. Marsh, USAF, Commander, Air Force Systems Command, has noted, "Our national strategy of deterrence . . . will increasingly require that our combat forces have improved and survivable surveillance, communications and force management capabilities."[1]

The Department of Defense is developing new and improved communication technologies to provide the capabilities described by Gen. Marsh. One of those technologies is laser communications, sometimes regarded as futuristic or "Buck Rogers." But, in reality, it is a most realistic option. The Defense Department has two major research and development (R&D) programs in laser communications, one managed by the Air Force and one by the Navy.

---

*This article originally appeared in the September 1981 issue of **SIGNAL**.*

Each operates under radically different requirements and system concepts. For example, the Navy effort concentrates on a single specific application (communication to the nuclear submarine fleet) and uses high energy lasers at low data rates.[2] The Air Force program focuses on several generic applications (primarily spaceborne and airborne communications) and uses low energy lasers over a broad range of data rates.

This discussion deals primarily with the Air Force system concepts and technological progress. It gives particular attention to the relatively high level of maturity achieved in Air Force technologies in contrast to Navy efforts that have not yet passed laboratory stage.[3]

Background

The Office of Naval Research, the Naval Electronic Systems Command and the Directed Energy Office of the Defense Advanced Research Projects Agency (DARPA) jointly manage the Navy/DARPA Strategic Laser Communications Program through the Naval Ocean Systems Center at San Diego, California. The Air Force Space Division at Los Angeles, California, manages the Air Force Laser Communications Program. These two independent programs focus their efforts on the widely divergent Navy and Air Force applications.

Despite their lack of technical congruence, the two offices have maintained close contact since their inception to ensure maximum crossflow of information. Cooperative efforts have ranged from discussion of concepts and sharing of data to planning a joint Air Force-Navy space laser communications test (since abandoned). Both programs began as low-level technology efforts, but over the past decade, their horizons have broadened.

The Navy/DARPA program is directed toward a single application—survivable communication with fleet ballistic missile submarines (SSBNs) operating at design depths and speeds. The Navy is developing two conceptual systems oriented toward this goal: the first system uses a spaceborne laser, and the second uses an earth based laser whose beam is reflected from space based mirrors.

Since the application requires only low data rates (communication bandwidth) capable of penetrating clouds are well as the sea, Navy efforts have focused on two types of lasers with low pulse rates but relatively high power: a mercury bromide dissociation laser for the space based system concept and a rare gas-halide laser for space based and earth based concepts. The Navy expects to have laboratory models of these technological linchpins in operation within a few years.[4]

Air Force efforts focus not on a single application but on a long-term technology goal: to ensure the availability of appropriate laser system technology to resolve communication problems expected to arise during the next two decades. The Air Force program is more diverse than the Navy program and is directed toward three principal applications: low data rate (approximately 100 bits per second) systems for teletype/telemetry; moderate data rate (10^4 to 10^7 bits per second) systems for voice and computer processed data; and high data rate (10^8 bits per second or greater) systems for extremely wideband transmission.

Spinoffs from the high data rate technology are expected to provide a survivable communication cross-link for satellite programs by the late 1980s. The low data rate technology offers not only a possible follow-on alternative (late 1990s) to present systems but also a capability for highly jam resistant aircraft/satellite and satellite/satellite telemetry.

The high data rate technology offers a survivable, wideband multichannel cross-link currently under discussion for the next generation of the Defense Satellite Communication System (DSCSIV). The Air Force is constantly evaluating many other applications, both near and far term.

The feasibility and utility of laser communications have been discussed since the laser was first introduced in 1960, but, like most concepts that evolve from advancing technology, the exact origin of laser communication is difficult to trace. The Air Force Avionics Laboratory was actively investigating laser communication system concepts by 1965, and, in 1970, Air Force R&D leaders foresaw a need for extremely wideband communication systems and established the Air Force Laser Communication Program (then called Project 405B) with a goal of creating a one gigabit (1,000,000,000 bits per second) laser communication system for space applications by 1980.

That goal expanded communication bandwidth approximately two orders of magnitude beyond any other system then envisioned. After 10 years of dedicated effort and numerous budgetary restraints, the Air Force successfully demonstrated a one gigabit system at White Sands Missile Range, New Mexico, in 1980. Furthermore, it has either demonstrated, or will demonstrate within the next two years, many spinoffs from technologies developed during that 10 year effort.

System Concepts

Since its inception, the Air Force program has concentrated primarily on high data rate (one gigabit per second) laser communication technology. Although the Air Force program office envisions no immediate specific applications for this technology, one can well imagine its value, especially for wideband, multichannel space cross-links such as may be required by DSCSIV. The high data rate system concept employs a closed, or reciprocal, link concept, that is, the high data rate receiver includes a laser beacon that provides a pointing target for the high data rate transmitter laser. Once the transmitter terminal has located this beacon through a multistep acquisition process,[5] the transmitter continues to track the beacon, points the high data rate beam and begins communication.

The selected high data rate communication laser is a mode-locked, frequency-doubled neodynmium: yttrium aluminum garnet (Nd:YAG) laser that emits 500 million pulses per second. Data is encoded onto these pulses at 1,000 million bits per second by a pulse quaternary modulation technique. The beacon, also a Nd:YAG laser, emits at a relatively slow rate of approximately

3,000 pulses per second, which can be modulated to transmit 20,000 bits of data per second.[6,7]

The Air Force has thoroughly and successfully demonstrated the feasibility of this high data rate system concept. After eight years of extensive laboratory tests and demonstrations, the Air Force Space Division exhaustively tested a prototype system in a quasi-operational environment at White Sands Missile Range from August through December 1980. The high data rate transmitter was installed aboard a C-135 aircraft, and the high data rate receiver was installed in a refurbished ground tracking station.

As the aircraft orbited the ground site at varying distances and altitudes, the Air Force test team repeatedly acquired and communicated data between the aircraft and the ground. Of course, this test was not conducted in the most probable operational environment (in space between satellites) or at probable operational distances (tens of thousands of kilometers). However, laser powers were reduced somewhat to simulate the effects of distance; atmospheric disturbances clearly constituted a "worst case" space condition; and the consensus was that the desert and internal operational aircraft environments are considerably "dirtier" and a "rougher ride" than any satellite used today.

Although data analysis will continue into 1981, the Air Force has demonstrated conclusively that lasers are now technically feasible for use in space communication systems and that the system technology is ready for application.

Lower Data Rate Concepts

As confidence and interest in laser communications technology increased during 1979, the Air Force investigated whether the high data rate laser system might be "slowed down" (in data rate) to accommodate the lower data rates (approximately one million bits per second or "moderate" data rates) that might be required by a survivable communication cross-link for a satellite system. This investigation accelerated development of a significantly smaller and lighter Nd:YAG laser and then led to a frequency tuned gallium arsenide (GaAs) semiconductor laser concept.

The selected system concept is operationally similar to the high data rate concept in that it is a reciprocal system, that is, each receiver includes a beacon laser. But, unlike the high data rate concept, which permits only a one-way flow of primary data, these satellites and their communications subsystem must be interchangeable, that is, there must be a potential for data flow in either direction between any two satellites. Consequently, this moderate data rate concept employs two lasers, each at a different frequency; one laser is the primary data transmitter and the second serves as a beacon. Thus, it not only achieves interchangeability and minimizes complexity but also allows a limited degree of redundancy.

The Air Force has developed and evaluated laboratory models of each laser and has selected the Nd:YAG laser for both functions (transmitter and beacon) in this concept. Fortunately, all other components are essentially direct spinoffs from the high data rate technology. Thus, one can estimate with a high degree of confidence the probable physical characteristics of this system. For example, the operational weight of this communication subsystem could be as low as 100 pounds, less than one half of the weight of a comparable radio frequency (RF) system.[8]

Power consumption of the system will be slightly less than, but comparable to, an RF system. The physical volume of the laser system will be significantly less than the volume of an RF system, although this advantage lies primarily in antenna diameter: less than three inches for the laser system but more than 12 inches for the RF system. Since these characteristics tend to minimize changes in satellite design, the Air Force could begin using the moderate data rate laser communication subsystem by the late 1980s.

Multiple User Access

As acquisition techniques for the one gigabit per second system were refined, the Air Force developed a low data rate device known as a multiple access receiver that greatly increased the allowable initial position uncertainties between two communication satellites.[9] This device consists essentially of a small grid of many optical detectors with a relatively wide aggregate field-of-view (compared to the field-of-view for a single optical detector). Capable of very fine resolution, this receiver can quickly recognize and pinpoint any incoming beacon/initiating signal, and it can readily discriminate among many simultaneous incoming signals. These acquisition beacon signals (not to be confused with the communication, or pointing, beacon laser mentioned earlier) have a low data rate, only some 100 bits per second.

The Air Force quickly realized that these characteristics were essentially congruent with the characteristics of a fleet-to-point, or report-back, communication system, that is, a system that allows many diverse units scattered over a large area to report their status (or other information) back to a single location. For example, the Air Force Satellite Communication System (AFSATCOM) now being deployed is such a system.[10]

This low data rate system concept envisions a small (approximately 60 pounds and less than one cubic foot) laser transmitter installed in the fleet (for example, one in each vehicle of the nuclear strategic forces) and several multiple access receivers located in appropriate space orbits. The receivers would each be linked, possibly through a relay satellite, to the fleet's home base.

Although it is relatively simple and uses existing components almost exclusively, this receiver system concept is not used currently except in its role as an acquisition subsystem. To demonstrate that application, the Air Force has fabricated a prototype multiple access receiver, which is being integrated into the Teal Ruby/P80-1 spacecraft scheduled for launch in 1983. Although the Air Force has no immediate plans to develop a laser report-back system (the AFSATCOM system is expected to last at least into the late 1980s), it is prepared with both concept and technology to develop such a system.

Technical Issues

Air Force system concepts raise several significant technical issues: some issues represent significant communication advantages; some suggest possible operational limitations; and others are simply illusions or misunderstandings. The most obvious advantages of laser systems derive from their high transmission frequencies (10^{12}-10^{14}Hz), short pulse characteristics and narrow beamwidths. The first consequence of a high frequency is a reduction in antenna size, which leads directly to savings in weight and volume. These savings are vital not only on spacecraft but also on today's highly complex aircraft.

A second consequence of high frequencies is increased communication bandwidth. Some heterodyne modulation techniques offer potential bandwidths in excess of 10 gigahertz. Unfortunately, heterodyne modulation technology has not yet approached this potential. Consequently, the Air Force relies not on the high frequency carrier/bandwidth characteristic but on the short pulse width, high pulse rate capability of lasers. As noted earlier, the high pulse rate capability has been used to demonstrate one gigabit per second communication in a quasi-operational environment, and engineers are investigating pulse modulation techniques in excess of 10 gigabits per second.

The short pulse widths contribute not only to the high rate of communication but also to the security, or jam resistance, of laser communication systems. In the simplest sense, short pulses provide jam resistance through sheer brute force. That is, by concentrating all the transmitted energy in a small fraction of a given interval, the effective peak power of a beam is multiplied manyfold, thereby significantly enhancing the system's signal-to-noise ratio, which contributes directly to jam resistance.

In addition to power enhancement, the short pulses permit use of sophisticated encoding techniques. This contribution is most obvious in such concepts as the low data rate acquisition subsystem and low data rate report-back.

In both concepts, the communication receiver must constantly view a large area; thus, a potential jammer might insert a false signal with relative ease. However, the extremely narrow and accurately timed laser pulses allow division of time into many narrow time slots, or bins, thereby permitting use of simple but highly jam resistant encoding techniques.

Such "time bin" encoding techniques are not so directly applicable to high speed systems; for these systems, the narrow transmitter beamwidths and narrow receiver fields-of-view provide the requisite jam resistance. For typical high speed receivers, a jammer would need to be within a cone only two miles across at 40,000 kilometers from the receiver. In either case — high data rate or low data rate, wide field-of-view or narrow field-of-view — laser technology leads directly to highly jam resistant, lightweight, small communication systems.

Laser communication systems have one serious operational disadvantage, at least for any application involving terrestrial based terminals: namely, the unavoidable physical fact that narrow laser pulses are drastically stretched in passing through clouds. Although the Navy has collected data apparently indicating that some low data rate signals will successfully penetrate cloud cover, the same data clearly indicates a need for operational alternatives for high data rate systems.

Two possible alternatives are dispersed earth based stations and air based relay platforms. Dispersal of ground stations will ensure at least one cloud-free space-to-ground path, whereas an airborne relay platform can convert the optical signal to a radio frequency signal that can "punch through" the clouds. Other solutions are possible, but any explicit choice must be based on specific operational requirements. Of course, no such problem exists for satellite-to-satellite communication links.

Misleading Myths

Three particularly misleading technical issues frequently arise, usually because of misunderstanding of operational concepts or inadequate knowledge of proven technology. The first issue is the possibility that narrow laser beams cannot be acquired or, once acquired, cannot maintain the required pointing accuracy. Most misunderstandings about the feasibility of locating, or acquiring, narrow laser beams in the vastness of space seem to overlook two vital points. First, successful acquisition for any system usually depends more on a wide receiver field-of-view than on a wide beam. Secondly, a system frequently may use relatively wide beams and fields-of-view for acquisition and then narrow them for communication.

The Air Force high data rate laser system is just such a system. It begins with a wide field-of-view (± 2.2 degrees) greater than any probable uncertainty in satellite position and a relatively wide acquisition beam. During the beam acquisition phase, both the field-of-view and the acquisition beam eventually narrow to less than 100 microradians (less than 1.75×10^{-6} degrees) before initiating high data rate communications. Thus, an acquisition process for a laser system, however complex, does not substantially differ from acquisition processes in other communication systems.

Another technical issue that generates misunderstanding is the feasibility of pointing narrow beams. Such misunderstandings seem to overlook the relative predictability of satellite orbits. The distance between satellites may be very great; orbital velocities may be very high; laser beams may be very narrow; and, in fact, a satellite may be expected to move a distance of several beamwidths (perhaps four to five beamwidths for two synchronous satellites) between the moment a beacon emits a signal and the moment a high data rate beam returns.

Clearly, a narrow beam may miss its receiver (hence, no communication) unless the transmitter points the beam toward the location where the receiver (beacon) will be, rather than where it was. But the transmitter can predict that location in a straightforward calculation based on continuous updates from the beacon signal itself. Thus it becomes straightforward, if not actually easy, for the high data rate beam to intercept the receiver, just as a hunter "leads" a bird on the wing or as a quarterback "leads" his receiver.

Various techniques can be used to obtain updates from the beacon signal; however, selection of a specific technique is based on the degree of desired operational flexibility. Finally, the Air Force tests at White Sands Missile range in late 1980 have demonstrated the feasibility of this point-ahead concept.

A final misleading technical issue simply implies that laser communications technology is too futuristic and unproved for practical use during the next decade. Although such a contention may be true for some specific concepts (and this author is not convinced), such a contention about the Air Force's system technology again overlooks the actual demonstrations of prototype hardware, not only in the laboratory but in the quasi-operational environment of White Sands Missile Range. A close examination of the data can only support the conclusion that much laser communications technology has reached maturity.

Joint Service Effort

Although the Air Force and the Navy have cooperated in the past when their interests overlapped, areas of cooperation have usually been the collection of data rather than any commonality of operational applications or requirements. A recent suggestion for application was the use of a low data rate broadcast system concept, such as the Navy's system, to provide point-to-fleet communication to both aircraft and submarines.

Coupled with the multiple access receiver system, this blend of laser technologies possibly offers a survivable two-way communication system unparalleled by currently envisioned radio frequency concepts. The technologies would be identical; only a slight shift in laser frequency might be necessary to separate the Air Force and the Navy signals and to avoid interference. Like the low data rate report-back concept, this concept has no extant operational requirement. However, communication system planners might consider not only the operational benefits but also the potential R&D savings that might accrue from such a joint development effort.

The Air Force and the Navy are pursuing new solutions to future communications needs — needs that may be best satisfied through laser communication technology. Radically different operational requirements dictate radically different technological approaches, but the Air Force and the Navy share developmental data whenever possible.

The Air Force has field tested its first prototype system and the Navy will soon demonstrate its first laboratory model; progress on all approaches seems to indicate eventual and even near term success. Proven system concepts have demonstrated clearly the versatility of laser communications technology. However, the very nature of the concepts suggests that lasers are not the perfect answer to all needs for all time. But accurate concept evaluations reveal the necessity to weigh carefully, on a case by case basis, not only the significant benefits of laser systems but also the possible concurrent advances in other communication technologies.

Ultimately, one may expect to find that laser and radio frequency systems are complementary rather than competitive, just as HF, VHF, UHF, EHF, etc., are functionally complementary. Recent Air Force tests have demonstrated that key laser communication technologies are essentially mature. Those technologies now await effective exploitation by today's planners of military communications.

Footnotes

[1]LTGen. Robert T. Marsh, "Air Force C^3 Technologies," *SIGNAL*, Vol. 35 (September 1980), pp. 32-36.

[2]Cdr. Thomas F. Wiener, "Strategic Laser Communications," *SIGNAL*, Vol.35 (September 1980), pp. 43-48.

[3]An article by Cdr. Thomas F. Wiener in the September 1980 issue of *SIGNAL* presented a detailed description of the Navy's laser communication efforts. This article mentions the Navy program only as a basis for comparison.

[4]*Ibid.*

[5]Lt. R. S. Mason and Lt. G. M. Lanham, "Free Beam Laser Communications System Technology Development," Proceedings of the Laser Institute of America (LIA) Electrooptical/ Laser '80 Conference and Exposition, Boston, MA, November 1980.

[6]The technical details and many technical advances achieved in developing this system are thoroughly discussed elsewhere. Footnote 7 lists an excellent technical overview that contains an extensive bibliography.

[7]S. B. Fishbein, et. al., "Laser Communications," *Satellite Communications*, (May 1979).

[8]"High Data Rate RF Satellite Communication Link Analysis," Technical Report AFAL-TR-75-24, October 1975, Wright-Patterson Air Force Base, OH, p. 110.

[9]Dennis L. Meredith, "LASER-COM Speeds Unjammable Messages Through Space on Light Beams," *Popular Science*, Vol. 215 (November 1979), pp. 86-88.

[10]"Air Force C^3 Technologies," p. 33.

Part V: Electronic Warfare

Editor's Introduction

The earliest example of electronic jamming of which the editor is aware occured in 1901 during the America Cup international yacht races. One of three fledging radio companies competing in providing news reports on the races successfully jammed the other two and coded the jamming in such a way that its correspondent was able to follow the progress of the races. What is now called electronic warfare came into real prominence during World War II in what Winston Churchill vividly characterized as "The Wizard War." Following that war, EW fell into a period of relative neglect. However, the Soviet Union did not neglect EW or, to use their term, Radio Electronic Combat (REC). In the United States the technology and military interest was largely kept alive by a few, for which we all owe a debt of gratitude to organizations like the Association of Old Crows.

A good introduction for the articles in this part is contained in the article "Electronic Countermeasures Technology" by Lee Meadows. It provides a good overview of the various techniques used in electronic countermeasures (ECM), including self-protection, escort, standoff and expendable systems.

Gen. Alton D. Slay, USAF (Ret.), is noted for the innovative management improvements he brought to the military systems acquisition process. His article "Electronic Warfare Integration and Planning" stresses the importance of integrating EW into the total combat aircraft system design as are the engine, radar, communications equipment, etc. He notes that EW systems have too often been treated as an afterthought and added on to the weapon system. Recognition of this fact has led to new acquisition management methods which are used to make EW an integral part of the weapon system planning process.

MGen. Larson's article about lessons learned about command, control and communication countermeasures (C^3CM) notes that electronic combat is finally emerging as an important element of modern warfare. It contains his observations about areas needing improvement based upon the numerous opportunities he has had to observe the attempted use of C^3CM. He sees three major areas requiring improvement before C^3CM becomes a true battlefield capability. They are connectivity, training and the establishment of sound doctrine.

Another more critical and thought provoking assessment of the U.S. EW posture is contained in "Electronics: Just Another Element of Warfare." Alan Campen also finds a dearth of doctrine, operational concepts, tactics, procedures and exercises necessary to the support of combat operations. He contrasts the paths taken by the world's two major military superpowers and concludes that "the United States, which would seem to have the greatest dependency on its electronics, may have the least capability to use them in combat." He believes that the United States needs a clear, consistent concept that integrates EW into the overall military campaign.

"Command and Control Warfare: Seizing the Initiative" by Stanley and Noggle has a similar theme, though more operationally oriented. The functions required to orchestrate active command and control warfare operations are seen to be the same as those involved in the command and control of any kind of a tactical battle employing all kinds of weapons. The article describes concepts which will help our understanding of how we can make command and control warfare work for us as a significant part of overall combat capability.

A somewhat contrarian view, hence one deserving particularly careful thought, is posed by Michael Thurbon in "Interoperability and EW." He argues that satisfying the need for interoperability (in the broader sense) means that major military systems of all kinds are becoming increasingly diffuse. Therefore, the advantage that ECM once had is shifting to favor electric counter countermeasures (ECCM). Further, for the foreseeable future, the cost of successful jamming will rise at a faster rate than the cost of introducing the relevant counter countermeasures. He sees the balance shifting to favor the weapon and communications systems and that the area that any jammer can hope to dominate is collapsing.

Dr. Ince describes the various communications media and systems being used in NATO as part of the NATO Integrated Communications System (NICS). Their vulnerabilities to ECM and possible improvements to make them more jam resistant are discussed. He concludes that NICS design must incorporate a mix of various media which have varying resistance to ECM so as to reduce the threat to the whole, plus the ability to reroute critical traffic over unaffected media.

One of the major problems confronting EW has been how to conduct realistic exercises. MGen. Marks, in this article on "EW In Operational Testing," tells us that operational test and evaluation of weapons systems and their employment concepts require that the EW environment be taken into account. He goes on to describe the organization and mission of the Joint Electronic Warfare Center, how operational tests are conducted, and the problems encountered in doing so. He points to the increased awareness of the importance of EW testing in today's operational testing community but also notes some significant problem areas.

"A New Airborne ECM System For The 1990s and Beyond" by Major Stiles is interesting for three reasons: It is a joint Air Force/Navy program; it describes the state-of-the-art in airborne ECM; and it briefly describes an acquisition strategy designed to reduce technical risk.

Electronic Countermeasures Technology: An Overview

Lee A. Meadows

Modern military forces rely increasingly on complex, high technology weapons. Many of these are controlled by information obtained by electromagnetic radiations. The effectiveness of these weapons is counteracted by a class of systems known as electronic countermeasures (ECM).

The history of warfare is filled with examples of the following sequence: One combatant introduces a new weapon and achieves an initial success by virtue of the element of suprise. The opposing combatant immediately initiates countermeasures (CM) tactics to nullify the effectiveness of the surprise weapon. The original innovator then responds to the countermeasure to restore his weapon effectiveness. Thus, there is a continuous reaction between the new weapon and the countermeasure to the weapon. Success will reside with the side that can introduce effective new weapons and can rapidly initiate countermeasures to nullify the opponents' weapons.

Electronic countermeasures originated shortly after wireless communications became an element in the military command and control doctrine. However, the earliest major use of electronic countermeasures occurred during WW II. When British antiaircraft guns and fighter planes began to inflict serious losses on the Luftwaffe day bombing raids, the Germans switched to bombing at night. They provided navigation data for their bombers by using

**This article originally appeared in the March 1978 issue of SIGNAL.*

radio frequency beams that would intersect over the intended target area. By flying along these radio beams the German bombers navigated to their targets and dropped their bombs under cover of darkness. The British countermeasure to this new tactic was a deceptive transmitting system (called BROMIDE) that effectively altered the navigation data by introducing false beams and caused the German bombers to drop their bombs harmlessly in the English Channel.

Later in WW II, when the Allies, as well as the Germans, developed radar systems for locating aircraft, both sides discovered that radar would be thwarted by dropping bundles of aluminum foil strips from aircraft. These became known as "chaff" or "window," and proved extremely effective in defeating the early radar systems of that period.

Experience gained in the 1970s in Vietnam and the Middle East led to a continued growth in quantity, variety and quality of electronically aided weapons. Some of these emerged as "surprise" weapons in the most recent Middle East conflict. The effective use of "surprise" weapons or weapons with unexpected capabilities posed new defensive performance requirements and led to new countermeasure tactics.

The mass deployment in the Middle East of the new Soviet supplied SA-6 short-range antiaircraft missile system, the SA-7 man-portable antiaircraft missile weapon, Styx antiship missile, and the wire control link, man-portable, anitank missile, presented a formidable challenge to the countermeasures community.

These four weapons use the semiactive microwave, infrared, active microwave seeker and optical tracking techniques respectively. Each has electromagnetic characteristics that are uniquely different from prior known equipment. The microwave ban SA-6 radar is the first operational example of a hostile radar applying doppler techniques that are especially effective against low flying aircraft. The doppler radar is also much less affected by the conventional noise jamming techniques that are effective against the multitude of low duty cycle microwave pulse radars that formerly made up the hostile radar inventory.

The recent Middle East conflict exposed Israel's preoccupation with its air strike capability at the expense of ECM capabilities. It clearly demonstrated the need for an ECM inventory appropriate to the threat.

By contrast, the Israelis remembered the sinking of their destroyer, Elat, in 1967, by a Russian Styx missile. Based on this experience, they equipped their gunboats with rapid-blooming chaff which effectively screened them from the weapon system's radar. As a result, they scored several successes during the October 1973 war against Syrian operated missile carrying ships.

The element of surprise will always be with us. However, the advantage derived from its use by an adversary will be effectively nullified if we heed the lessons learned:

1. Design system architecture with inherent quick reaction flexibility and versatility;
2. Rapidly implement the findings of the intelligence community in the field; and

3. Continue to recognize countermeasures as an integral part of the overall force structure.

The relationship of hard kill threat systems to own force platforms and their soft kill countermeasure assets are summarized in Table 1. The passive Electronic Support Measures (ESM) provide warning of threat presence by detection of both intentional and unintentional threat radiations within the radio frequences (HF, VHF, UHF, SHF and EHF) and electrooptical infrared, visual and ultraviolet portions of the electromagnetic spectrum. The ESM in combination with STELTH are designed to achieve detection and identification of threat system electromagnetic emissions and direction well outside the threat system's target detection envelope. ESM provides a valuable passive means for self-protection by:

1. Implementing evasion, cover and deception tactics and
2. Setting up essential electronic countermeasures (ECM) parameters both prior to penetrating the threat lethal envelope.

HARD KILL THREATS & WEAPON CATEGORIES	OWN FORCE PLATFORMS	SOFT KILL COUNTERMEASURES ASSETS			
		INTEGRAL	EXPENDABLE	ESCORT	STANDOFF
AAA, SAM, AAM	Strategic Aircraft	RF/EO ESM Internal ECM Stelth	Decoy Chaff Flare	Decoy	None
ASM, SSM	Tanks & APCS	RF/EO ESM Stelth	Chaff RF/IR Decoy Smoke	Decoy Airborne ECM (RPV)	Airborne ECM Man Pack ECM
ALCM, SLCM	Ships & LCA's	RF/EO ESM Internal ECM Stelth	RF/IR Decoy Chaff Smoke	Decoy Airborne ECM	Airborne ECM
AAA, SAM, AAM, A-AGUN	Tactical Aircraft	RF/EO ESM Pod ESM Internal ECM Stelth	Decoy Chaff Flare	Airborne ECM (RPV)	Airborne ECM
Guns, Mortars, Missiles	Man Portable	Man Pack ESM & ECM	Decoy Flare Smoke	Airborne ECM (RPV)	Airborne ECM & ECM

Table 1. Own Force Platforms and Countermeasures vs. Hard Kill Target Categories.

The self-defense integral countermeasures assets of own force platforms are frequently complemented with expendable, escort and standoff countermeasures to maximize the platforms' survival. The coordinated application of the four key CM assets enjoys a synergistic impact on platform survivability.

Chaff remains a valuable screen and track break CM technique against non-coherent radars. Flares are an effective tract break CM against IR guided weapons, and smoke is effective against visual spectrum guidance systems. Coherent processing, multicolor discriminates and multimode (RF-IF-Visual) seekers demand added sophistication in response time, dispensing tactics, color fidelity and cloud densities.

The expendable and escort decoy asset is manifested in the track break (false target) and saturation (multiple false target) role where the fidelity of electronomagnetic signature is key to the own force platform survivability.

A new generation of lower power active expendables shows promise as rocket or remotely piloted vehicle (RPV) delivered jammers dispensed around threat radar and communication sites. This tactic complements the escort and standoff missions.

The airborne and manpack ECM in the escort and standoff roles is designed to deny threat system acquisition and track of companion own force platforms throughout their engagement.

Radar Countermeasures

Radar controlled guns and missiles depend for their effectiveness on the reception of reflections from the skin of their targets. The basic purpose of ECM is to introduce signals into the radar system that will deceive or confuse the system and its operators' ability to acquire and track targets and thereby degrade the performance of the radar system so that it is unable to perform its intended purpose. The ECM system designer, in order to accomplish his task, needs some knowledge of the radio frequency (RF) characteristics and operational tactics of the threat radar system.

The user of the ECM system needs a radar order of battle showing type and geographic position of hostile weapons to properly plan his mission. Thus, electronic intelligence gathering is a prerequisite for effective use of electronic countermeasures. Given this knowledge of enemy radar controlled weapon systems, the ECM system designer will configure an ECM architecture that will permit effective evasion, deception and/or jamming of the hostile weapon network.

The typical countermeasure engagement sequence consists of the following steps:

1. Receive signals from the RF environment and determine significant parameters for processing;
2. Sort the received signals so that they can be associated with specific emitters;
3. Identify and prioritize all incident emitters;
4. Display the identified hostile emitters and decide on the countermeasure actions required; and
5. Activate the selected countermeasures against those emitters that represent the greatest threat to the friendly force.

The ECM tactic will consist of one or more of the following: self-protection, escort, standoff and expendable. These tactics are described below.

Self-Protection Systems

In this type of countermeasures system, each weapon platform carries sufficient countermeasures equipment for its own self-defense.

Self-protection countermeasures typically concentrate on degrading or nullifying the effectiveness of those hostile systems used for the aiming of guns, the guidance of missiles, or active fuzing. The self-protection countermeasures "bag of tricks" includes track break countermeasures using velocity, range and angle deception repeaters and noise jammers with deceptive range and angle track break modulations. Expendables in the form of smoke, chaff, flares and low power jammers are also used to achieve track break countermeasures for self-protection.

Escort Systems

This CM tactic uses a dedicated platform to carry high power jammers in formation with the friendly strike force and provides a cooperative jamming "shield" in support of the entire force.

The typical escort jamming system consists of a versatile RF signal source (exciter) driving a high power, 100 percent duty cycle amplifier which feeds a high gain steerable antenna. The exciter may be either preset on the basis of pre-mission intelligence information or tuned to detected signals, or both. Various modulation modes in the form of noise or false targets are imposed on the exciter output to match the countermeasures to the mission requirements.

Standoff Systems

Standoff CM tactics use very high powered jammers that direct their jamming at hostile radar systems in support of tactical strike forces. They operate from platforms located just outside the effective range of the hostile weapons.

The standoff jamming system typically contains the same components and incorporates the same modes of operation as the escort jammers. Since the standoff platform is required to operate at greater ranges from the threat systems, high transmitter powers are required to achieve the same level of jamming signal into the hostile radar receiver.

In addition, the situation geometry may place the standoff jammer in the hostile radar antenna sidelobe, which significantly increases the jammer power requirement. The Navy's EA-6A and EA-6B have been used as standoff jamming platforms with ALQ-76 and ALQ-99 transmitters, respectively and are currently deployed with the Fleet.

Expendable Systems

An aircraft or ship ejects radar reflectors or small jamming transmitters on a precursor basis for self-defense or to saturate an area prior to or during a mission. Expendables are configured either as dispensed units or propelled units.

Dispensed countermeasures in the form of chaff, flares or active jammers are used by tactical forces, on a programmed basis, for effectively breaking track of passive radiation seekers (RF, IR and optical) incorporating monopulse and lobe-on-receive-only angle trackers. In this mode, the dispensed device decoys the hostile weapon away from its intended target. Strategic forces may use the dispensed expendables in large quantities to provide a continuous ECM screen for the entire task force.

Propelled expendables are typically forward fired rockets or drones that contain flares, repeaters, chaff or jammer payloads. They are frequently used as decoys to draw hostile fire as a preferential target or to confuse or saturate hostile weapon networks to gain time to evade, degrade or destroy the threat systems.

Outlook for the Future

Potential hostile forces will probably continue to deploy lethal weapon systems in large quantities that operate in new "surprise" portions of the electromagnetic spectrum and that will incorporate contemporary countermeasures technology and tactics.

The trend of threat radar design is toward increased electronic counter countermeasure (ECCM) capability and operation at higher microwave frequencies. This trend is forcing the ECM system designer to use directional antennas to achieve the higher effective ECM power that will be required and to use coupled output devices to achieve the required wide bandwidth and high duty cycle. In addition, the increasingly dense threat environment will require spatial discrimination of the threats to aid signal sorting and minimize beaconing. These considerations have led to the development of the multibeam technology.

The key to success of future ECM systems will lie in the dynamic management of their resources to counter these threats. This management involves analysis of real time on-board sensor data for emitter identification; emitter location and threat identification; priority assessment; technique selection; situation display to operator (if required); system output management in terms of direction, frequency, time and power; effectiveness assessment; and the coordination of tactics between platforms. The complexity of this task strongly favors the performance of the management function by a digital computer. A successful dynamic countermeausre management system will also incorporate, as an inherent part of its design, the flexibility to be programmed with software and to be readily adapted in the field to provide rapid reaction to surprise threats or threat tactics. Systems based on these requirements are currently under development.

An additional factor that will dramatically influence future countermeasures systems is the relization that contemporary warfare is bounded as much by cost-effectiveness as it is by technology. The design-to-cost philosophy, which is currently evolving, challenges the system designers to optimize their system architecture to achieve the best acceptable level of performance for a predetermined life cycle cost value. The countermeasure system designer now has the challenge to develop cost-effective systems for countering the electromagnetic characteristics of a larger number of lethal weapons of widely different characteristics.

Electronic Warfare Integration and Planning

Alton D. Slay

There are many who believe that the traditional "Old Crow" of electronic warfare (EW) fame may really be a Pterodactyl—a creature so out of the mainstream of evolution that it is becoming extinct. Not so!! This tough old bird is much too smart to let that happen. Crows are, by nature, gregarious with their own species but have the ability to adapt to new environments and to become friendly with other species when it is to their benefit to do so. Our Traditional Old Crow (TOC) has started his evolutionary process by emulating his feathered namesake. He is becoming very friendly with various other avionics, electric and command, control and communications (C^3) birds—because it is to the benefit of the EW business and, in turn, our combat capability, for him to do so. Once the process is complete, our TOC will be an NOC (New Old Crow).

This evolutionary process has been predictable for some time and is now as inevitable as death and taxes. The critical forcing function is technology itself—technology used by the enemy to foil our combat goals and technology used by us to build combat systems. Centrally important is the fact that these two sides of technology are forcing total integration of electronic warfare system planning, hardware and software—the whole nine yards—into our overall weapon system planning, development and acquisition process. Since I

**This article originally appeared in the March 1979 issue of SIGNAL.*

so fundamentally believe this and am driving my command to this concept, a few paragraphs of explanation will follow.

What is EW?

To be sure that my writing and your reading are in sync, I will describe my version of the answer to the question. Figure 1 shows the most sensible breakdown of electronic warfare into its component parts.

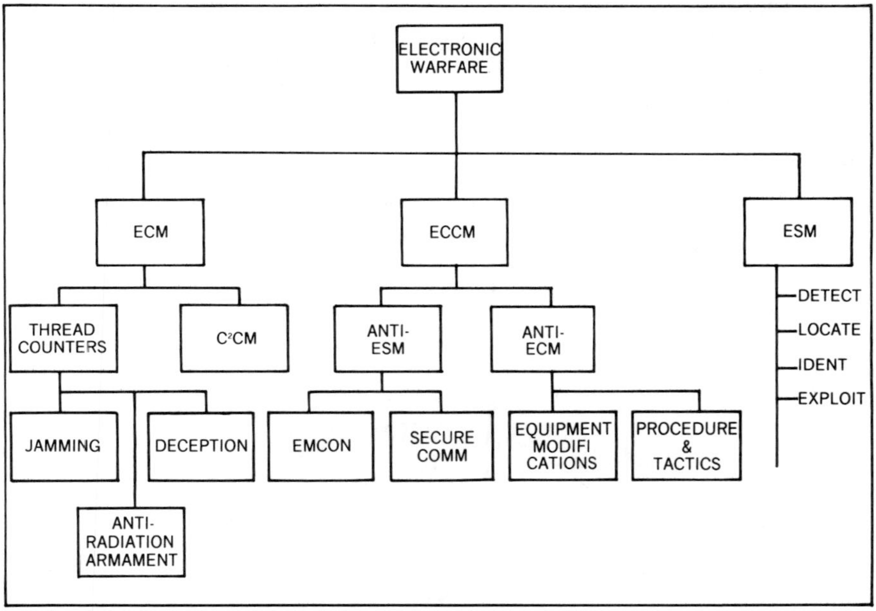

Figure 1.

Electronic countermeasures (ECM) are what is conventionally thought of when EW is discussed. This is where our past primary commitment in equipment has been—jammers, chaff, flares and antiradiation missiles (ARM). Now the thinking in this area is evolving from countermeasures against the enemy's lethal defenses to include countermeasures against the enemy's own command and control (C^2) systems, since his C^2 systems are as critical to his combat operations as ours are to us.

Electronic counter countermeasures (ECCM) are focused on permitting our own combat forces to operate effectively in an unfriendly countermeasures environment. Here we may be trying to counter his attempts to exploit our electromagnetic radiation by antielectronic support measures (anti-ESM) techniques such as emission control (EMCON) or secure (encrypted) communications. We may also try to counter his jamming techniques by antijam modification to our equipment, or even tactics, such as extremely low level flying operations. ECCM is getting a great deal of attention with the Air Force today.

The last major category, electronic support measures (ESM) involves actions taken to search for, intercept, locate, record and analyze radiated electromagnetic energy for the purpose of exploiting such radiation in support of military operations. The classical equipment associated with this area are radar homing and warning receivers (RHAW), but many new techniques are in development. There is obviously a close tie between ESM and other functions, such as reconnaissance, surveillance and intelligence.

An examination of Figure 1 shows that EW may also fall in the domains of other important functional or mission areas. The Venn diagram of Figure 2 illustrates the real overlays of EW with avionics and command, control and communications. Since, of course, avionics and C^3 are very critical to the Air Force today and in the future, there is extensive planning going on in these areas. The major challenge is the successful integration and marriage of these major plans and capabilities.

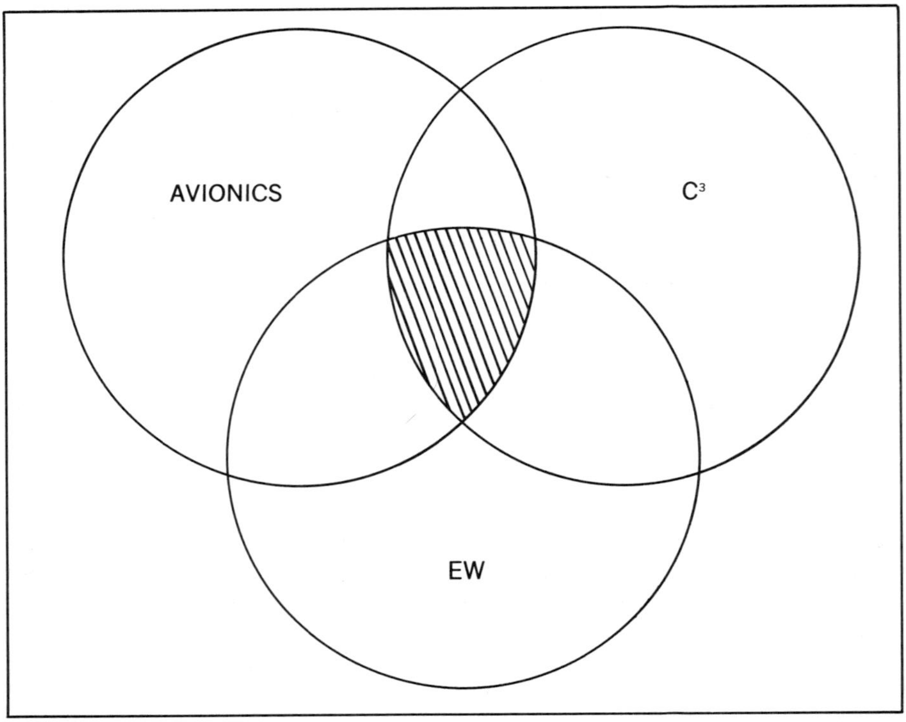

Figure 2.

The Problem and the Issues

Anyone remotely connected with the defense planning business knows that control and use of the electromagnetic spectrum can be decisive in a future war. At least 14 years ago, Soviet generals were writing definitive articles about the criticality of waging electronic war. Recently, declassified accounts of the Second World War show how critical the success of electronic warfare was to the outcome.

Recent writings and testimony to the Congress by many high officials of the Department of Defense (DOD) further substantiate the growing threat to our forces from an ever increasing and sophisticated enemy commitment to Electronic Warfare. Certainly, the recent experiences of the Vietnam War and the Mideast War emphasize the decisive importance of EW.

But is the system reponding adequately to this recognized decisive importance? In my opinion, it is not.

It still bothers me, after many years in this business, to see the way EW systems are generated and fielded. Why do we usually treat EW as an afterthought? Why is EW added on to an aircraft *after* it is fielded, which is always the most expensive and inefficient way? We certainly do not buy engines or most avionics that way. With the combat electronics environment in which we must operate today, it seems clear that an EW suite is a critical part of any combat airplane and, in fact, should be as much an integrated part of the total weapon system design as the engine, radar, communications equipment and so forth. "Should be" but "has not been" is the problem.

All of the complexities inherent in the study of EW boil down to the two major issues of combat effectiveness and cost-effectiveness:

1. How do we plan for, manage and use the combat capabilities of electronic warfare in the best operational sense?

2. How do we develop, buy and support the necessary high-capability things associated with electronic warfare at an affordable cost, with a maximum return on investment and in the context of other acquisitions of combat capability?

Part of the problem is related to how we plan for EW. These two major issues are really *interdependent* in a fundamental way—the EW things we buy, use and support are a result of the planning processes that identify the need for these things. In the past, the area of EW planning has been neglected. We have too long looked upon EW as some arcane art which is best left to those qualified practitioners, the so-called EW community.

The Acquisition Process

The EW community has a long history of significant accomplishment. It has developed—and is still developing—tremendously capable equipment and tactics. The Old Crows mentioned at the beginning of this article *are* that community—a select group of industries, operators and technicians that have been fighting the electronic war for many years. A number of them are in our EW Systems Program Office at Wright-Patterson Air Force Base (AFB)—in fact, they are found in many organizations throughout the Air Force Systems Command (AFSC). Also, in almost every industrial and government laboratory there is an EW division—indeed, there are whole companies dedicated to EW.

There are analogs in other functional areas such as avionics and propulsion laboratories and Systems Program Offices (SPOs). But somehow, they are different from EW. Avionics and propulsion are integrated with the prime airframers; they are considered an integral, integrated part of the weapon systems—considered from the outset at the design of every major weapon system. Not so with EW. Only the B-1 had an integrated defensive avionics suite and even that lagged the development of offensive avionics. Even today, our most modern airplane flying, the F-16, does not have ECM gear as integral equipment. We are still trying to figure out what we want as an ECM suite long after the rest of the avionics has been developed, tested and fielded.

Upon examination of the architecture of EW suites, it is found that they are usually separate from the rest of the avionics. They have their own dedicated computers, internal software and dedicated controls and displays—sometimes not even connected to the main avionics suite by the ubiquitous system multiplex bus. Only in the laboratories at the exploratory development stage do we see true integrated avionics/EW concepts, similar to those found in navigation and force control radar functions such as the Integrated Tactical Electronic Warfare System (ITEWS) concept at the Air Force Avionics Lab.

We had an acquisition technique which was used regularly during the Southeast Asia conflict period called quick reaction capability (QRC). QRC programs short cut conventional development and procurement procedures to get new equipments fielded quickly in response to the day-to-day demands of fighting a war. The Old Crows did themselves proud during the QRC days, as they countered surface-to-air missile (SAM) threats in support of major tactical operations like LINEBACKER strikes over North Vietnam. In fact, the QRC syndrome is central to the way we have handled EW from the time it was invented. Why? Because its technology, operational procedures and contractor base was unique and the operational need was for fast reaction development and acquisition.

One of the troubles with the QRC syndrome is that the long-term effects are painful. Programs done under QRC procedures are generally expensive. You often do not have time to shop around, compete or haggle over prices. Also, the press to reduce acquisition time under QRC usually sacrifices the "illities"—reliability, maintainability, testability, affordability and so on. Long, risk reduction test programs are not *de rigeur* under QRC. Another problem is that the QRC approach is just not adequate to cope with the high technology and extreme sophistication required today and tomorrow in our EW suites.

The question is "Can we use the more structured and integrated architecture system approach which is being used successfully for other elements of avionics/electronics/C^3 today?" I believe we can and must, if we are to do the EW job effectively.

The Integrating Forces

The microelectronic revolution is having the same dramatic impact on EW it has had on our other combat capabilities. Digital reprogrammable EW systems, using solid state radio frequency (RF) front ends are the "normal" configuration we see today. I see a great convergence of system characteristics, such as processing throughput, data rates, software program sizes and circuit densities in modern EW systems with other avionics subsystems such as solid state, programmable fire control radars.

If you examine the concepts under consideration by the logistics community, such as their EW reprogramming centers at Warner Robins Air Logistics Center (WRALC), you see a fundamental similarity between the complex of host computers, hot benches, simulators and software support systems with the architecture of the developing weapon system avionics integrated support facilities (AISF) such as the F-16 AISF at Ogden Air Logistics Center (ALC). The principal difference at an EW facility is an extensive capability devoted to threat systems analysis, validation and simulation, which basically drives the rest of the facility.

Our challenge in this area is to carefully search out the areas of architectural commonality and look for interfaces and protocols that can be standardized or at least made interoperable with other interfacing systems, whether they be aircraft, avionics or C^3.

Another strong integrating force can be management attention. I am trying to focus management attention within the acquisition community toward that end. By now most of you are aware of the existence of our newest AFSC/Air Force Logistics Command (AFLC) organization, the Deputy for Avionics Control (DAC), located at Wright-Patterson AFB. The DAC's main charter is to rigidly discipline the avionics acquisition process and to do so under a master plan, touching all weapon systems and applying across the full spectrum of the life cycle.

I view EW as being totally under the DAC's purview. I have directed the DAC to work toward full integration of EW into our system with the rest of avionics from now on. The first major emphasis will be to ensure that EW system standards and protocols are avionics standards and protocols, such as the use of MILSTD 1553 Multiplex buses and the use of avionics higher order language such as J-73 for EW system programming.

The EW Master Plan will be a *subset* of the Air Force Avionics Master Plan. The Annual Avionics Conference this year, which is really the forum for forming the DAC's activities for the coming year, had a strong contingent from the EW community as one of its panels.

Now that I have broached the idea of planning, which I regard as a thrid integrating force, it is time to return to that first issue I mentioned earlier—how do we plan for, manage and use the combat capabilities of EW in the best operational sense. We must understand how EW planning fits in the context of the rest of our Air Force planning—how do we set priorities and that bottom line—the budget.

Planning—EW and Otherwise

Ad hocery is dead. There is not enough money, time or people for it in this day and age. In June of last year, I charged my planners with the task of putting

AFSC "up front" in planning. Project Vanguard is the result—with EW as an integral part.

We have divided AFSC's "business" into roughly 40 planning areas. Figure 3 shows the three basic *types* of plans that make up the 40. All AFSC programs fit into one or more of these planning areas. Mission Area Plans correspond to the DOD and Congressional mission areas considered under A-109 type of mission area budget formulation. Major Force Element plans correspond to the major weapon systems indentified as budget elements. Functional Plans tie together technology related programs where cross-mission are or cross-weapon system considerations apply, such as standardization or reliability.

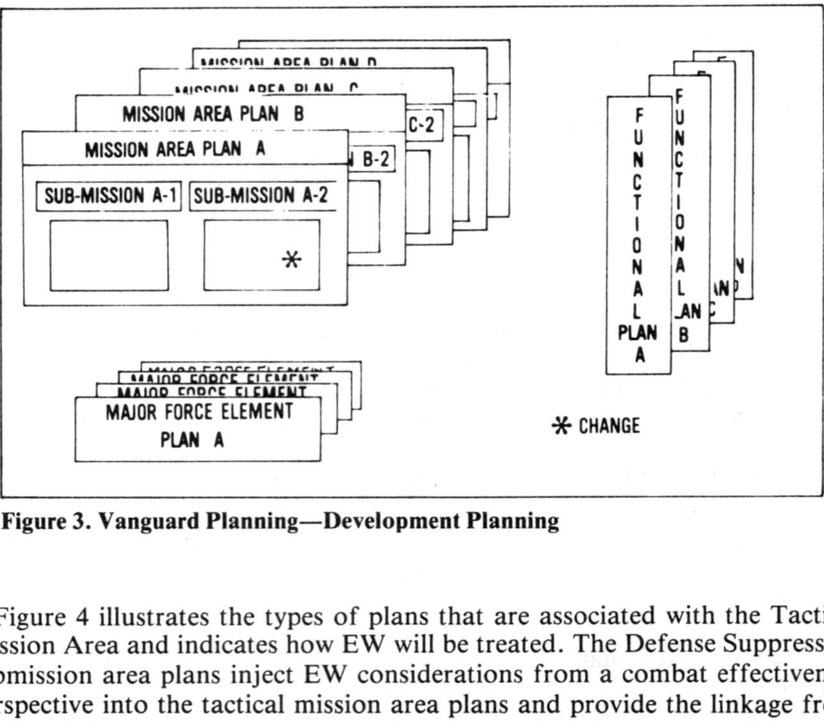

Figure 3. Vanguard Planning—Development Planning

Figure 4 illustrates the types of plans that are associated with the Tactical Mission Area and indicates how EW will be treated. The Defense Suppression submission area plans inject EW considerations from a combat effectiveness perspective into the tactical mission area plans and provide the linkage from the mission area into the associated Major Force Elements, such as F-4G Wild Weasel aircraft or airborne relay vehicles. Figure 5 shows how this coupling is accomplished. The example is a preliminary defense suppression planning outline showing how the major EW Program Elements couple with the major Force Elements.

Of course, EW will also be treated functionally and will generate a Master Plan, which is to be a major input into the Avionics Functional plan. The purpose of the functional plans is to tie together programs that are function-

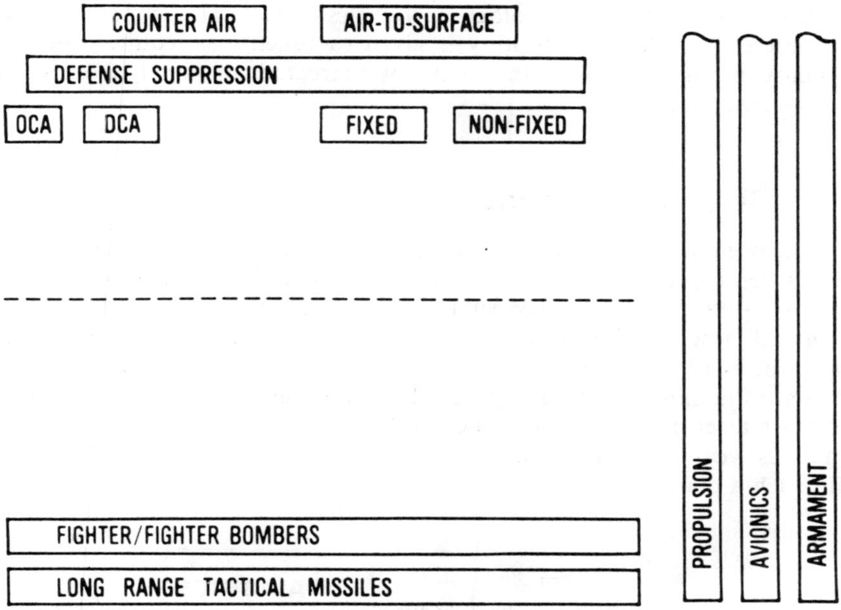

Figure 4.

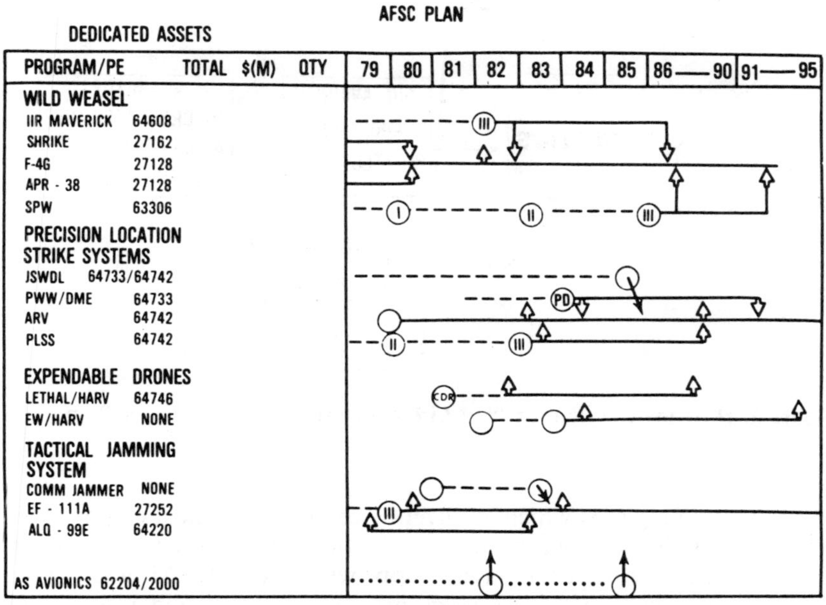

Figure 5. Defense Supression

ally related, perhaps by technology or logistics considerations, and to ensure that necessary budgetary visibility is given to functional requirements, like reliability or maintainability, that may not be treated adequately from a strict mission area perspective.

The EW Master Plan

At the beginning of this article some time was spent defining the elements of EW, as represented in Figure 1. The EW Master Plan will have this same structure—that is the major EW subplans will be ECM, ECCM and ESM. Our major task now is to integrate a multiplicity of previously disconnected planning efforts, fill in the gaps and produce a coherent whole that fits in the Vanguard Planning structure. Figure 6 illustrates the EW plan's linkages to the avionics master plan. As I mentioned earlier, this was a major item of discussion at last year's Avionics Conference which actually produced a roadmap to build such a plan.

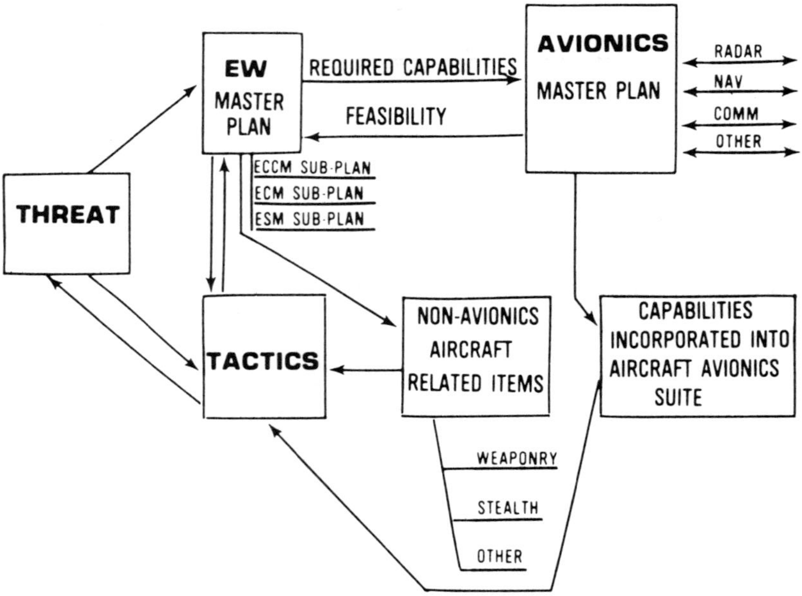

Figure 6: How the EW Master Plan Links with the Avionics Master Plan

The EW planning area which is currently receiving the most emphasis and is therefore in the best shape, is the ECCM Master Plan. Plan objectives have been established, terms of reference approved and a USAF-wide planning group formed. Two workshops have already been held and responsibilities assigned to various field organizations. It is planned that this ECCM Master

Plan will be the Air Force's response to DOD Directive 4600.3 which requires that counter countermeasures (CCM) be considered for all major weapon systems prior to Defense System Acquisition Review Council (DSARC) approval and will, in fact, apply to all USAF systems, even those below DSARC thresholds. This plan will be administered by the HQ AFSC Deputy Chief of Staff, Development Plans, with the Electronic Systems Division (ESD) as the primary Product Division.

Less emphasis has been placed on our ECM and ESM planning. We have made a start in ECM architectural planning but much remains to be done. ESM planning is very preliminary, having only been looked at by a single laboratory. Considerably more effort will certainly be forthcoming in the ESM planning area in the near future.

The most important planning effort of recent years in tactical C^3 is the Tactical Air Forces Integrated Information Systems (TAFIIS) Master Plan (which will, incidentally, be the major component of the Vanguard C^3 Functional Plan). This plan, which is a joint user-developer definition of an Architecture for Tactical C^3, breaks C^3 down into 12 major architectural components. One of the major subarchitectures, Force Management, identifies EW as a special branch.

Figure 7 is an extract from the TAFIIS Master Plan, showing the Tactical EW Roadmap. Every numbered or lettered node is accompanied by a precise description of required actions and decisions in the bulk of the TAFIIS document. This roadmap shows how to move through four successively more effective (and costly) paths to achieve the objectives shown on the right side of the roadmap. Figure 8 illustrates how the various roadmap paths rate in goal attainment. We need to integrate this planning perspective into our overall EW planning and to keep the EW-C^3 interfaces identified.

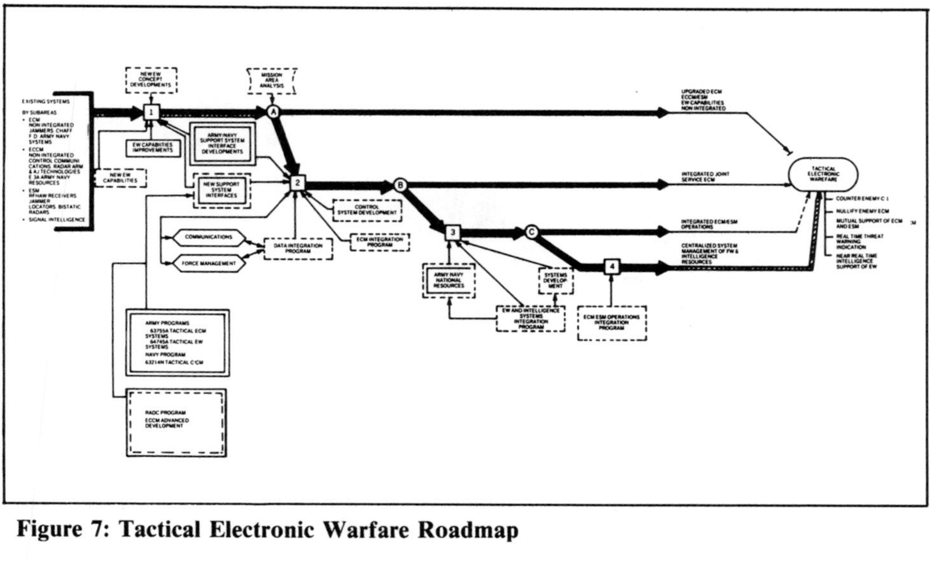

Figure 7: Tactical Electronic Warfare Roadmap

OPTION NO.	OPTION TITLES	COUNTER ENEMY C2	NULLIFY ENEMY ECM	MUTUAL SUPPORT OF ECM AND ESM	REAL TIME THREAT WARNING INDICATION	NEAR REAL TIME INTELLIGENCE SUPPORT OF EW
		GOAL CAPABILITY				
I	UPGRADED ECM/ECCM/ESM/ EW CAPABILITIES	3	0	0	1	1
II	INTEGRATED JOINT SERVICE ECM	3	0	1	2	3
III	INTEGRATED ECM/ESM OPERATIONS	4	0	5	3	4
IV	CENTRALIZED SYSTEM	5	5	5	5	5

(UNCLASSIFIED)

(U) LEGEND:
0 – No Capability
1 = Little Capability
2 = Some Capability
3 = Substantial Capability
4 = Less Than Full Capability
5 = Full Capability

Figure 8.

A Clear-cut Task

What I have tried to illustrate is that with all the complexities of EW, including the instabilities created by unforeseen threat changes, there is a structure which can lead stability through its integration with the structures of the rest of the Air Force. The task is not going to be easy for any of us. Integration is never easy, but it is the rule of the survival of the fittest.

The Old Crow is still alive and healthy.

C^3CM: Lessons Learned—Where Do We Go from Here?

Doyle E. Larson

Electronic combat is finally emerging as an important element of modern warfare. The conflicts in the South Atlantic and Lebanon's Bekaa Valley propelled this developing military strategy to the forefront in the popular press. But where are we today and where do we go from here?

In both my joint and U.S. Air Force positions, I have had numerous opportunities to observe the attempted use of command, control and communications countermeasures (C^3CM) in field operations. I am convinced it is a potentially excellent strategy, but there are many areas which demand improvement if we are ever to employ C^3CM effectively.

The major lessons learned from these observations center around three areas:

1. Connectivity (in the sense of interoperability and the communications necessary to support it);
2. Training (realistic field exercises and staff command post exercises); and
3. Establishment of sound doctrine (putting concepts into practice).

In my opinion, improvements in these areas seem to offer the greatest opportunity for return on our invested efforts.

This article originally appeared in the April 1983 issue of* ***SIGNAL.*

Connectivity

What is connectivity? It is the network that ties the C^3CM players together. Connectivity is the communications and data links required to transmit timely information simultaneously to multiple battle managers, exchange data base information and issue battle execution orders to numerous lethal and disruptive elements simultanously and in near real time. It also implies a considerable amount of mutual support coordination and cooperation among services and among nations.

Five troublesome connectivity problems impact on our ability to effectively implement the C^3CM strategy. In fact, these identifiable problem areas prevent us from effectively using the assets we already have:

- Organization;
- Intelligence support;
- Cross-service jamming support;
- Frequency deconfliction; and
- Communications.

These are key problems, ones we should try to resolve swiftly. Let us take a closer look at each one of them.

Organization

Current organizational arrangements for fighting the air-land battle do not support C^3CM operations well.

For example, the several staff sections within our battle organizations which have a piece of the C^3CM pie often perform their respective jobs in an autonomous manner. Operators, too often, are only knowledgeable of the lethal aspects of C^3CM, and even though they are responsible for the entire C^3CM strategy, including operations security and deception, they tend to be woefully ignorant of these functions. Operations people seldom talk to intelligence, so the intelligence people go about their business gathering information which may or may not meet operational needs. Furthermore, by the time it is screened, validated and delivered, the data is often too late anyway. The communications officer, who should have a big role in all aspects of C^3CM, especially defensive, seldom gets involved at all.

Only rarely does someone take full responsibility for using C^3CM strategy and begin directing the coordinated activities of everyone involved. We need to pull all the players together under one battle manager in order to make C^3CM work.

This individual must be an operator—the operations officer, the chief of staff or the commander. It must be someone who is in a position to monitor all activities and has the responsibility and authority for directing the full C^3CM effort—both lethal and non-lethal.

Intelligence Support

Meanwhile, where is a requirement for simultaneous, timely intelligence exchange between all intelligence sources and multiple battle managers, each getting information tailored to his battle needs.

Currently, there is practically no way to exchange information and intelligence between these entities on a near real time basis. For example, information arriving at the Allied Tactical Operation Center (ATOC) is not fed to the Army, other than on an exception basis, and Army real time steerage of the airborne intelligence collection is practically non-existent.

To solve this problem, secure jam resistant communications must be established from all intelligence assets to the Allied Tactical Operations Center, corps and division tactical operations centers, and to the field reporting centers and posts actually controlling the forces.

This tailored information flow is begging for technological, particularly software, solutions. Timely information about the enemy is the cornerstone of successful C^3CM operations.

Although 100 percent of the required intelligence data will probably never be acquired, some strides have been made in the collection of such information. The world is full of data bases of all sizes, shapes and descriptions. Still, none of them are adequate for C^3CM. They do not contain information about low level echelons (battalion or regiment). Likewise, no provision is made for timely data transfer.

Technological advances in information processing and retrieval are required to overcome this shortfall. Digital signal processing currently enhances data gathered from multiple national and tactical resources for display at remote centers. However, this process currently takes too long and is not available for the commander who must know quickly what is happening in the vicinity of his forces. Precise location data is rarely available. When the commander selects the destruction option, he must know where to shoot or where to send the unconventional warfare team.

Cross-Service Jamming Support

Immediate cross-Service tasking for jamming support is also a problem. Immediate tasking of Air Force resources relies on the tactical air control system which the Air Force uses to control aircraft. The Army uses the system to request immediate air support, primarily close air support. Current procedures call for using this same system for requesting immediate cross-Service jamming support.

Experience has shown that the tactical air control system is the only channel currently available for requesting this support and its responsiveness and timeliness is not sufficient to handle immediate tasking. Requests originating with aircraft commanders or ground units in the division area of operation must complete for usage of the air request net and may go as far as the ATOC before tasking back down the chain to the appropriate jammer is completed.

No solution to this problem is readily apparent. However, a first step would be to ensure that all elements required for C^3CM possess compatible communications systems.

Deconflicting Frequency

Deconfliction of frequency usage for jamming between ground and air forces and between allied forces continues to be an observed problem during air-land exercises. This issue will become acute as more airborne jammers with more power come into use. It is possible that jamming generated by one service will indeed interfere more with the C^3 systems of another friendly unit than it will with those of the enemy.

To overcome this, procedures must be worked out to keep unintended interference to a minimum. Frequency managers supporting C^3CM activities need to be given the tools, including automatic data processing support, so that efficient, near real time frequency management can be accomplished. Most importantly, channels of communications must be established between users of the spectrum so that conflicts can be quickly resolved.

The control of operating frequencies should be automated to the point of changing without user action, as the situation dictates. These changes could be the result of a desire to exploit or gather intelligence on certain frequencies. Other frequencies could be highly susceptible to enemy jamming. Additional frequency changes might be required to avoid self-interference. The days of controlling and executing frequency allocation by handwritten logs are long gone.

Communications

Finally, our last connectivity problem: communications. The criticality of good communications to military operations cannot be overemphasized. The commander depends upon intelligence transmitted by communications systems to formulate his plan. He also depends on communications for the connectivity required to control and maneuver his forces.

Despite the known Soviet jamming threat, none of our military services today possesses adequate jam resistant, secure communications. In every recent exercise involving Electronic Warfare or C^3CM, lack of communications has been cited as the glaring deficiency.

Technology is available to redress this deficiency. Frequency hopping radios are ready to be deployed to provide jam resistant secure communications. However, these equipments have not been procured in sufficient quantity to provide connectivity such as from the Army ground commander to the airborne forward air controller.

Operations with allied nations add to the communications problem because of the many layers of staffs involved, the incompatibility of communications gear and a host of associated frequency management problems. The root of the problem for the United States lies in the Department of Defense (DOD) programming system itself.

Weapons systems are the big dollar items. Program managers plan for 100 B-1s and direct communications programmers to plan for supporting communications. During the funding process, the communications get squeezed out of the service budget, along with other support systems, and we end up with combat forces we cannot properly employ.

In summing up the connectivity issue, it is fair to say that most of our combined exercises have shown that much work needs to be done. Without increased connectivity, we cannot properly use the forces we already have.

Training

Turning now to training. Where should we be heading? First of all, I believe the ultimate goal of training is to mentally prepare the armed forces to cope with the unusual and the unexpected, to give poise in a situation where all else is in disarray.

Realistic exercises, getting as close as we can short of actual combat, represent the best opportunity for doing this. Classroom instruction is important, but it is by repeatedly exercising our war procedures that we are able to pinpoint problem areas such as I discussed previously. Realistic training, conducted often, will lead to solutions and better wartime procedures.

During the fall of 1982, people from the Joint Electronic Warfare Center and Electronic Security Command participated in Northern Wedding, Cold Fire, Carbine Forces and Display Determination. Each of these exercises provided valuable lessons which smart managers need to observe closely.

Our current intelligence collection systems can quickly saturate key decision makers with the sheer volume of information. It is critically important that the commander is not swamped with data. We can prevent this by training personnel to selectively provide the information each commander needs to accomplish his specific task: focused, tailor-made, battle information.

We found, too, that in our C^{3}CM training it is important to concentrate on teaching how to think, not what to think. The training must seek to advance general principles rather than inflexible procedures. The C^{3}CM strategy requires that we continuously refine our assessment of enemy actions and modify our own plans to best advantage. It is a dynamic situation.

This cannot be readily accomplished by a regimented environment where every eventuality is covered by published standard operating procedures. People must be taught to stay flexible and be prepared to create a new plan when things do not go as expected. The battle manager, as the quarterback, must be ready to "call an audible."

We have committed some errors in getting ourselves organized for C^{3}CM. We now know everyone must use the strategy: commanders, intelligence people, operators, planners and communicators. Previously, C^{3}CM had been accomplished by a small group of individuals conducting their own "little war" in a separate cell. Although this was adequate for a time, we found that by involving personnel of all disciplines in planning and executing C^{3}CM, much better results were attained.

Combined exercises have also pointed out needs for compatible terminology and communications equipment. Prior to integrating the forces of different services and nations for the purpose of conducting realistic combined operations, little consideration was given to the fact that we could not talk to each other. Once some of these required communication links were established, we found that we did not understand what each other was saying, even though we spoke the same language. Now, at least partially due to the C^3CM strategy, we are being forced to consider acquisition of compatible communications equipment, and staffs are beginning to talk with each other using common terms that everyone understands.

All of this is simply to point out that we are making progress in applying C^3CM and that exercises are the key training tool that will put us over the hump. We must, however, ensure all commanders continue to include C^3CM objectives in exercises and that these objectives are realistically pursued. Using canned scenarios to fill exercise squares is a useless drill.

Doctrine

My third subject is the difficulty of establishing doctrine acceptable to all the players. To focus on this problem, let us look at the composition of the four services in a joint command: Navy, Air Force, Amphibious Forces and Army. Comparing the number of movable entities subordinate to each service three star commander, we find the Navy units number in the tens, the Air Force in the hundreds, the Amphibious Forces in the thousands and the Army in the tens of thousands.

Continuing with the comparison, besides a smaller number of moveable subordinates, the Naval and Air Force commanders deal with officer subordinates, whereas the Amphibious Forces and Army subordinates include all ranks.

When communications are compared, we find the Amphibious Forces and the Army are operating amid the clutter of the earth's surface. Their communication systems tend to be portable which puts certain inherent difficulties into play regarding command and control.

More importantly, when comparisons are made concerning the information available about the movable subordinates, the Navy and the Air Force commanders have, by and large, fairly precise information about their subordinates and through secure communications can direct their efforts quite easily.

Consequently, in terms of C^3, the Navy and the Air Force commanders have the greatest tactical flexibility.

This leads to certain styles of command, or doctrinal differences. The Navy and the Air Force commanders tend to bring all information into one place to make the fundamental decisions and thus use their tactical flexibility to the best advantage. A Navy or Air Force approach must stress centralized control, while an Amphibious Force or Army commander (not having this capability) may have developed doctrine stressing decentralized control in combat operations.

If we were to go beyond this short study of the different styles of command, we would see other differences among the Services which impact on our task of establishing doctrine acceptable to all the forces involved.

For example, an analysis of the way sea, air and land forces fight their individual battles leads us to some generalizations on differences in engagements. Although naval forces may have few major encounters with the enemy during a war, each encounter may last for days. Air forces engage the enemy frequently for short time spans and land forces are normally in constant contact.

Additionally, if we consider deception, we would see that different types of forces view its use in different ways and direct their efforts toward different enemy targets. Naval forces use jamming for defensive purposes; land forces use it offensively, and the Air Force applies jamming both offensively and defensively. This is not to say one idea is right and another wrong, but to show there are differences and they must be reckoned with when formulating C^{3}CM concepts and doctrine.

Doctrine normally lags behind technology and innovative thought. It does not precede technology or experience; it formally documents it. Military doctrine is predominantly the outgrowth of exercise or combat experience. We have simply not had the combat experience necessary to put forth a final, battle-tested C^{3}CM doctrine. Further, we do not have all the equipment, particularly the communications, jamming and deception gear, that we need to use C^{3}CM as we would like.

In spite of the difficulties we face, there remains a pressing need to develop concepts for the implementation of C^{3}CM. Even if we cannot define a final doctrine for combat, our current tactical doctrine is sufficient today to begin the coordinated execution of the C^{3}CM concept. And even though we may not have all the equipment we need, we can and must get started now by educating and training our commanders and staffs on how to think C^{3}CM and on the methods to use to fight smart.

We can test our C^{3}CM techniques and ideas during training exercises and then go back and refine our concepts. Eventually, our smart battle managers will overcome problems related to differences in perceptions on how to employ C^{3}CM. We know, too, that the C^{3}CM strategy is compatible with current and planned weapons systems—it is not dependent on a particular piece of equipment.

All that is required is for us to continue hammering out and refining combined concepts using available resources. Doctrine will follow in due course.

Summary

Within the Services, there is an emerging sense of hope for the effective use of C^{3}CM. We have made good progress in integrating C^{3}CM into battle plans. We know there are areas in connectivity, training and exercises and doctrine development that require additional attention. But, the potential for C^{3}CM is excellent, and service and national leadership have recognized that potential.

Industry will provide some technological improvements, but much hard work by our staffs and middle managers remains to be done before C^3CM becomes a battlefield capability.

Electronics: Just Another Element of Warfare

Alan D. Campen

A brilliant Pacific sunrise bathed the beaches of Guam and sent slivers through shell holes in the wall of what had once been the home of a family of some means. Rivulets of rainwater from the usual dawn shower filtered through the torn roof and coursed over a poncho which sheltered the radio but not its operator. CINPAC ADVON was operational on the heights overlooking Agana and Navy Radio Station NPN was back on the air after a two and one half year interruption.

As I took over the watch, my nervous and unnecessary key adjustments drew a barrage of unofficial and irreverent "Z" signals concerning my technical aptitude as net controller and the state of my equipment. That was almost a ritual, but this morning was different. Buried in the cheerful cacophony was a strange signal with a rhythm and tone which did not fit. The "fist" just did not sound accidental. It was an amateurish attempt at what I was later to discover was "imitative deception." When that effort failed to disrupt net discipline, he locked his key down and tried to jam.

Such was the author's initiation to electronic warfare in 1944.

Manipulation of the electromagnetic spectrum to gain military advantage is not a new concept, although the attention that electronic warfare (EW), or radio electronic warfare (REW) or command, control and communications countermeasures (C^3CM) is now receiving would seem to indicate otherwise. In fact, the history of EW can be traced back in time to at least 1904 and the Russo-Japanese War. The British used radio codes to help deceive Turkey during the Third Battle of Gaza in 1917.

This article originally appeared in the April 1981 issue of **SIGNAL.*

The Russians were almost fatal victims of communications exploitation during the First World War and again in the early days of World War II due to German attacks against wire line communications. The Luftwaffe even employed a bomblet designed to snag overhead lines and then used artillery to discourage repair. Even the United States employed EW as part of an integrated effort at tactical deception from June 1944 to June 1945 in Europe.

It would seem then that history would be of some use in developing a capability to conduct or resist EW operations. Certainly the technologies are new, but the problems of intergrating that technology into the overall military campaign cannot be all that different now than when Clausewitz first called attention to the "friction" in warfare which makes "the apparently easy so difficult." One might expect, therefore, to discover a continuum of EW development, beginning with World War I. In the case of the Soviet Union, this does seem to be true. The Soviets have drawn heavily from bitter lessons in the Great Patriotic War and developed a highly structured doctrine and capability to use electronics to enhance the effectiveness of weapons as well as to protect their own means for troop control.

In the United States, one does find evidence of a continuum in the development of technology to detect and counter electronics (the ECM part of EW) and at the national level to monitor and analyze signals. But there are few signs of the necessary complementary development of doctrine, operational concepts, tactics, procedures and exercises which are essential if modern technology is to be employed in a useful and non-disruptive manner to support combat operations. Could it be that this is our legacy from South Vietnam and its benign electromagnetic environment, or of Korea and WWII where EW was not generally regarded as a major factor in the outcome?

Whatever the reasons, two major military powers have taken different paths with markedly different results in regards to EW. The United States, which would seem to have the greatest dependency on its electronics, may have the least capability to use them in combat. An issue of *SIGNAL,* devoted to EW, is an appropriate place to examine a few of the non-technical aspects of EW in this seeming paradox.

Learning from History

The October 1973 "surrogate" war provided a unique opportunity to compare opposing technologies, operational concepts and resultant capabilities. The Israelis, who took an understandably hard-nosed approach to the evaluation, found the East to lead in system philosophy and design with crude, robust but reliable components. The West, not surprisingly, was found to excel in technology; a virtue of questionable utility in combat.

That the Soviet Union should have a mature and tested EW and command, control and communications (C^3) countermeasure system philosophy should not be surprising. The Soviets take their history quite seriously and the history of EW has not dealt with them kindly. Indeed, few nations have been so sorely victimized by one or more of the disruptive elements we now collectively call counter C^3. The defeat and surrender of the Russian fleet in the Straits of Tsuchima during the Russo-Jappanese War (1904-1905) may have been caused, in no small way, by failure to jam a sighting report. However, there is nothing ambiguous about the near disasterous results of loss of Soviet troop control during the early months of World War II, when German attacks on telephone switches and open wire lines virtually severed communications at the division, Army and Front level.

Soviet statistics report 40 percent of their communications installation inoperative in 1941, however, their reaction was immediate, it was effective and it provided the fundamentals of what has now become widely known as Radio Elecronic Combat (REC). Just how well the Soviets profited from the 1941 humiliation was manifest in the spring of 1944 and Operation Bagration which was to mark the end of the German army on the Eastern Front.

In the January 1980 issue of *Military Review,* Maj. Barney F. Slayton, USA, cites Bagration as a case study of Soviet ability to integrate all aspects of radioelectronic warfare into their tactics. Further, Maj. Joseph C. Arnold, in the July 1977 *Military Review,* draws attention to similarities between Bagration and current Soviet military thinking in Europe.

The United States could, if it chose, also draw usefully upon the lore of EW in WWII. The short but colorful history of the 23rd Headquarters, Special Troops, from June 1944 to June 1945, covers some 21 operations of increasing complexity and effectiveness, involving radio and sonic deception, decoys and special effects. While the technologies have radically altered in these past four decades, the practical matters of integration with tactics have changed little. But, lacking the Soviet reverence for history and sensitized to criticism of constantly girding for past wars, we seem not to have worked to integrate our superior technology.

Instead, the United States seems to have opted for quick-reaction, black box, ad hoc responses to cold war and Vietnam demands or, indeed, no action at all in some areas of EW. While eletronics are clearly essential, it leaves doctrinal, organizational and procedural issues to chance; a dicey proposition in the short but violent combat scenario now favored for a NATO conflict. Indeed, it is disquieting to note how frequently the question of a need for operational concepts surfaces at EW/C^{3}CM symposia and conferences. We appear to be fielding equipments with only the most superficial understanding of the means for their successful integration into the chosen tactics for combat.

Hardware in Search of Operational Concepts

As one looks back over the years to the origins of EW, there are the similarities and differences which may be useful distinctions to draw on as we contemplate what lies ahead in the 80s. It is particularly important that the strengths and weaknesses of EW be accurately assessed as an aroused and impatient public and a new administration search for quick solutions to the serious mismatch between our military posture and the strengths it needs to secure national interests.

If history teaches anything, electronics will again be asked to perform miracles: to leverage inferior capabilities, to provide quick answers, to become a force multiplier. It is important to understand just how far electronics can carry the load, alone. Only a few years have passed since we last threw billions of electronics dollars at a problem in Vietnam which had defied conventional military solutions and exhausted the nation's will. There is little argument that sensor technology was advanced by several generations and many wholly fortuitous benefits were derived, perhaps justifying the cost. For example, the conceptualization of an Electronic Battlefield was conceived in Vietnam, but did the McNamara Wall or Igloo White systems really help achieve military objectives? More to the point, how might those electronic marvels have fared against a sophisticated electronic counterattack? History shows how quickly the British countered "Headache" with "Aspirin," or "Benito" with "Domino" and "Benjamin."[1]

Perhaps it is inevitable that a nation with untempered faith in technology and impatient for simple, expedient solutions will place disproportionate emphasis on electronics to leverage a smaller, more affordable combat force. Again, history has some lessons. In a speech in Washington to the National Security Industrial Association (NSIA), Gen. Robert T. Marsh, USAF, said his problem was not "technology," but "management of technology." Put another way, we often find technical solutions searching for problems with the resulting high probability of expensive mismatches. Making black boxes work in complex systems involving humans is usually far more difficult than initially envisioned; evidence abounds.

Gen. Westmoreland introduced the concept of the automated battlefield in a speech in 1969, saying he felt it was achievable in no more than 10 years. The magnitude of such a task was badly underestimated. Building individual electronic control systems is one thing; getting them to play on an integrated battlefield—as they must to be effective—is a step which has so far eluded virtually every attempt. The February 1981 issue of *Armed Forces Journal*[2] contains one perspective of what went wrong with Army efforts at tactical automation. It should be noted that the Air Force and the Navy have had similar difficulties in systematizing electronics to support command and control.

The East and West do things differently. The Soviets adapt technology to doctrine with feet firmly planted in tradition and a speculative eye to the future. The United States, on the other hand, too often harkens to the "technologic imperative," builds and fields things before trying to discover why.[3] The leisurely pace of WWII and Vietnam abetted this concept of trial and error. A North Atlantic Treaty Organization (NATO) scenario may not be so permissive. But, just how important is C^3 anyway?

Who Needs C^3?

When dueling was an elegant and honorable method of resolving differences, it was the custom to divide the variables of weapons, timing and terrain constraints more or less equitably between the principals. This permitted a modicum of preplanning of stategies on both sides. Such conditions no longer prevail. U.S. policy of "countervailance" (in a nuclear sense), NATO MC 14/3 strategy of "flexible response" or our new policy of "conscious ambiguity" towards terrorism are all reactive in nature with our actions to be developed, ad hoc, in real time to constitute an appropriate and calculated response to an offending act.

The objectives are twofold: to deter or to prevail should deterrence fail. To satisfy the former objective, one must convince an opponent of the certainty of response. To achieve the latter, one must mask the character and timing of that response. This becomes difficult when the opponent makes all the critical choices about weapons, tactics, timing and terrain.

It becomes almost overwhelming when one adds the additional burdens of a purely defensive strategy and a numerically inferior combat force. The implications for a C^3 structure to successfully underwrite such policies are staggering and may explain why the Soviets regard our C^3 force multiplier more as an Achilles Heel and fair game for their REC capability.

Recent studies of REC and C^3CM have examined respective C^3 dependencies and vulnerabilities in NATO and the Warsaw Pact. The vulnerabilities are generally agreed and need not be recounted here. The C^3 dependencies, however, are not so straightforward; it all depends. What it depends upon are how much preplanning has been accomplished and how much control is needed to maintain the planned action.

Our current policy permits extensive and specifically directed planning by opponents while limiting our planning to the development of countless options to be triggered by a hopefully timely and accurate perception of events. Given such a scenario, it can be argued that the opposing commander is only minimally dependent upon his C^3 system, even granting his proclivity for tight, top down control and lack of initiative by lower echelon commanders.

Our reactive strategy, on the other hand, demands perfect C^3 all the time. He will remain relatively independent of C^3 so long as his game plan unfolds on schedule. This raises the logical question of what is needed to disrupt his plan and force C^3 back into the combat equation.

No Substitute for Firepower

Neither deterrence nor war winning capabilities can be completely and accurately assessed without considering the perspective of the enemy. History and current analysis forcefully suggest that the Soviet Union will not likely initiate or continue to press an attack against any adversary without reasonable assurance of a favorable correlation of forces. What this favorable ratio might be is subject to continuing study. A lower limit is probably 3:1 and the upper limit might be the incredible 10: 1 tank ratio seen in Operation Bagration. The question to ponder is the extent to which EW or counter C^3 can either deter by creating unacceptable uncertainties about favorable force correlations, or, having failed to deter, can then create an unfavorable correlation by delaying, diverting or destroying forces.

In an article in the November-December 1980 issue of *Naval War College Review,* LTC Allan A. Myer, USA, looks at correlation of forces through the Soviet prism. He notes some serious difficulties facing Soviet planners in what he calls "wandering values," which are important factors that can only be employed in a qualitative sense in the otherwise scientific Marxist-Leninist dialectic. "Marxist-Leninst methodology is the Soviet prism, and it really does not matter if it is right or wrong. That can be left to philosophers."

One of the illusive unquantifiables described by LTC Myer is called "Adaptability—The Art of Putting It All Together." It reflects a Soviet concern about a growing gap between theory and practice. There are many facets to adaptability, but a recurring one is command and control which to the Soviets means the qualitative features of the organizational structure and character and the degree of troop control.

If troop control is cause for concern now, it can only become more difficult in the future as their combined arms force becomes more complex with automation of planning and fire control, forward reaching tactical and naval weapons and increased mobility. Certainly any action by NATO to protect C^3 systems and to acquire an active counter C^3 capability cannot but add uncertainty to a "scientific" process already encumbered with "wandering values." To this extent, EW and counter C^3 can add to deterrence by furthering doubts about favorable correlation force. But, should deterrence fail and NATO be attacked by a Warsaw Pact force combining surprise, mass, concentrated fire power, shock and rapid exploitation, NATO will face a problem whose solution lies in firepower supported by electronics.

The dawn of lethal electrons may lie nearer than most suspect, but until it arrives, electronic warfare must be seen only as an important but complementary element to firepower. If firepower disrupts the enemy plan, conditions will be ripe for an integrated and precisely focused counter C^3 strike which can cause the break in troop control the Soviets experienced in 1941 and continue to fear.

The Key is Integration

Many articles have been written about REC, now that this subject has come out of the closet. Most either describe REC, discuss our vulnerabilities to it or debate the merits of the United States developing a similar doctrine and capability. Usually noted, but seldom given the emphasis deserved, is the matter of integration. REC is more than electronic boxes to search, listen or disrupt. REC is part of the integrated combat operation and even part of a weapon system should target destruction be decided. REC is part of a concept of combat which integrates intelligence, maneuver and combat units. It derives its strengths and lends its support through synergistic interaction which is fostered by integration. Further, it is a part of a combined arms unit which has already integrated all combat units.

This observation is no precis for the United States adopting combined arms as an organizational concept. The Air Force, the Marines and the Navy have persuasive reasons for current organization and support concepts which may, in fact, be most appropriate for the policy of flexible response. The reason for emphasis on integration is that it cannot be solely achieved through black boxes or ad hoc actions after a conflict begins. Integration involves rules and procedures, training and, above all, realistic, regular exercising. It involves the things the U.S. military has found the most difficult to do in peacetime in the past. Integration into the loose coalition which is NATO simply adds another order of magnitude to the urgency and to the degree of difficulty, which is serious enough when only U.S. forces are involved.

Summary

In a scenario worthy of Gulliver for its absurdity, two frustrated Goliaths warily maneuver in a miniature Third World becoming ever less pliant to their ambitions, policies or prowess. Further, their mutual anxieties over energy resources hazard a confrontation that neither may wish but which neither may be able to avoid. Such a confrontation will involve electronics as an essential element of combat and the effectiveness of that contribution will be measured by the extent of the successful integration of electronics into the battle strategy. This integration may reflect the respective historical perspectives of the necessity for and the means of EW employment.

Paradoxically, the party whose strategy places the utmost demands upon C^3 can hardly be labeled the front runner in the EW contest. Why? Perhaps because the corporate memory cannot conjure up, as the Soviets clearly can, a military campaign which turned on the issue of troop control. Instead, to modify the thoughts of Dr. Robert Hermann,[4] we seem implicitly to assume a "free signalling environment."

Perhaps our historical perspective reminds us only of past conflicts where the luxuries of time, boundless resources, technical superiority and unbridled ingenuity permitted us to recover from seemingly fatal errors in prewar perception and planning. Are the stark realities of a "come as you are" war at all compatible with the time needed to integrate superior technology into the total battle management process, as it must to be effective? If muddling through is again to be our concept, one must ask what sort of a contribution that makes to underwriting a believable policy of deterrence and of war winning capability if deterrence fails.

Footnotes

[1]An Historical Survey of Counter C^3, The BDM Corporation, March 30, 1979.
[2]"Tactical Automation: The Achilles Heel of the U.S. Army?"
[3]"Are Our Weapons Too Complex?," OPED, *Washington Post,* February 22, 1981.
[4]*SIGNAL* Magazine, July 1980.

The views expressed in this article are those of the author and not necessarily of the Department of Defense or The BDM Corporation.

Command and Control Warfare: Seizing the Initiative

Verl R. Stanley and Phillip L. Noggle

Command and control warfare (C^2W) encompasses both lethal and non-lethal means for attacking the enemy's command, control and communications (C^3) structure while protecting our own. We use this term in preference to the recognized Department of Defense (DOD) term, C^3 countermeasures (C^3CM) in order to avoid a reactive connotation of the term "counter-measures." Effective execution of C^2W requires an active, rather than a reactive, U.S. command and control concept which will assure that these potentially powerful weapons are employed to achieve battle objectives, not just to counter enemy moves.

In parallel with the evolution of strategy, tactics and doctrines for C^2W, distributed, decentralized decision processes and a "pull" type intelligence support system must be developed. Information on demand is the key to keeping abreast of battle dynamics as well as avoiding information overload. Planning and force employment for C^2W depend upon correlation of key intelligence and operational information. Local, selected source correlation is as critical to C^2W force employment as central multisource correlation is for higher echelon C^2.

A New Means of Gaining "High Ground"

Throughout history, local superiority has been gained by seizing high ground before it was heavily defended. During the Middle Ages, in the plains of Europe, a small artificial hill with a stockade fence around it, called the "Motte," enabled a commander to create his own "high ground." In addition to improving his ability to see the enemy, it was also effective in enhancing the

This article originally appeared in the April 1981 issue of* ***SIGNAL.*

commander's primitive weapons and decreasing the effectiveness of the weapons and mobility of his opponents. The speed of advance of men and horses was significantly reduced, as was the kinetic energy of human propelled projectiles. The demoralizing psychological effect was probably as important as the effects of gravity since the attackers recognized that they faced an exceedingly clever foe.

How can this principle of local superiority be achieved on the modern battlefield? The weapons systems are far more complex than in the Middle Ages. What is the innovation that will enable us to obtain the equivalent of high ground or to outflank our opponents? It is generally accepted that if we can attain effective use of our present technology, involving coordinated planning and synchronized use of sensors, fire power, maneuver elements and electromagnetic energy, we will be able to gain the equivalent of creating our own high ground. The question thus becomes, how can we accomplish the coordinated planning and precise execution to enable us to effectively employ all of the weapons that can be made available? This is command and control, and ours must be superior to that of the enemy.

C^2W includes concepts, strategy, tactics and military operations required to gain and maintain superiority in command and control of forces. This includes capabilities and tactics both to destroy/degrade enemy C^2 capability while improving and protecting/preserving friendly C^2 capability. This superiority does not necessarily mean overall superiority, but rather localized superiority at places, and for periods of time, selected by a battlefield commander to enable him to defeat the enemy.

An example commonly used is the North African campaign of World War II, where the British 8th Army maintained superior force levels generally, but Rommel gained many victories through actions that assured him localized superiority. Rommel did not accomplish this through countermeasures in reaction to British action, but by commanding his forces in light of the information (intelligence and his own force status) available to him in such a way as to bring about his purposes and objectives of battle. By seizing the initiatives with an active concept, his command and control was superior to that of his adversaries.

Active Command and Control Concepts

To identify the vital components of an active concept for orchestrating C^2W operations, a list of required functions was made. (See Figure 1.) It is significant that this list describes the functions involved in the command and control of any kind of tactical battle employing all kinds of weapons. This, of course, indicates that command and control of C^2W operations is integral to command and control of the overall batle and is equally dependent for its success upon information to accurately assess the situation. While previously a matter of debate, a consensus now supports the proposition that C^2W is the province and responsibility of tactical commanders and their battle staffs, not that of specialists. While specialists are required as integral parts of the staffs supporting commanders, the responsibility rests with the commander.

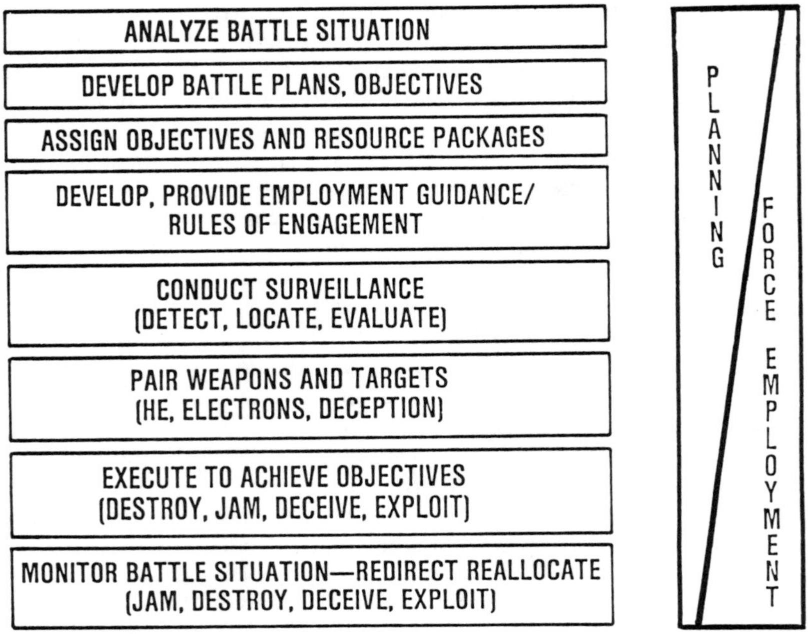

Figure 1. Active C^2W Battle Management

C^2W operations must be directed by the C^2 structure that directs other kinds of operations, since C^2W presents additional tactics/weapons options as a part of other operations. Figure 2 depicts relationships for air missions (e.g., counter air and interdiction) and land operations (e.g., fire support and maneuver).

Exercise Experience

The Air Force Electronic Security Command (ESC), during the past few years, has drawn upon its strong intelligence capabilities and from this base has supported a number of exercises in an attempt to introduce and to assist in developing operational implementation of C^2W, both offensive and protective. Success in these endeavors, however, has not approached that attributed to Rommel. A review of unclassified comments of participants in those exercises reveals deficiencies which might be summarized as follows:

1. Orchestration did not really happen.
2. Simply "overlaying" C^2W on existing structures does not work in practice. C^2W needs to be integrated with other operations.
3. Required connectivity and decision aids are not presently available.
4. Information systems (both intelligence and combat information) were not selective — a truckload of data was simply dumped on commanders.
5. Decentralization of decision making and of information access is needed. Too much was held at too high a level at the expense of timely action.

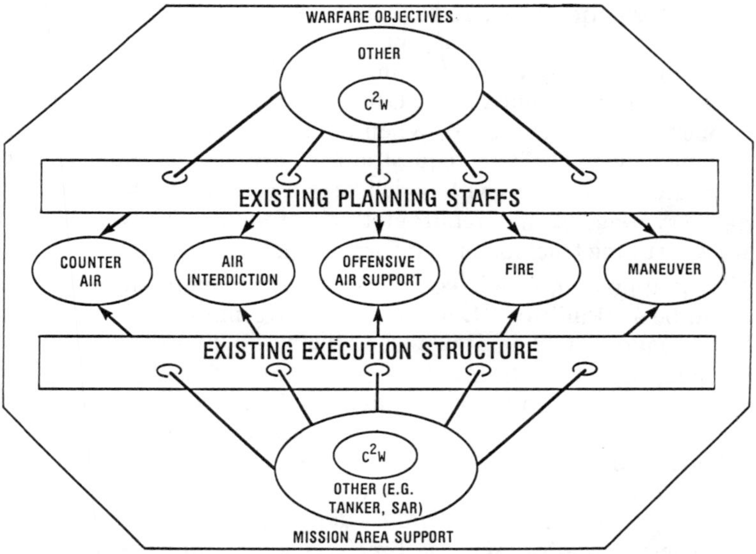

Figure 2. C^2W Integrates with Existing Command and Control Structures

Strategy, Tactics, Procedure and System Evolution

In light of the above analysis of required functions and a review of exercise experience together with frequent discussions with tactical operators, no fundamental changes in basic command and control structures and doctrine are requred to intergrate C^2W operations. Appropriate strategy, tactics and procedures (to include decentralization of execution decisions) do, however, need to be developed and applied. In connection with this, both operational and intelligence information systems, communications interconnectivity and analysis and decision aids must be developed in order to realize the full potential of C^2W opportunities.

In the meantime, C^2W concepts, elementary tactics and procedural changes can be implemented now to more effectively employ existing C^2W capabilities. Each improved capability, whether technical or operation, could then be implemented over time in a controlled, evolutionary process. The initial step in this evolutionary process to integrate and orchestrate C^2W operations is to more fully involve commanders and battle staffs in the process. The assistance of specialists as a part of the battle staff would be helpful, rather than in separate structures attempting to employ C^2W as an overlay of the battle staff actions.

Selective Information Systems

Another conclusion that can be made is that all command levels require a tailoring of information flow. Though an oft used term, it is rarely realized in actuality. One reason for this may be the "Push" concept for information systems currently employed, wherein information is transmitted to commanders based upon predetermined need. Usually that determination is based upon advance queries to commanders. When queried in advance, a commander's pre-engagement response is to request almost everything, because he cannot fully anticipate the situation with which he will actually be confronted. Thus, receiving everything, he may realize nothing of value because he cannot sort through the resulting truckloads of information.

A different approach, which would also alleviate communications net overload, would be a "Pull" or "Demand" type information system from which needed information could be drawn when the battle situation could reasonably be anticipated, as well as during the actual course of the battle.

Such "Pull" type information (dated by hours at best) could then be supplemented with near real time information on a similar demand basis through battle staff access to sensor/surveillance system control and processing elements for dynamic query and tasking. Intelligence and operational information systems employing "Pull" design concepts would then provide both the general situation and the timely, selected information actually required for C^2W orchestration, as well as for overall battle management. Obviously, a "Pull" type of information system requires assured interconnectivity but should substantially reduce communications volume.

Distributed C^2: Decentralized Decision for Survivability

Earlier, the need to protect friendly command and control from attack was mentioned. This is an integral requirement if we are to gain and maintain C^2 superiority even on a localized basis. Various forms of distributed C^2 which address this question of survivability are being pursued by each of the Services. Elements of the Army's Air-Land Battle 2000 concept and the Air Force's work toward Future Tactical Air Control Systems are two rather well publicized examples.

In considering the U.S. C^2 architecture into which C^2W operations must fit, we need to think about distributed architectures and the need evidenced in exercise experience cited earlier for decentralizing the decision process. Such decentralization would be dysfunctional without corollary decentralization of access to intelligence and operational information required to enable desisions. Since there is coherence in all of these C^2 vectors, sound doctrinal concepts need not, and should not, be violated in the process. Once such time proven doctrinal concept essential to effective employment of air power is centralized control coupled with decentralized execution.

Centralized Planning and Command

Functional analyses of the Tactical Air Control System, for example, indicate a strong, continuing requirement for centralized control of air forces to include the functions of:

1. Situation Analysis—What is going on?
2. Battle Planning—What do I want to happen?
3. Assign Objectives—Who does what to make it happen?
4. Apportionment-Assignment of Resource Packages—Weapons plus eyes and ears are necessary to achieve assigned objectives.
5. Employment Guidance and Rules of Engagement—To assure that decentralized execution level decisions are made in the context of the overall battle plan.
6. Follow-up Action—To make the battle plan as dynamic as it must be.

These planning and command functions are essential in an active approach to command and control. Properly accomplished, they provide the means to direct execution level battle staffs to act and to employ resources in accordance with the overall battle plan. Time sharing of "eyes and ears" (critical resources such as JSTARS, COMPASS CALL, COMFY CHALLENGE, etc.), as well as frequency management/protection, deception and changes in the situation also must be outputs of these functions. The planning level battle staffs must take care of essential joint air-land coordination (to include frequency management and deception planning) and become integrated in operations orders.

Concepts Underlie Architectures

The above discussion leads to some general concepts about effective future architectures. Figure 3 illustrates a generic C^2 structure and major C^2W functions to include planning and force employment levels. The essential planning level functions, discussed above, are required, as shown, to allow effective force employment decisions. Resource packages, including weapons and "eyes and ears" (some on a time sharing basis), must be employed by the force employment level battle staff. The force employment guidance, with access to both node finders and control of node breakers, provides for decentralized execution level decision making for offensive air support, counter air and interdiction air missions, and for maneuver and fire support land operations. Decentralization of force employment decision making in this context would contribute significantly to the realization of near real time battle capabilities, given anticipated technological improvement in sensors, weapons and weapons direction capability.

Of great importance to the success of an operational concept encompassing these procedural changes are the information cycle (Figure 3), comprised of both intelligence (information about the enemy) and operations reporting (information about own/friendly forces). The long information cycle supports planning levels in preplanning and in the dynamic update of plans and orders. This cycle includes the multisource correlation process which is vital to understanding the battle situation in order to plan and select objectives.

Overall situational information must be provided via this cycle to force employment levels.

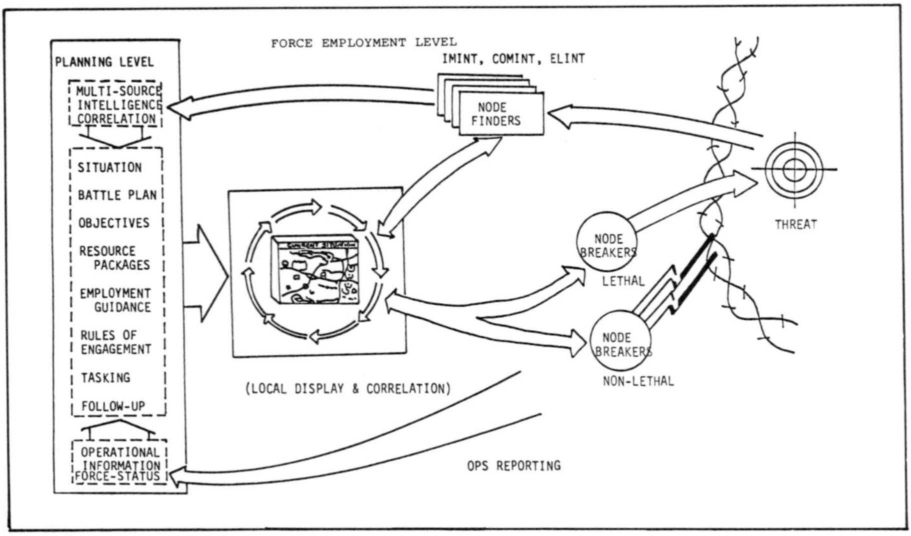

Figure 3. Generic C^2 Structure and Functions

Short cycles support force employment battle staffs, providing specific types of information are "pulled" by the battle staff and correlated with primary sensor data upon which action is largely based (e.g., air surveillance radar for air defense operations). This local, limited sources correlation function is as critical a requirement for force employment as the central multisource correlation is to overall command and control at higher echelons.

The concepts described herein have been considered in the contexts of air missions (close air support, defensive counter air, offensive counter air and interdiction); individual combat resources (COMPASS CALL, COMFY CHALLENGE, EF-111 RF-4G and RPV/DRONES); and a Central European battle scenario (developed by the Joint Electronic Warfare Center) and have been found to be appropriate.

C^2W Provides Added Fighting Options

Command and control warfare offers options to fight differently from what we are accustomed to through heavy dependence upon intelligence in ways that will allow us to gain the best possible results from limited forces. The concepts summarized in Figure 4, contribute substantially to understanding how we can make command and control warfare work for us as a significant part of overall combat capability.

The improvements suggested really apply to the total C^2 process. If these improvements were made they would facilitate the integration of all command

- ACTIVE APPROACH TO C^2W
- C^2W MUST BE INTEGRAL TO ALL C^2
- START NOW WITH CONTROLLED EVOLUTION
- INFORMATION SYSTEMS MUST BE SELECTIVE
- DISTRIBUTED C^2 - DECENTRALIZED DECISIONS
- CENTRALIZED COMMAND REQUIRED
- CONCEPTS UNDERLIE ARCHITECTURES
- C^2W PROVIDES ADDED FIGHTING OPTIONS

Figure 4. Concept Summary

and control warfare operations. The successful military campaigns of history combined an active concept of employment of force and innovative use of intelligence and available technology to create an advantage, such as high ground or force concentration. It appears that C^2W offers a potential twentieth century equivalent of creating your own high ground.

Footnotes

[1]*Air Force 2000: Air Power Entering the 21st Century,* Department of the Air Force, 1982.

[2]*Air-Land Battle 2000,* U.S. Army Training and Doctrine Command, 1982.

[3]*Ground Attack Control Capability Concept Analysis,* E-Systems Center for Advanced Planning and Analysis, August 12, 1982.

[4]*Integrated Planning and Force Employment of C^2 Warfare,* E-Systems Center for Advanced Planning and Analysis, February 1982.

Interoperability and EW

Michael T. Thurbon

In a BBC broadcast in 1965, Gen. William Westmoreland was quoted as saying:

> We are on the threshold of an entirely new battlefield concept. On the battlefield of the future, enemy forces will be located, tracked and targetted almost instantaneously through the use of data links, computer assisted intelligence evaluation and automatic fire control. With first round kill probabilities almost a certainty and with surveillance devices that can continually track the enemy, the need for large forces to fix the opposition physically will be less important. With cooperative effort no more than 10 years should separate us from the automated battlefield. [1]

More than 10 years have now passed since that prediction was made and, while it is apparent that the time scale implied was rather optimistic and that one might register some reservations concerning the degree of automation desirable, it is quite clear that the fundamental concepts proposed are correct. It is also clear that, although Gen. Westmoreland was concerned with the land battle, his remarks apply with equal force to Naval and Air Forces operating independently or in concert with their sister services.

Embedded within Gen. Westmoreland's view of the battlefield of the future lies the crucial element of interoperability. His statement is revealing because, while it makes no specific reference to this element, it contains all the factors which make interoperability a key to the effectiveness of forces in the future. It binds together the essential components of location, tracking, targetting and intelligence—the factors defining the threat. It concerns itself with fire control and it indicates the desired end product of improved interoperability; that is,

---

This article originally appeared in the April 1980 issue of* ***SIGNAL.*

the more effective use of our weapons and manpower. On the technical side, it stresses the importance of communications and the power of computers.

It is my aim to examine the characteristics of interoperability and to explore very briefly from the perspective of electronic warfare, some of the possible consequences arising from increased interoperability.

The Fusion of C^3 with Weapons

Before looking at the underlying qualities of interoperability represented by the factors revealed in the quotation, it is useful to trace the fusion of communications, command and control with weapons, a characteristic which appears to be of increasing significance in modern military systems. Historically, a clear dividing line has been drawn between command, control and communications systems on the one hand and weapons systems of all kinds on the other.

Until comparatively recently, in most Services communications and weapon systems have been the responsibility of separate groups possessing different training and specialized expertise. Where such conditions existed, either little or no interoperability was required or, it was desired, only limited success was possible. This state of affairs was perfectly satisfactory so long as communications were primarily voice or, at best, capable of passing only low speed data, and weapons had very severely restricted capacity to absorb and process data.

It was also adequate where forces from a small number of nations were cooperating who spoke a common language sufficiently well enough to exchange small quantities of information. In particular, there was formerly only a strictly limited dependence on communications to enable individual weapons to be used. Thus, there was little need to bring together in intimate contact communications and weapons. This was a situation which until quite recently changed only slowly but which is now evolving at an increasingly rapid pace.

This evolution is especially clear in the case of air defense where, to make efficient use of the potential of modern aircraft and their weapons, it has been necessary to move from simple voice communications between individual aircraft and between aircraft and ground sensors to high capacity data links exchanging large quantities of information in real time on a wide variety of aircraft and weapon parameters.

This dependence of weapons on communications is being highlighted by a number of developments.

Dependence on Communications

The first of these developments is the increasing cost and effectiveness of all weapons. The kill probability of conventional weapons is rising rapidly. At the same time, their cost is increasing at a comparable or even faster rate. For example, a recent analysis[2] has shown that in the United Kingdom the cost of successive generations of certain fairly typical weapons has risen by a factor ranging from two for a tank to eight for an antiaircraft system. More significantly, perhaps, the annual increase in the cost of most weapons far exceeds the annual increase in gross domestic in all the major Western nations.

Thus, although the total destructive power available to a commander is increasing, the total number of weapons in his stocks is likely to diminish at a time when the number of targets he may have to engage is also increasing. To ensure the most effective use of his resources, the commander will need to maintain firm control at least over the selection of types and priorities of targets if not over the selection of individual targets and on munitions expenditure.

The importance of this last point has been heavily underlined by the experience in the Middle East since 1967. As the intensity of conflict has become progressively greater, the consumption of warlike stores has risen even faster, exposing forces to the danger of a rapid and total loss of fire power. To exercise the necessary control to handle these situations will demand progressively more efficient means of command and control.

The dependence on communications also reflects the increasing number and performance of potential targets, especially those faced by the North Atlantic Treaty Organization (NATO) forces and the effectiveness of the weapons. The increasing effectiveness of weapons is a general phenomenon and is certainly not limited to Western systems. These factors in turn mirror the changing character of modern warfare which is becoming more complex, more precise and more dependent on electronics.

To cope with this situation requires a multiplicity of different types of sensors dispersed throughout a considerable volume of space. Only in this way is it likely to be possible to maintain a more or less continuous track on high performance targets which is essential if they are to be engaged successfully and economically. This in turn requires very rapid transfer of large amounts of information between sensors pools of intelligence data, demand and control authorities and the individual weapons.

The third development (which both demands interoperability and contributes to its achievement) is the employment of digital processing in almost every element in the military system from the individual weapon to the Command and National Headquarters. The computational power now becoming available provides the potential for better and more timely decisions both locally in the weapon itself and systemwide.

This ability to absorb and process information on the local and general environment, if properly integrated with, and supported by, adequate communications, effectively equips each weapon in the system with the sensors of some, or all, of the others, enormously enhancing the potential of the individual weapon system and the system as a whole. Again, the general trend can be seen in the air defense context where, for example, the relevant data collected by an Airborne Warning and Control System (AWACS) aircraft can be transferred as required to the fire control system of associated fighters and information on fighter and weapon status can be displayed in the control platform.

These advantages only accrue, of course, if the data is immediately available and immediately useful to each element in the system. Thus, either a single, common, format must be employed or individual participants must be designed to accept and transform data for their own and others' use. Thus, too, it seems that the old distinction between command and control aspects on the one hand and the weapon system on the other is becoming more and more blurred.

Defining Interoperability

The need for interoperability is frequently discussed and widely recognized but the general definitions of the term are not particularly revealing. In the United Kingdom for example, it is defined as:

> The ability of systems, units or forces to provide services to and accept services from other systems, units or forces and to use those services so exchanged to enable them to operate effectively together.[3]

Elsewhere, for instance in Sir John Hackett's thought provoking book, *The Third World War,*[4] interoperability seems to be equated with standardization and certainly interoperability involves a degree of standardization. However, discussion of the subject usually concentrates on the requirement to design the relevant characteristics into the system; for example the appropriate performance of the interface units. Standing back a little from this approach and examining interoperability in the light of some of our own studies connected with the general air defense problem, the following key features emerge.

Interoperability has, of course, no independent existence: it is an essential property of a complex system. This property possesses only three distinct components. The first of these components is the ability to transform data (that is meaning) into a commonly understood and usable form. The second component is the ability to transform information into a common form suitable for transmission and reception (engineering transformation) and the third component is control.

The transformation of meaning involves the translation of information into a form every recipient can understand and use. This is most clearly seen by means of examples. Perhaps the simplest is that of the problem of multiple coordinate systems, the solution to which may require conversion or standardization. Whichever solution is chosen must result in information in a commonly usable form which may involve translation at the element generating the information or by the recipient or by both.

A second example is that of track numbering. If local numbering methods are employed, a translation process must be available to enable the individual tracks to be identified and, if necessary, discussed by some or all of the participants in the overall system.

The second component of interoperability, we suggest, is the transformation of information into a common form suitable for transmission and reception: in other words, those processes necessary to ensure that signals can be received, demodulated and the waveforms and bit streams correctly processed; communications engineering.

The third component of interoperability is what we have termed control.

This term is used to embrace those functions essential to the viability of the system as a whole. Some of these functions may be specific to a particular system. More generally, however, control involves the protocols and techniques that enable the units, having been provided with the appropriate technical capabilities, to be assembled into a system at a specific time and for a specific tactical purpose. Control enables the system to be generated or closed down, to fuse with one or more other suitably equipped systems to form a new, enlarged system or to fragment into a number of smaller self-contained systems as the situation demands. The control property of interoperability also enables the individual unit to join an established system at any time and to leave it temporarily or permanently.

Diffuseness

Within this brief examination of the nature of interoperability can, again, be seen the importance of communications. What is also apparent from all of the foregoing is that major military systems of all kinds are becoming increasingly diffuse. This diffuseness applies geographically and in the use elements make of the electromagnetic spectrum and it also applies in the sense that the individual elements are themselves becoming more and more interdependent. The analysis of the performance of systems attempting to draw boundaries around the elements is becoming an increasingly difficult and artificial exercise.

A glance at some of the weapon systems now in, entering or planned to enter service highlights this trend. At the lowest level, it is apparent in the use of designated precision guided munitions where the illuminator may be far removed from the weapon launch point and where communications and weapons form a single integrated, indivisible system. In this case, too, a single designator may provide illumination for a variable number of launch points, say guns, or several designators may work with a single weapon.

This trend to diffuseness is also apparent in the renewed interest now being taken in bistatic radar systems. Here one or more transmitters may provide illumination for many remote receivers. Such a system is dispersed through a considerable volume, sensors and transmitters are firmly netted together and, if more than one transmitter is available, the system is relatively invulnerable to loss of individual elements.

The trend can also be seen in the numerous systems employing remote intelligence sensors working to points at which fusion and filtering can take place. In communications, a good example is provided by the use of packet switching networks. Thus, the performance of the total system is likely to become much greater and its influence more far reaching as it embraces larger numbers of disparate elements.

At the same time, the dependence of the overall system on the performance, and indeed the existence of, individual subsystems is likely to become less crucial. The state of affairs to which we are moving seems to be one in which relevant information from each element is always available to every other element and in a form that conforms to the essential properties of interoperability described above. At the same time, when viewed from outside by a potential enemy the overall system will present few uniquely identifiable vital points.

Clearly, in the near future, the systems that are most closely to conform to this ideal are those planned to use Joint Tactical Information Distribution System (JTIDS) or one of its similar competitors. If the interoperability characteristic of control is properly included and suitable net management techniques are devised, any system employing JTIDS should be self-healing and possess no vital points, at least in terms of communications.

As an aside, it is worth remembering that if interoperability is to be achieved, the demands on those responsible for ensuring that it is achieved are becoming more and more severe. Any failure on their part will result in systems that are unable to provide the operational capability that the Services intended and that the taxpayer has paid for.

The formal means established to ensure that interoperability is achieved must, like interoperability itself, possess the means of management and control able to spread their influence across service, contractor and, ideally, across national boundaries. The increasing complexity and diffuseness of military systems means that larger numbers of contractors are inevitably involved, that numerous operational requirements must be scrutinized to ensure mutual compatibility and that a growing number of international standards are complied with. These requirements can only be met if a suitable organization is established equipped with the necessary tools to process the mass of data generated.

These requirements also suggest the need for a unified approach to be adopted to establish and control interoperability throughout the life of the system from specification, design and development through testing and trials and in-service modifications.

Balance Between ECM and ECCM

At this point, it is appropriate to gather up the second strand in the title and examine certain facets of electronic warfare and the possible influence of interoperability and the resulting diffuseness of military systems on electronic warfare (EW). The aspects of EW covered are primarily those concerned with communications because of the powerful involvement of communications with interoperability. However, a number of the remarks are also relevant to the non-communications field. What I hope to show is that a significant shift is occurring within EW between electronic countermeasures (ECM) and electronic counter countermeasures (ECCM) and that this imbalance will be further distorted as a result of the increasing diffuseness of systems discussed above.

Fundamentally, the ECM problem hinges on bandwidth, and in communications this means, basically, the developing ability of designers to employ wideband spread spectrum techniques. These techniques have two advantages of great significance to electronic warfare. They provide processing gain and they generate signals that are extremely difficult to detect. To overcome the effects of processing gain, the jammer must either seek out and then exploit specific weaknesses in the design of the victim or it must find some means of increasing the jamming flux at the target receiver.

If the first course of action is selected, assuming an exploitable weakness can be discovered in time, a jammer can be produced which is economical in the use of power but which may have to be technically very complex and very expensive and, being specific, may have a very limited life. If the second approach is chosen, i.e., increasing the jamming flux, the jammer may elect to increase its effective radiated power over the widened bandwidth.

This solution has the merit of being independent of detailed intelligence but it is expensive and, especially if wide area coverage is required, will almost certainly be limited by the availability of primary power. If the effective radiated power cannot be increased sufficiently, the jammer must move closer to its potential victims. If it does close the range it increases its own vulnerability to physical attack. At the same time, the area a given jammer can dominate is also reduced. Thus, the cost of covering by jamming a given operational area is likely to rise very steeply. A partial solution to the dilemma is for the jamming system to be designed to concentrate its attack against selected key participants, a point we shall return to in a moment.

The detection problem introduced by wideband signals is particularly severe for vehicle mounted EW systems where wide apertures and long base lines cannot easily be made available. Regardless of which method of spreading the signal bandwidth is chosen by the designer of the communications system and regardless of the EW platform, the EW receiver will have to be complex and consequently expensive, especially when the signal environment is dense.

In addition, if wideband low peak power signals are employed, the EW receiver is again likely to have to move closer in with consequent increase in the vulnerability of its platform to attack. However, in the context of communications, a second receiver associated problem arises from the lack of channelization inherent in spread spectrum systems. This loss of identifiable channels makes interpretation of the spectrum extremely difficult. As a result, the ability to isolate and interpret critical activity and units is reduced, at least in the tactical context. This in turn limits the ability to concentrate the jamming attack in time, space and frequency against crucial components of the systems.

The shift in balance between ECM and ECCM, to the advantage of the weapon or communication designer, is further emphasized by the power of modern signal processing techniques. A simple, but good example here is the use of self-adaptive interference cancellation loops which are cheap even when the cost of any additional antennas is taken into account. These loops can be extremely effective in reducing the effects of most kinds of jamming, having the additional advantage that their employment at any instant is by no means obvious to the enemy.

The difficulty of jamming communications might in the past have been received with some resignation, if only because of its contribution to answering the question, "Do we jam or do we listen?" In the future, this particular question may be much less relevant as more circuits, both voice and data, become powerfully encrypted. This is unfortunate as it swings the emphasis onto jamming at precisely the time that the ability to jam is coming under greater strain.

One could, of course, take this examination of the balance between ECM and ECCM much further. It would be interesting for example to explore the effects of high speed time division multiple access techniques on jammer design and tactics. All we have attempted to do, however, is to select one or two fairly obvious trends and suggest a number of qualitative results. The overall effect seems to be quite clear though. For the foreseeable future, the cost of successful jamming will rise at a faster rate than the corresponding cost of introducing the relevant counter countermeasures. The balance is moving to the weapon and communication systems and the area any jammer can hope to dominate is collapsing.

Effects of Interoperability on EW

So finally, we can tie the threads together and see what, if any, are the effects of interoperability on this EW problem.

If one looks at the problem through the eyes of the jammer designer faced with the general trends against him or her that we have outlined, it appears that his or her best strategy must be to concentrate his or her limited resources to the maximum possible extent. This seems the most hopeful approach to restoring, at least to a limited degree, the EW balance. The means by which this concentration can be achieved depend to some extent on the jamming methods chosen, (and in the longer term these will have to be mainly wideband noise) but basically the designer has four options. He or she can control his or her use of jamming in time, frequency, space and in priority *vis-a-vis* the threats.

Concentration in time and frequency, while being extremely important in terms of power management, may not be greatly influenced by the increased diffuseness of the system the ECM is attacking. Concentration in space and in target priority do seem to be more sensitive to this effect.

To be successful, the jammer should attempt to select a key point, preferably a vital point, and concentrate the jamming effort against this point so that any disruption caused at this one victim is spread indirectly over a wider area than the jammer could otherwise influence.

To achieve this desirable state of affairs, it must first be possible to identify and isolate the appropriate target, and we have already seen that the detection and identification problem is becoming more tricky as a result of the changes in waveform design. This situation will be seriously worsened as systems become more and more diffuse and the number of uniquely identifiable participants falls.

At the same time, of course, the relative independence of such diffuse systems to the performance of individual elements again limits the effect of the jamming. Thus the ECM designer and operator are presented with a dichotomy.

The technical trends in weapons and communications of all kinds are limiting the capability of jamming, forcing concentration against specific targets and the confining of the jamming energy into narrow beams, while the increased diffuseness induced by heightened interoperability is pulling strongly in the opposite direction, accelerating the rate at which the area any jamming can dominate is collapsing.

Footnotes

[1]"The Electronic War," A BBC Broadcast on Radio 4 on February 2, 1975.

[2]"Budgeting for Defense," by Sir Arthur Hockaday, *Journal of the Royal United Services Institute for Defense Studies,* December 1979, page 6.

[3]Joint Services Publication 101.

[4]"The Third World War," by Sir John Hackett et al.; Sidgewick & Jackson, London 1979.

EW and NATO Communications

A. Nejat Ince

Communications are a vital cohesive force in an organization as territorially fragmented as the North Atlantic Treaty Organization (NATO). A mixture of different systems, collectively referred to as NATO communications, they provide the means for political consultation between the NATO Allies in times of tension and crisis and for effective command and control of the NATO assigned forces in the event of a coventional or nuclear war.[1,2]

In this article, we shall be examining the make-up of NATO's communications and how secure they are from interference by a potential enemy. At the present time, NATO communications which tie together headquarters, weapons and sensors throughout the NATO territories consist of NATO owned systems and of facilities leased from the various national civil administrations and military agencies.

The NATO satellite communications (SATCOM) system has, at present, 12 large static ground terminals but will be expanded in the near future to some 20 large static and transportable terminals. Two satellites—NATO IIIA and IIIB—are in service. The description of the satellites and the ground segment are to be found in Footnotes 3,4,5 and 6. The system uses Frequency Division Multiplexer/Frequency Modulation/Frequency Division Multiplex Access (FDM/FM/FDMA) modulation but is scheduled to be converted into digital working using mod/phase-shift keying (PSK)/FDMA modulation.

The Allied Command Europe (ACE) High system[7] consists of 49 over-the-horizon (O/H) and 41 line-of-sight (LOS) links. This system currently carries the majority of the long haul voice and telegraph communications for ACE. Digitalization of the system is being considered.

A large basically LOS radio relay system, called CIP-67, is being implemented in Central Europe. This will provide thousands of circuits for NATO

*This article originally appeared in the March 1978 issue of **SIGNAL**.

use. Initially, the system will be analog but later will be converted into digital working.

In 1970, NATO decided to integrate all its communications into a single system called the NATO Integrated Communications System, or NICS for short, and to set up a Management Agency (NICSMA) to implement the system. The basis for NICSMA's initial planning and implementation, which is presently underway, was the system concept developed by Strategic Headquarters Allied Powers Europe (SHAPE) Technical Center (STC) and set forth in its report TR-86 (3), entitled, "A proposal for a NATO Integrated Communications System." The original concept is, at present, being revised[8,9] to take into account new developments, particularly in the area of digitalization, for application to the subsequent stages of NICS implementation.

NICS Architecture

NICS is required to support clear, narrow and wideband secure voice, message and data services, to enhance survivability, flexibility and communication security and to provide improved interoperability with national civil and military strategic and tactical systems.

To meet these objectives economically, NICS is planned as an automatically switched, common-user grid network. The switches, located outside target areas and interconnected via independent transmission paths, are stored program controlled. The routing principle ensures that a high probability exists that a route can be found for a call, even when the network is damaged, jammed and severely overloaded. Priority is given to high precedence calls, with low precedence calls being pre-empted if necessary.

Communications security is provided by the application of a combination of link-by-link and end-to-end encryption techniques. The network connectivity is supported by a mix of NATO owned cable, LOS, O/H and satellite systems and leased post, telegraph and telephone (PTT) facilities. High frequency (HF), XHF and airborne platform overlays are also being considered as a means of enhancing network survivability during extreme stress situations. The general characteristics of NICS are illustrated in Figure 1.

The achievement of the goals stated above in a cost-effective manner—including the provision of adequate communications capacity under all conceivable threat situations—requires very careful planning and a good balance between the survivability of headquarters, weapons and sensor systems on the one hand, and that of the associated communications systems on the other. The issue survivability raises questions of "against what threat?" and "how much of survivability?"

We shall not discuss here how these problems are handled quantitatively in the design of practical systems. Suffice it to say that in NICS a diversity of communications media and techniques are used to take account of different features of range, capacity and mobility, and of their susceptibilities to electronic warfare and physical threat including electromagnetic pulse (EMP) and sabotage.

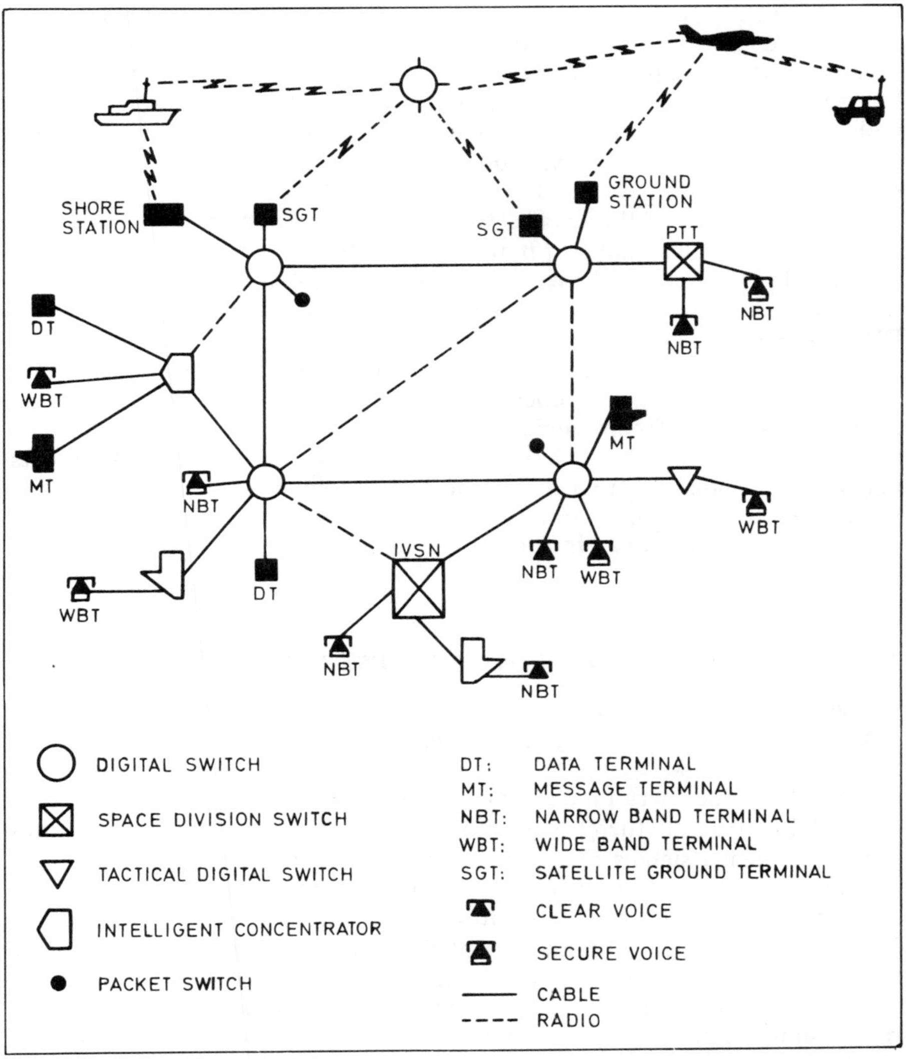

Figure 1. Future NICS Architecture

The use of radio which makes NICS susceptible to electronic warfare (EW) is necessary on the grounds of cost and flexibility and for the provision of communications to mobile users. Moreover, because of the fragmented nature of the NATO geography, we must use radio, sometimes as the only means of communications but often in combination with cable facilities and other radio systems.

NATO SATCOM plays an important role in NICS by providing reliable long-distance internodal links, connections from mobile/transportable and isolated users into the switched network and direct communications between users co-located with terminals. SATCOM is, of course, ideal for broadcast purposes. NICS also uses cable systems, since these can provide jamming free circuits, and radio relay and troposcatter systems, which can work over a wide range of frequencies providing several channels and spanning large distances. For access into the network and also for direct communications between mobile users on land, at sea and in the air, use is made of thin-line radio systems utilizing various propagation modes in addition to the systems already mentioned.

In the remainder of this article, we shall look at the characteristics of the radio systems which may form part of NICS, or indeed of any similar network, briefly describe the nature of the EW threat and the vulnerability of these systems to that threat and discuss methods by which their resistance to EW can be improved.

The degree of actual protection incorporated into each radio system depends, of course, on the number of users relying on the system and on the alternative means of communications available to them, i.e., on network topology, connectivity and design. These features of the network and its performance under different threat and traffic scenarios, including EW, can be studied using computer simulations of the whole network, supported by computer and laboratory simulations of individual communications systems.

Characteristics of Radio Systems

The properties of the various radio systems which are, or could be, used in NATO communications are given in Table 1.

Most of the links listed in Table 1 have been used and reported upon extensively in the literature, except perhaps meteor-burst and field aligned scatter links. We shall therefore briefly describe them here. Meteor-burst links rely on bouncing signals off the ionized trails left by meteors in the height range 80 to 120 km. The signals are intermittent and peaky (duration of the order of one sec) and necessitate storage of the data and its transmission at high burst rates when a suitable oriented meteor trail is present between the two terminals to obtain a continuous but lower rate throughput.

The attraction as far as military applications are concerned is the possibility of communications over distances of up to 2000 km which cannot easily be located or jammed. Another attraction is that in northern latitudes meteor-burst communication is inherently less susceptible to blackouts than HF communication.

Meteor reflections and their application have been investigated at STC over a period of several years. These studies led to the development of a system in the mid-1960s, named COMET, which has significant advantages in performance over earlier systems. The description and application of the STC system and its interception and jamming properties are to be found in Footnotes 10 and 11.

Another system not apparently used so far is the field aligned scatter system. Heating of the ionosphere by a powerful transmitter operating at the plasma frequency of the F-layer can cause irregularities consisting of filaments

Type of System	PROPERTIES (NO EW) RANGE	CAPACITY	(a) AVAILABILITY (b) EW FEATURES
VLF Links (3-30 KHz)	Typically 4000 km (groundwave)	1-4 telegraphy channels	(a) High (b) Limited frequency range and agility/spread
LF Links (30-300 KHz)	(1) 900 km in summer (groundwave) (2) 1700 km in winter (groundwave)	(1) 4-6 telegraphy channels (groundwave) (2) 1 telegraphy channel (skywave)	
HF Links (3-30 MHz)	(1) 500 km over sea (groundwave) (2) Any range (skywave) but 3000 km Single-hop	(1) 1 secure voice or medium speed data channel (groundwave) (2) 1 telegraphy to 1 voice or medium speed data channel (skywave) (worse at night)	(a) High (groundwave) for telegraph and analog voice but lower (skywave) for secure voice or medium speed data. (b) Potential for frequency agility/spread and other ECCM
Extended HF Links (30-100 MHz)	2000 km (meteor-scatter propagation	(1) Burst capacity Mbits/sec (2) Average capacity several telegraphy channels	(a) High (b) Inherent resistance to EW
Field aligned scatter links (VHF-UHF)	A few 1000s of kms	1 voice channel or equivalent telegraphy channels	
Troposcatter hop (UHF-SHF)	Several 100s of kms	Up to 120 analog digital voice channels	(a) High (b) Tandem links and fixed operating parameters
LOS hop (UHF-SHF)	10s of kms	A large number of channels	
SATCOM Links (7-8 GHz)	Any range within satellite coverage region	100s of analog and digital voice circuits	(a) High (b) Good frequency range and agility/spread and other ECCM

Table 1.

of increased ionization (the increase is of the order of one percent) aligned with the earth's magnetic field. The irregularities are capable of scattering incident electronmagnetic radiation. Such field aligned scattering (FAS) is highly aspect sensitive so that communication via FAS is limited to certain zones having widths of some tens of kms.[12]

Electronic Warfare

Electronic warfare (EW) is defined as that division of the military use of electronics involving actions taken to prevent or reduce an enemy's effective use of radiated electromagnetic energy, and actions taken to ensure our own effective use of radiated electromagnetic energy. EW arises, therefore, in the context of radio communications and is related to frequency spectrum utilization which, apart from EW, also involves electromagnetic compatibility (self-inflicted EW) and the effects of natural phenomena such as precipitation and solar activity and of nuclear weapons on propagation media.

The two elements of EW are electronic countermeasures (ECM) and electronic counter countermeasures (ECCM). The ECM can be affected in passive and active modes.

1. *Passive ECM*: By the interception of radio signals, useful intelligence information may be obtained in one of two ways. Communication intelligence (COMMINT) may be obtained directly from their content if messages are unprotected by cryptographic equipment. Information may be built up on levels and patterns of communications activity. This is referred to as electronic intelligence (ELINT) and may enable the interceptor to perform more effective jamming.

Direction finding (DF) may be employed to determine the location of headquarters or force elements. This is of particular importance in the case of mobile operations involving high value units such as ships.

Characterization of communications traffic in terms of, for example, radiated power levels, modulation types, frequencies polarization, identification of keying transients and other unique properties (fingerprints) of transmitter waveforms, can be used by a potential enemy to aid the design of active ECM systems and also, possibly, of beam-riding weapons.

2. *Active ECM*: Jamming can be defined as the means by which communications are interfered with in an intentional manner by an enemy; it may be of an overt or covert type. In overt jamming, no attempt is made to disguise the fact that jamming is being attempted.

Covert jamming may take the form of "spoof" signals which appear to be legitimate but which deceive or confuse the receiver. An example of this may be the intrusion of any enemy transmitter into a clear voice net. Alternatively, the jamming signals may be such that they adversely affect the performance of the communication system without it being immediately evident that jamming is taking place, e.g., by attacking a synchronised system using delayed replicas of the wanted signal. Operationally, this will have the effect of increasing the time before countermeasures are applied.

Vulnerability of Radio Systems to Passive ECM

1. *General*: All radio systems may be protected against COMMINT by encryption. To reduce the vulnerability of radio systems to electronic intelligence (ELINT) and DF requires that detection by the interceptor be made difficult. Generally speaking, this may be achieved by using one or a combina-

tion of the following techniques, depending on the propagation mode and type of traffic involved:

- Using lowest possible transmitter power;
- Not emitting in directions which are favorable to an interceptor (use of directive antennas);
- Emitting in short bursts and at unexpected times;
- Changing the frequency of transmission (frequency agility);
- Use of spread spectrum modulation (SSM) to reduce signal power density below the level of the interceptor receiver; and
- Use of radio systems in which signals arrive at the receiver via a reflection or refractive media (man-made or natural) which are aspect sensitive

2. *SATCOM Links*: Satellite communication and control downlinks can be monitored directly from Warsaw Pact (WP) territory and from international waters. Uplink monitoring is also possible from the region in the vicinity of the ground terminals via sidelobe radiation. The main uplink beam can also be intercepted from an appropriately positioned aircraft or surveillance satellite.

The communications traffic and the telemetry signals from the satellite can be encrypted, and thus the content will be secure; however, if the signals are not masked, downlink monitoring may provide an enemy with data on the essential system parameters and hence allow him to implement his active ECM systems more effectively.

3. *O/H Links*: Because of the large transmitter powers used to overcome high scatter/diffraction losses in O/H links, interception from long range with ground based and airborne receivers becomes possible. Link encryption must, therefore, be used to provide traffic flow and communications security.

4. *LOS Links*: To a lesser degree, the comments made for troposcatter links also apply in the case of LOS links.

5. *HF Radio*: The state-of-the-art in the collection and analysis of HF data and in direction finding techniques is very advanced. Although communication security can be provided for telegraphy by encryption of the signal, reliable secure voice is difficult to achieve mainly because of multipath. Burst communications, frequency agility and the use of spread spectrum techniques, as we shall discuss below, can be used to reduce the vulnerability of HF links against interception.

6. *VLF and LF Broadcast Transmission*: The interceptions of both very low frequency (VLF) and low frequency (LF) broadcast transmissions does not present a significant problem for a potential enemy. However, the telegraphy traffic normally carried by such transmissions is always encrypted and thus little useful information can be obtained from this type of monitoring.

7. *Meteor-burst Links*: Meteor-burst links, when encrypted, possess perhaps the highest degree of resistance to interception. Tests conducted by STC have shown that the percentage of traffic that may be intercepted can be extremely low (of the order of one percent) provided that the interceptor is not within, say, 200 km of either terminal of the meteor-burst system.[11]

Vulnerability To Active ECM

The effectiveness of jamming depends on the parameters of the jamming system, including its relative location, as well as on the characteristics of the communication system. Most important of the jammer characteristics are:

1. Radiated power level;
2. CW, pulse, noise, repeater modes of emission;
3. Frequency agility;
4. Antenna gain, directivity and steerability;
5. Bandwidth;
6. Capability and responsiveness of coordinated interception and tracking facilities (e.g., analysis time); and
7. Ability to adapt in response to changes in target communication signal parameters.

Equally important is the proximity of the jammer to the communication system receiver (or of an intercept or DF station to the communication system transmitter). For example, this may dictate the type of propagation involved, ie., LOS, groundwave, skywave or scatter, and hence the characteristics of wanted and unwanted signals. If these characteristics, such as fading, can be considered as independent, then this may afford a method of discriminating in favor of the wanted signal at the expense of the unwanted.

Another factor influenced by the disposition of ECM and communication system terminals is propagation time delay, assuming that the jammer has access to coordinated intercept and tracking facilities and that the parameters of the communication system can be changed (e.g., frequency-agile pulse mode).

The degree to which the communications system performance may be degraded as a result of jamming will, of course, depend also on its characteristics in the following aspects:

1. Transmitter power;
2. Modulation type;
3. Bandwidth;
4. Coding for error protection (error detection and correction);
5. Degree of frequency agility;
6. Synchronous or non-synchronous operation;
7. Receiver design (filters and synchronization circuitry);
8. Antenna gain, directivity, steerability and ability to form nulls in direction of a jamming signal;
9. Use of diversity processing;
10. Capability and responsiveness of coordinated surveillance and analysis facilities for on-line threat assessment; and
11. Ability to adapt signal format and data rate in response to changes in the nature of the jamming threat.

Items (1) to (11) above are concerned with the characteristics of particular NATO communications media. In addition, one must also consider the relationships between the various NATO media and the organization of the overall communication network incorporating these media.

Overall Communication Systems Considerations

The important factors to consider under this heading are:

1. The provision of the correct mix of communications media and techniques to maximize resistance to the ECM threat at an acceptable cost.

2. The alternative routing and control strategies which must be available to combat the non-avilability of certain links—either through physical destruction or ECM activity. Here it is important to consider the provision of robust and secure system control links to allow adaptive operation and reconfiguration. In practice, the quantity of system control data to be passed may be relatively small, requiring only low capacity links. Furthermore, it may not be necessary or desirable for control of a given communication medium to be affected by a control link using that same propagation medium.

3. The ability to pass the minimum essential operational traffic in the face of the maximum ECM threat. This may require the application of adaptive on-line data compression techniques to ensure that the data to be passed are matched as nearly as possible to the available channel capacity. It must be borne in mind, too, that some of this capacity will be needed to provide redundancy to combat the ECM threat.

In optimizing the performance of NATO communications in an ECM environment, little can be done to influence the ECM equipment characteristics. However, a reliable estimate of the potential EW threat is essential for the development of optimum ECC strategies and techniques.

Specific Techniques for ECM Resistance

We shall now review briefly specific techniques which are being, or could be, applied to NATO or other communication systems to enhance their ECM resistance.

1. *SATCOM Links*: ECM against SATCOM systems can be of basically two types: those directed against the satellite itself and those directed against a specific ground terminal. A third category, directed against all ground terminals of a system from a satellite, could be postulated, but we shall not consider this possibility any further here.

ECM can also be divided into two categories in another way: ECM directed against the control of the SATCOM system or its elements (e.g., beacon tracking, satellite and ground segment control, system timing) and jamming directed against the communications signals themselves. The command link used for maintaining the position and health of the satellite and any communication link used for the control of the entire SATCOM system [13] can be encrypted against spoofing and may even be protected against ECM. The SATCOM control system should, of course, be designed not to depend heavily on a centralized control. The satelliteborne which may be employed by ground terminals for automatic tracking and also for central timing should and is protected against ECM by on-board filtering.[5]

ECM against ground terminals may be applied from local ground based and seaborne jammers as well as from aircraft. The use of antijamming (AJ) modulation, discussed below, coupled with the means and procedures for automatic indications and location of such jammers may help to shorten the peroid during which communications to the particular terminal may be disrupted.

As far as the jamming of the satellite is concerned, the following equipment techniques can be used singly or in combination to provide protection of the communications links accessing the satellite:

- Spread spectrum modems;
- Increased effective satellite power for the ground terminals (SGT);
- Satellite receive antenna discrimination;
- Processing transponder; and
- Increased effective power in the satellite;

The last four ECCM techniques are self-evident so we shall, very briefly, talk about the first one.

Spread spectrum techniques aim to achieve high resistance to interference by spreading the information over a very wide bandwidth. This is achieved by adding the input data stream with a high speed coded bit-stream. The resulting signal, when applied to a modulator, covers a wide bandwidth (typically many megacycles). Basically, the jamming resistance provided is determined by the ratio of the RF transmission bandwidth to the information bandwidth and can be as high as a factor of a thousand or more.

We, at STC, have designed, built and tested modems which employ spread spectrum techniques not only for satellite application, but also for use in the frequency bands of LF, HF and VHF/UHF to obtain low density output signals with resistance against interference and impulsive noise, to negate or circumvent degradation due to multipath and to effect message multiplexing, as well as precise range measurement.[4]

The performance of whatever ECCM equipment/techniques are to be adopted for a given SATCOM system and against ECM threat should be studied using computer/laboratory simulation methods.[4]

2. *O/H Links*: These links are inherently vulnerable to ECM by virtue of their fixed operating parameters and immobility. ECCM techniques that may be employed include jammer detectors and locators, metal fences around stations, as well as bypassing the affected station by means of chaff.[16]

3. *HF Radio Links*: The properties of the ionospheric propagation medium are such as to offer a considerable potential for ECM resistance. The ECCM techniques which could be applied to HF systems include the following:

- Adaptation of the spread spectrum encoding technique for the HF environment to combat the effects of high level interfering signals or jammers and also to overcome multipath time dispersion by selecting a single mode from a multipath set. We have, at STC, developed a modem of this type which is, at present, undergoing tests.[15]
- Variable rate transmission to take account of time varying properties of the HF channel capacity. This capacity can typically vary rapidly between say 100 bits/sec and several kbits/sec, depending on whether fading effects cause the wanted signal to be much stronger than interfering signals or, alternatively,

cause one mode of a multipath set to be dominant. A burst-mode frequency-agile transmission system would be suitable in these circumstances and would also make use of the communicator's time advantage over the jammer.

- When spare transmission capacity is available, to use it for providing masking transmissions.
- The use of real time channel evaluation (RTCE) to select operating frequencies suitable for the communication link and unfavorable for the jammer's propagation path.

4. *Meteor-burst Links*: Meteor-burst links operating in XHF band possess inherent antijam properties which make jamming from beyond the horizon very inefficient. Investigations carried out at STC show that to jam a meteor-burst link operating at 40 MHz for 90 percent of the time the jammer would be at a disadvantage relative to the communications transmitter in the power ratio of 25 to 1.[11] If the operating frequency were raised to say 80 MHz, the advantage for the link would be in the ratio of almost 100 to 1.

5. *VLF and LF Links*: Unless protected, VLF and LF links have very low resistance to ECM. The extremely stable nature of these propagation media as evidenced by their use for navigational and position location transmissions, could also be exploited in ECM-resistant signal design and processing schemes. For example, full advantage could be taken of the ability to integrate a received signal effectively over long periods in say a very low rate data transmission mode.

Conclusion

It is axiomatic that a determined enemy can jam any radio system that may be devised for use on its own. The ECCM equipments/techniques and procedures discussed above can increase the cost and difficulty of ECM considerably. The appropriate use of redundant and diverse transmission media not susceptible to the same threat in a switched network together with an intelligent routing and precedence system, is the general prescription for the design of a survivable communications network such as NICS.

Footnotes

[1]W. B. LaBerge, "Cooperative Improvement of NATO Communications," *SIGNAL*, February 1977.

[2]L. T. Seith, "C^2 In Allied Command Europe," *SIGNAL*, February 1977.

[3]Staff of Communications Division, "A Proposal for a NATO Integrated Communications System NICS," *SHAPE Technical Center Report TR-86,* February 1971.

[4]A. N. Ince, "Design, Testing and Operation of X-band Satellite Communications System," *IEEE Trans. on Communications,* Volume COM-22, No. 9, September 1974.

[5]E. T. Bobak and R. G. Clabaugh, "NATO Phase III Satellite Design," Easton 77 Conference Proceedings.

[6]A. N. Ince et al., "Specifications of Technical Parameters of the Ground Segment of NATO SATCOM System," *SHAPE Technical Center Report,* CR-25, January 1973.

[7]A. N. Ince et al., "A Review of Scatter Communications Systems," AGARD Proceedings on "Aspects of Electro-Magnetic Scattering in Radio Communications," Cambridge, Mass., October 1977.

[8]Staff of Communications Division, "An Evolutionary Concept for the Enhancement of NICS including Progressive Digitalisation," *SHAPE Technical Center Report,* March 1977.

[9]L. K. Wentz, "Planning the Future NICS," EAS-CON-77 Conference Proceedings.

[10]P. J. Bartholomé and I. M. Vogt, "COMET—A New Meteor-Burst System Incorporating ARQ and Diversity Reception," *IEEE Trans. Communications Technology,* Volume COM-16, April 1968.

[11]A. N. Ince, "Interception of Signals Transmitted via Meteor Trails," AGARD Conference Proceedings on "Aspects of Electro-Magnetic Scattering in Radio Communications," Cambridge, Mass., October 1977.

[12]W. F. Utlaut, "Ionspheric Modification Induced by High Power HF Transmitters," *Proc. IEEE* Volume 63, No. 7, July 1975.

[13]A. N. Ince et al., "Power Control Algorithms for Satellite Communications," *IEEE Trans.,* Volume COM-24, 1976.

[14]A. N. Ince, "Code Division Multiplexing for Satellite Systems," Proc. NATO Advance Study Institute on "Communications Systems and Random Process Theory," published by A. W. Sijthoff, Leyden, Netherlands, 1977.

[15]M. Darnell, "New HF Transmission Techniques" *ibid.*

[16]A. N. Ince, I. M. Vogt and H. Gössl, "Design of Chaff and Chaff-Supported Communications Systems," *IEEE Trans.* Volume COM-24, No. 8, August 1976.

EW in Operational Testing

John B. Marks

This is a time of rapid change for the communications-electronics community. It seems as if new equipment and systems are coming on-line every month, and the state of all the electronic arts is advancing almost too fast for the operations people. The pages of *SIGNAL* are filled with advertisements from manufacturers who have developed better and better means of shooting, moving and communicating. With so many choices available, and so much equipment being developed, the role played by the operational testing organizations becomes increasingly important.

Operational test and evaluation requires extremely careful planning and thoroughness to ensure new items meet the critical tests of combat. The pressure is high, and the pace is rapid. There is a lot of "background noise" in which the voice of the electronic warfare (EW) interests can easily become last. That voice has become much louder in recent years, and there are a number of positive steps being taken to ensure that EW considerations are built into operational testing. One of the voices belongs to a relative newcomer to the community.

The Joint Electronic Warfare Center (JEWC) opened its doors on October 1, 1980. Functioning under the staff supervision of the Joint Chiefs of Staff (JCS) J3, the JEWC was given broad responsibilities for providing support to the Unified and Specified Commands, the military services and Department of Defense (DOD) agencies in all aspects of EW including command, control and communication countermeasures. As a part of these responsibilities, we pro-

This article originally appeared in the April 1984 issue of* ***SIGNAL.*

vide analytical and technical EW support to DOD activities conducting joint operational tests. Our charter allows us to respond to requests to support single service tests, provided we have the resources. In the past three years, the Joint Center has provided varying degrees of support to 44 test programs involving EW or command, control and communications countermeasures (C^3CM) related systems. Our personnel have traveled thousands of miles, and spent thousands of hours on the scene of tests, or participating in planning or post test conferences and discussions. We have a unique, and privileged viewpoint of joint and single service EW testing, and after I have acquainted you with our capabilities and some examples of the kind of support we provide, I will share some of our insights concerning the EW and C^3CM aspects of operational testing.

The Tests Division of the JEWC is composed of 15 personnel. Led by Cdr. Rolland Skiles, USN, these personnel are primarily in pay grades 03 and 04 with backgrounds in airborne EW, communications and electronic engineering, EW analysis, avionics, air control, air defense, naval surface warfare, operations research and intelligence. We also have two civilian electrical engineers employed on a full-time basis to provide both expertise and the long-term continuity that is not possible with military personnel.

When required, personnel from other divisions assist the Tests Division. The Tests Division provides support on the EW aspects of defense system tests by joint or service organizations. Our support extends to all levels of testing from development through operational and follow-on test activities. The nature of our support varies according to the project and is highly consumer oriented. Our involvement can range from one small, specific task to long-term continuing or recurring support activities that require extensive field deployments.

The JEWC does not have an organic instrumentation or data collection capability, but instead, uses equipment belonging to other military and DOD agencies and contract support when necessary. We do, however, have an interest in the equipment. In response to tasking from the JCS, the Joint EW Center compiled a comprehensive catalog of threat simulators.

The Joint Threat Simulator Handbook lists all simulators currently fielded and under development by each of the military services and was compiled from three separate catalogs previously maintained by the Army, Navy and Air Force. The Handbook will be updated annually. The current edition contains an overview of the technical characteristics, system description and locations of the simulators.

The 1984 edition will contain photographs of the simulators and have a much more readable format, with additional information such as telephone numbers of points of contact for each system. The Tests Division of the JEWC has responsibility for maintenance and update of the Handbook as well as maintaining contact with organizations that may be developing new simulator systems. However, the majority of our efforts are not directed to this kind of activity. Let me give you a few examples of what my people have done recently or are currently doing.

When the Tri-Service Working Group was tasked in early 1983 to perform operational test and evaluation of some equipment under the Foreign Military

Equipment Exploitation Program, we were asked to provide an individual to act as Test Director/Coordinator. Cdr. Skiles took on the task himself and spent five months at various test sites.

Additionally, we provided two personnel to give training and assistance to one of the single service portions of the overall test program. After the individual members of the Tri-Service Working Group publish their test results, the Joint EW Center will compile an Executive Summary of these results for the Tri-Service Working Group.

On a much larger sacle, we have been asked by the Army's Operational Test and Evaluation Agency (OTEA) to provide support to their follow-on evaluation (FOE) of the Patriot air defense missile system. Specifically, we were asked to analyze the survivability and vulnerability of the Patriot system regarding electronic, infrared and visual signature. Once we have obtained all of the system characteristics, we review the past test history for areas that might require further or renewed emphasis. Our analysts noted four areas of concern, and we so informed OTEA. We joined with the Office of Missile Electronic Warfare (OMEW) and OTEA in order to coordinate electronic signature analysis and active communications jamming during the FOE.

Concurrently, we arranged for the JEWC Tests Division to coordinate the use of F-4G Wild Weasel and RF-4C TEREC aircraft during the maneuver phase of the evaluation. As an internal action, the Test Division requested the JEWC Directorates of Studies, Analysis, Concepts and Doctrine to update and validate a study of the electronic threat that the Patriot system will face when initially deployed. This study is being done using threat information provided to the JEWC by the Defense Intelligence Agency. In addition, we will perform several other coordination and evaluation functions and will submit a final analysis report to OTEA upon completion of our assistance.

From the Joint EW Center's point of view, there are two types of operational tests. On the one hand, there are tests of equipment and systems such as Patriot or Joint Tactical Information Distribution System (JTIDS), which are hardware oriented. On the other hand, there are concepts tests, such as EW in Close Air Support (EW/CAS), or the command, control and communications countermeasures (C^3CM) Joint Test and Evaluation (C^3CM JT&E) which may use hardware, but are focused on the success or failure of tactics or ideas.

Each of these tests represents a set of unique challenges and problems for the planners and executors. In many instances, the Test Division of the JEWC has been closely working with these people, and we have gained a deep appreciation for their flexibility and dedication. In the course of planning and running a test of either equipment or concepts, a variety of problems may arise. These problems seem to be related to the situations of testing, and not so much to the composition of the testing teams. They do seem to differ between equipment and concept testing. I will address the problems separately, beginning with the testing of concepts.

We have noted four major problems in testing concepts such as EW/CAS or C^3CM:

1. Control of variables in live tests;
2. Difficulty in achieving realism;
3. Use of exercises as tests; and
4. Collecting accurate data.

Control of Variables

When testing EW tactics or concepts, there are a number of variables which impact the final performance assessment. Power levels of transmitters, weather, terrain, operator behavior, relative positions of participants and even the time of day can have an effect on the success of an EW engagement. This becomes even more difficult when aircraft are used as part of the test. Even though pilots thoroughly understand the need to fly at a precise speed or altitude, flight conditions may force deviation from the plan. The deviation may result in the loss of instrumentation coverage or make it impossible to judge whether or not a change in EW effectiveness was due to a preplanned change, or the unplanned variation.

In a test that simulates the dense EW environment we expect to encounter on a modern battlefield, the chances of avoiding random and damaging variables become very small. When this damage occurs, the Test Director is forced to choose between repeating an expensive portion of the test, or trying to do the best he can with the compromised data. Careful planning and prior review of previous tests to find "lessons learned" information will minimize the damage caused by failing to control as many variables as possible during live testing.

Difficulty in Achieving Realism

This problem is faced by an organization trying to duplicate real work situations, whether for training or testing. Duplicating the real dangers of a battlefield or obtaining large quantities of enemy equipment is out of the question. This brings up the cost trade-off between extensive simulation and comprehensiveness of the test. EW threat simulators are expensive to build and operate, and it is not cost-effective to assemble large numbers of military personnel and equipment for the sole purpose of conducting EW testing over a long period of time.

Limiting the assets used in order to minimize expenses tends to limit realism. Five aircraft, conducting three attacks each on simulated air defense positions, is not the same as engaging those positions with 15 aircraft at once. Failing to force participants to perform within real world constraints, such as reload times or availability of missiles, is a correctable, but often seen flaw which compromises realism in testing. Once again, careful planning and intelligent consideration of trade-offs is required to ensure that the test closely reflects reality.

Exercise as Tests

In view of the cost and realism constraints facing the Test Director, the ideal solution would be to use one or more of the regularly scheduled military exercises as portions of a test of EW tactics or concepts. In general, however, each time this has been done the results were less than satisfactory.

This is due mainly to the differing goals of the two events. The purpose of a military exercise is the training of the participating forces. The purpose of a test is essentially data collection under controlled conditions. The participants in the exercise cannot be expected to put the same emphasis on complying with the requirements of the test as they would on doing their best in the exercise.

This situation introduces unexpected variables in the test and usually produces unusable data. Although using an exercise as a data collection opportunity appears to be an excellent idea, in the experience of the Joint EW Center, it has only a minimal chance of succeeding, even with the most meticulous planning and instrumentation.

Accurate Data Collection

The collection and reduction of test data is central to any test effort, be it a test of equipment or concepts. The significant difference is that it is much easier to devise instrumentation for equipment tests than for testing intangibles such as EW effectiveness. Part of the difficulty lies in the establishment of measures of effectiveness (MOE), which leads one to determine what data must be collected during the test. Presupposing that the MOE are correct, then the problem of collecting data must be addressed.

In many cases regarding EW or C^3CM effectiveness, data consists of such things as response time, or messages missed, which do not lend themselves to automated collection. This requires the use of human data collectors, who must be trained to do the job. Unless the criteria for data elements are rigorously defined, the results may reflect subjective estimates on the part of the collector and therefore produce an inaccurate picture of what actually occurred.

We have found a tendency for the larger problems of testing, such as equipment status or aircraft availability, to overshadow the importance of proper training for data collectors. This training is one of the support activities for which the personnel of the Tests Division of the JEWC are often tasked.

Although the Joint EW Center was commissioned well after the EW/CAS test began, we were fortunate in having several of our personnel closely associated with the tests up until their completion in 1982. We have been involved in support of the current C^3CM Joint Test since before the Joint Test Force (JTF) was established.

In response to tasking from the Technical Advisory Board, the Joint Center did extensive research preparation for the creation of the JTF. The project encompassed previous tests, exercises and studies; current C^3CM concepts and scenarios; computer models and a list of all available military hardware and electronic systems of all the Services. After compiling all of this information, we prepared a report with recommendations for use by the C^3CM JTF. The recommendations in the report were intended to assist the JTF Director in avoiding or minimizing the problem I have addressed here.

When the Joint Electronic Warfare Center was first established, our original intentions were to provide operational test directors with assistance in the selection of instrumentation and range capabilities that they would need to meet the requirements of the particular test. We were willing, and to a degree able, to provide independent observation and evaluation of the electronic warfare effects that occurred during the operational testing of equipment or systems and to perform computer modeling, or other analytical services before, during or after the testing.

Most of all, we envisioned ourselves as having the opportunity to provide a repository of ideas and lessons learned that would allow us to raise the overall effectiveness of all Service, joint or DOD operational testing (at least in the area of EW effects or considerations) by carring the best ideas from one test to another. It was for just this kind of thing that the Joint EW Center was formed. What we have discovered is that the needs of the operational test agencies are much more extensive.

In our initial evaluations of several test plans that were submitted to us, we found that electronic warfare effects are either not considered or have been substantially underrated regarding the possible degradation of system performance. After providing assistance to a variety of test groups, we have begun to pinpoint where some of the problems lie. I will describe the major difficulties as generic examples only, pointing no fingers at any particular test or agency. The Joint EW Center works hard to avoid being anything but an advisory or assistance organization and has no desire to tell any Test Director or organization how to conduct their business. We do, as I mentioned earlier, enjoy a privileged overview of the entire spectrum of operational testing, and I believe that our insights are worth considering.

There are four significant problem areas:

1. Inadequate knowledge of EW on the planning staff;
2. Inadequate threat definition;
3. Conflicting goals between operations and procurement; and
4. Lack of sufficient time for testing.

Inadequate EW Knowledge

Operational testing is usually a temporary or short duration duty assignment for a planning staff position. Personnel are frequently assigned to the staff on the basis of availability not skill. Therefore, in many cases, personnel assigned to the planning staff have no knowledge of EW. Planning staff are normally disbanded following completion of their assigned tasking, making it unlikely that they will have the opportunity to see the results of their efforts or to learn anything from any problems or mistakes in the planning process. Therefore, no matter how well intentioned the planning staff is, there is a good chance that their plan will overlook one or more significant aspects of the proposed test, most frequently EW.

There is no easy solution to this situation, as staff selection is driven by other organizational requirements. When the Secretary of Defense directed the establishment of the Joint EW Center, it was to minimize the difficulties caused by such shortfalls. We have the requisite continuity and can be tasked by any test planning staff director to assist them right from the start in ensuring that any EW considerations are built into the plan.

Inadequate Threat Definition

The general threat definition for a particular piece of equipment is usually established during the procurement state prior to production. These definitions are based primarily on what we know at the time and are projected to the estimated initial operational deployment of the equipment, which may be as much as 10 or 12 years from the time the threat definition is done. Obtaining an updated threat definition is a responsibility of the procurement community.

Updating the threat definition is an expensive, time consuming process that can result in pointing out requirements for major, and possibly very expensive, changes to be made in the system being developed and is therefore seldom done. By the time the system is ready for operational testing there are too many hours and dollars invested in it for any change to the original threat definition to be at all welcome, and it is unlikely that the intelligence community would be willing to invest the time and effort to do a new definition when it is nearly certain that no appreciable change in the system would be made. While this situation is certainly less than desirable, it is a fact of life which continually confronts the operations testing community.

Conflicting Goals

The essence of the operational test is to exercise the new equipment in a wartime environment with service personnel and procedures. Any problems identified reflect directly upon the manufacturer and the Procurement and Development Agency, both of whom have already tested the equipment to their satisfaction. The goal of these communities is to provide the piece earlier, with the capabilities that were requested and the ability to deal with the threat as defined at that time. With rare exceptions, they do their jobs very well, and are justifiably pleased with their product.

On the other hand, the operations personnel of the Services are concerned with the current or even projected threats and look forward to testing the system under the most realistic conditions possible. Their goal is to deploy equipment as close to the current state-of-the-art as possible. Both groups are correct and are doing their jobs to the best of their ability. Unfortunately, they often end up in direct conflict over the operational testing of new equipment, and again, especially over the EW aspects of the test. This is due to the rapidly changing state of both EW threats and countermeasures capability.

There is no easy answer to this conflict, but the ongoing efforts being made to shorten the development and procurement cycle have the potential to minimize the problem.

Lack of Time for Testing

This is an end result of the problems I have already addressed. Even though the operational testing of new systems or equipment is designed to ensure that the new system will perform as expected when deployed, there is a tendency on the part of the operational organizations to want to deploy the system as soon as possible. After all, there is a definite operational need for all of the systems we have coming on line, and the field commanders can hardly be faulted for wanting the testing done as quickly as possible.

In this instance, the pressure reinforces the existing pressure from the procuring agencies, who already know that the equipment meets specifications and are eager to get it into production. It takes a firm stand by the Test Director to ensure that the system is properly tested within time and cost limitations. In the Joint EW Center, we are fortunate in being able to show the interested organizations the results of cutting too many corners and to help the Test Director devise as comprehensive a test as possible in the time allotted.

Conclusion

Lest I leave the impression that operational testing universally fails to take into account the effects of electronic warfare, let me point out that to the contrary, things have never looked better. Certainly, there is increased awareness of the importance of EW considerations in today's operational testing community. The fact that the Joint EW Center is contacted on a daily basis for advice, assistance or other support is proof of that. The fact that DOD would dedicate funds to test the concept of C^3CM in the current Joint C^3CM Test and Evaluation in order to determine the optimum blend of EW and lethal capabilities is encouraging.

The fact that all four military services are also conducting tests or studies to try to determine the best courses of action for future research, development and procurement of military equipment to support C^3CM is an extremely healthy sign. The fact that organizations such as the Armed Forces Communications and Electronics Association (AFCEA) are growing in strength is further proof that EW and C^3CM are no longer concepts for the future. We have entered a new age of EW and C^3CM emphasis in military thought and action, and the operational testing organizations are in the forefront of the change.

In short, the road has not been smooth, but there has been great progress made in leveling out the rough areas. Best of all, we are rapidly finding the answers and solutions to longstanding questions and problems, and I have no reason to believe that we will not continue the upward trend.

A New Airborne ECM System for the 1990s and Beyond

James V. Stiles

ASPJ—the answer to a fighter pilot's prayer? Well, maybe not, but it has to rank pretty high on the list of capabilities he wants with him when he rolls in on a target.

The Airborne Self-Protection Jammer (ASPJ), designated AN/ALQ-165, is an electronics countermeasures package designed to protect U.S. tactical aircraft against terminal threat defenses through the turn of the century. It will be mounted internally on the Navy's F-18, F-14, A-6 and EA-6B aircraft and on the Air Force's F-16. A pod version will be developed for the Navy's AV-8B and some of the components will be repackaged to provide a power management capability for the Air Force's AN/ALQ-131 Electronic Countermeasures Pod.

History

The ASPJ concept has its origins within the Navy's Dual Mode Traveling Wave Tube Jamming Program in 1969 and the Air Force's lightweight, low cost countermeasures effort in 1975. Joint Navy/Air Force development of an internally mounted self-protection jammer for advanced tactical aircraft was

---

**This article originally appeared in the April 1973 issue of SIGNAL.*

initiated as a result of Director of Defense Research and Engineering (DDR&E) direction in March 1975.

Subsequently, the Navy and Air Force agreed to pursue the definition of a common internal self-protection jamming set for the F-16 and F-18 as well as other aircraft. This agreement established the ASPJ Technical Coordinating Committee and an ASPJ Steering Committee in mid-June 1976 calling for a joint Navy/Air Force technical specification and management approach to the project. The Navy was designated the lead Service for the program in January 1977 and the Navy/Air Force Advanced Tactical Aircraft Protection System (ATAPS) joint program office (JPO) was established within the Naval Air Systems Command in December 1979.

The relationship of the JPO to the various organizations involved in the program is quite complex but can most easily be described in terms of the signatories to the memorandum of agreement among the Service organizations.

The Commander, Naval Air Systems Command; the Commander, Aeronautical Systems Division, Air Force Systems Command; and the Commander, Acquisition Logistics Division, Air Force Logistics Command have jointly signed a memorandum of agreement which outlines responsibilities and the program management relationships among all interested parties. They form the basic hierarchy above the joint program office and provide the foundation upon which the JPO derives its authority to task field support activities of both Services to assist in the development and testing of the equipment. In addition, 21 joint Service working groups have been formed to ensure that all aspects of the program (i.e., integration, testing, configuration, software, etc.) receive the proper degree of attention.

Why ASPJ?

The ASPJ will enhance the probability of survival of U.S. tactical aircraft in the face of the growing Soviet threat in radar terminally guided antiaircraft weapons. Depending upon the mission scenario these aircraft will be exposed to a complex mix of antiaircraft artillery, surface-to-air missiles and air-to-air missiles. This situation imposes a high risk to aircraft without electronic countermeasures (ECM) equipment. With ECM, the jamming substantially slows down and often completely prevents the enemy weapon system from being successfully launched.

The ASPJ uses two basic techniques for self-protecting the tactical aircraft, noise jamming and deception jamming. In noise jamming, the ASPJ will simply saturate, or overpower, the enemy radars. In the deception mode, the ASPJ will essentially project an electronic image of the tactical aircraft in some other bogus location, seducing enemy missiles and antiaircraft batteries into attacking a "ghost." The ASPJ will respond to a dense radar environment automatically handling multiple threats simultaneously.

All radiating electronics can be jammed to some extent. The role of the ASPJ equipment will be to detect and identify the presence of threatening radars in real time and, by combining noise jamming and deception with maneuver and defense suppression weapons tactics, degrade the effectiveness of the enemy.

The ASPJ will not really add any new capabilities to the repertoire of electronic countermeasures available to today's fighter pilot but it will provide substantially more of them in a smaller, updatable package. The basic ASPJ consists of five pieces of equipment including high band and low band transmitters, high band and low band receivers and a central processor unit. This configuration will be used primarily on the F-18. The common configuration has augmentation units consisting of an additional high band receiver and transmitter which will be used on the F-16 and other aircraft.

The ASPJ covers a broad frequency spectrum to provide real time, power managed jamming in a stand alone configuration or in response to inputs from a radar warning receiver. The system has dual central processors with distributed control features to accomplish special, time critical tasks and to provide built-in test capabilities.

The system is expected to experience a high degree of reliability with a low failure rate and low maintenance times. The built-in test capability will detect 98 percent of the system failures and will isolate those system failures to the failed Weapons Replaceable Assembly/Line Replaceable Unit (WRA/LRU) approximately 99 percent of the time. The system architecture has been developed so that the ASPJ is highly modular in nature, fully reprogrammable and contains room for future growth.

ASPJ Modularity

Modularity has been a design criteria in five different areas:

1. *Aircraft:* ASPJ utilizes tailored, blindmated racks to adapt unique aircraft mechanical/electrical requirements to the ASPJ units.

2. *Avionic:* ASPJ interfaces with multiple avionics data busses on the aircraft and has provisions to work with several different radar, expendable, radar warning receiver, missile and data link systems. Additionally, ASPJ accepts both direct and "smart" blanking in several formats.

3. *Equipment:* ASPJ contains seven pieces of equipment which can be assembled in different configurations depending upon the requirement. Each unit can be functionally upgraded/modified without affecting the whole system.

4. *Module:* ASPJ is divided into 56 separate plug-in modules which can be functionally replaced or upgraded as required. (Examples are exciters, microprocessors, filter assemblies, instantaneous frequency measurement module and the traveling wave tubes.)

5. *Software:* ASPJ software has a top down structure and is fully modularized with most modules averaging 200 words. Furthermore, operating software programs (i.e., Operational Flight Program) have been separated from threat and technique data (i.e., User Data File), allowing independent updates.

Unlike most EW systems today, the ASPJ will have electrically alterable programmable memory (EAPROM). This will facilitate changes in technology and the threat environment to be reflected in both the Operational Flight Program and the User Data File.

Finally, ASPJ has both the physical room for additional hybrids, components and circuits, and the memory capacity for additions or changes in the future.

Since electronic technology and user requirements change so fast, it is imperative that any ECM system fielded today be sufficiently flexible to adapt to these changes as they occur so that it does not become prematurely out-of-date. A Preplanned Product Improvement (P^3I) program has been established within ASPJ to prevent changing requirements and advances in technology which occur after the design is frozen from causing the gradual obsolescence of the equipment.

Generally the P^3I initiatives represent technology risks if incorporated immediately in full scale development. But when mature, they should result in a substantial operational capability. Most of the potential candidates are currently under advanced development. They include studies of new threats, new techniques, Very High Speed Integrated Circuitry (VHSIC) and other hardware improvements.

The P^3I program has addressed both near and long-term objectives. The near term objective is to insure that the ASPJ has the growth, reserves and interfaces to be easily modified when these technology initiatives mature. The long-term objective is to provide the resources, planning and system integration effort to efficiently insert the product improvement into ASPJ. The entire P^3I effort is closely coordinated between the Services to insure maximum advantage is taken of maturing technologies.

Acquistion Strategy

Historically, ECM equipment has been acquired for a very narrow objective, purchased in very limited quantities and often from a sole source contractor. Although expedient, this has resulted in a rather haphazard and expensive approach to buying ECM equipment. Congress has made it clear that the Department of Defense should pursue a more generic ECM suite with application across a variety of tactical aircraft in an effort to keep costs down. To implement this direction, it was necessary to address both technical and business aspects of the acquisition strategy. The technical aspects included the various system parametric requirements to meet the threat along with considerations of cost and risk. From this, the technical specification was developed.

Equally important were the business considerations. Since the contractor actually designs the equipment, it was necessary during the market analysis to compare the size of the industrial base and the production capability of each candidate contractor with the potential quantity and cost of the total procurement. Incentives and source selection criteria were chosen very carefully to provide program flexibility and to focus industrial interest on the program goals.

Four primary objectives were included in the acquisition strategy. First, design competition was seen as a way of optimizing the results of the early program investment. Second, since the design competition usually only produces a paper design, certain hardware risk reduction requirements were included, so that those areas of the design that were considered risky would actually be demonstrated with hardware and in a competitive environment. Third, a high initial production rate was considered essential due to the high number of outstanding requirements that needed to be satisfied. Finally, and probably most importantly, sustaining competition even in production was seen as a way of obtaining the best price for each initial and follow-on production lot.

Developing competition is a very effective way to keep costs down. With that in mind, competition was created from the system level down through the component level. The prime contractors have been required to develop multiple sources for components and subassemblies. This has the effect of reducing costs and forcing a wider involvement of the industrial base.

A plan was devised to split the production of ASPJ among existing companies—fostering competition but maintaining the broad industrial base. In this competitive environment, the high production rate required for this program can be achieved. Competitive production was established by creating a team of contractors where both members would have the capability to build the ASPJ to the same specification.

The contractor teams were to act as a single entity, providing a single proposal in a joint venture. They would mutually support each other through the system design, development, fabrication and test phases and then compete for portions of the production contract.

Graphically, the teaming concept is shown in Figure 1. After some initial industry studies were conducted, proposals were solicited from teams of contractors. Two teams were awarded identical contracts for the design phase of the development. The designs (developed in parallel to one set of specifications) were then compared, and one team (ITT/Westinghouse) was selected to

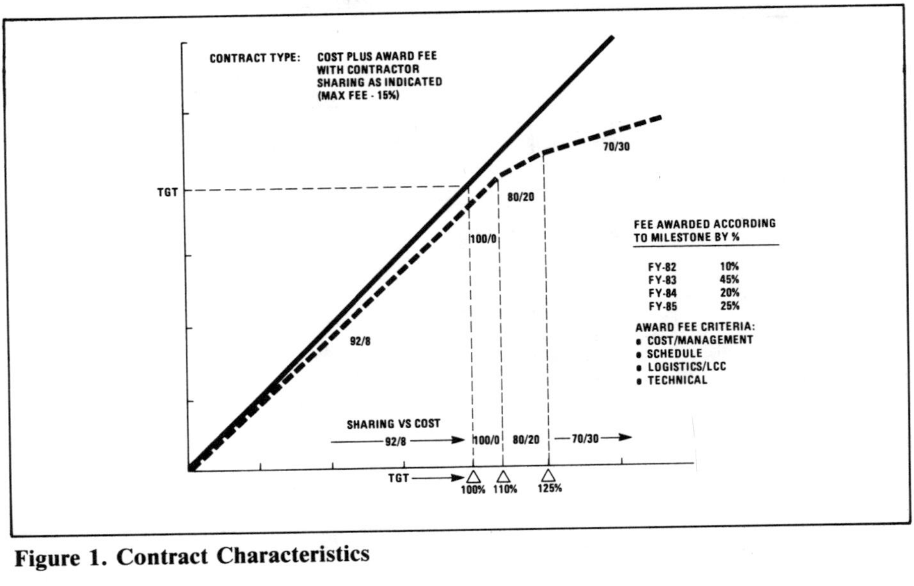

Figure 1. Contract Characteristics

continue as a joint venture through full-scale development, fabrication and test.

Upon completion of testing, the joint venture will be dissolved and ITT and Westinghouse will then compete as independent vendors for disproportionate shares of the production contract (e.g., 60 percent to the low bidder and 40 percent to the high bidder).

The advantages of this teaming concept are that each contractor will have the capability of producing the entire system, thereby providing a synergistic development capability which takes advantage of a wider range of available technology. This wider range of technology will result in a better system at a lower technical risk than would be produced by a single company. Additionally, it results in two fully qualified sources.

Recognizing the risks inherent in developing a complex piece of equipment which utilizes the most current technology, it was necessary for the development to be conducted under a cost type of contract. However, a rather unique cost sharing arrangement has been devised which was attractive to both the government and the contractor. Figure 2 depicts how this arrangement works.

The basic development contract was awarded for $80.5 million, but the contractor will pick up eight percent ($6.4 million) of the development costs. Government/contractor share ratios in the event of an overrun vary at the rate of 100/0 for the first 10 percent of overrun, 80/20 for 10 to 25 percent overrun and 70/30 for any overrun in excess of 25 percent. Further performance incentives of up to $12 million are available over a number of award fee periods.

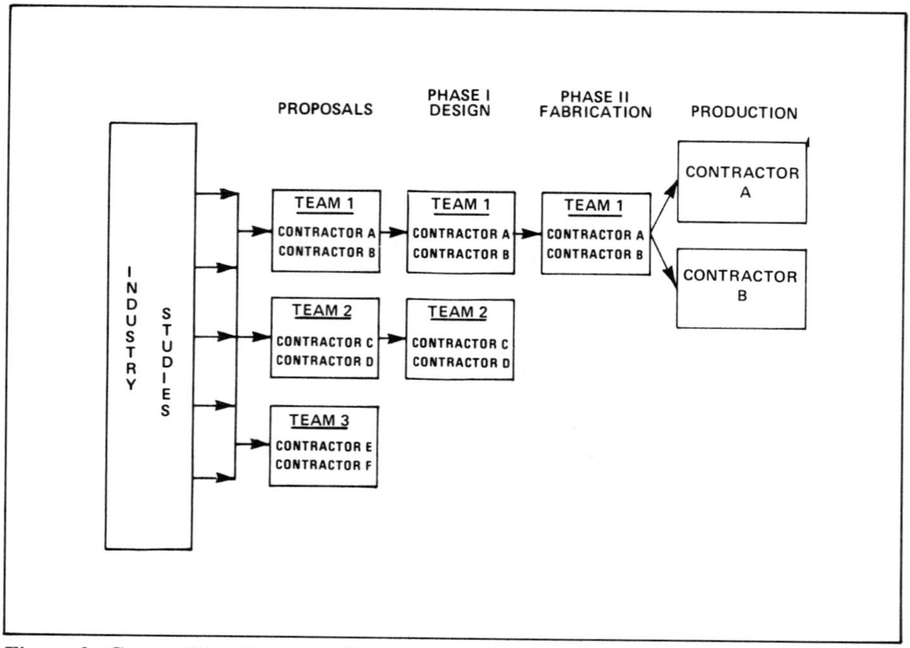

Figure 2. Competitive Teaming Concept

Current Program Status

The joint venture has successfully completed several major program reviews and the equipment is well into fabrication. The next major milestone will be the delivery of the first test units which should occur this fall. An extensive laboratory and field test program will follow with a limited production run to begin based on the successful completion of development testing. A full-scale production decision is expected in early 1986. At the quantities presently planned, production contracts should total $3-4 billion.

In summary, the ASPJ has been carefully designed to provide ECM protection for a variety of combat aircraft for many years to come. It is the primary airborne ECM system for the future and is being procured using a unique acquisition strategy. The program objectives, the wide degree of industry involvement and the level of competition have been tailored to satisfy the needs of the fighter pilot within the fiscal and programmatic constraints imposed by Congress and the state-of-the-art.

About the Authors

Colonel William E. Bettis, USAF (Ret.), served 33 years in the Air Force, including combat tours in WW II, Korea and Vietnam. He held various assignments in the communications-electronics field — Research and Development, Operational Commander, and Systems Acquisition. Since his retirement in 1975, Colonel Bettis has remained active in the communications-electronics area serving as a consultant to several industrial organizations.

Lieutenant Commander Michael M. Blume, USCG, is currently serving as Commanding Officer, Communication Station Miami. He was formerly on the Commandant's staff in the Plans and Policy Division of the Office of Command, Control, and Communications. He also served as Commanding Officer, U.S. Coast Guard LORAN Station Wake Island and the Operations Officer of the cutter Cherokee. He attended the Coast Guard Academy and the Naval Postgraduate School.

Dr. Frederick E. Bond is currently a Principal Engineer with Aerospace Corporation, engaged in satellite communication system architecture and planning activities. Earlier associations at Aerospace Corporation include the TACSAT and FLEETSATCOM satellite programs. From 1973 to 1977 he was a Deputy Director of the Defense Communication Agency, with primary responsibility for all MILSATCOM Systems Architecture. Dr. Bond is a Fellow of the IEEE, an Associate Fellow of AIAA, a member of Sigma Xi, and a retired Lieutenant Colonel in the U.S. Army Reserve.

Dr. Jon L. Boyes, Vice Admiral, USN, (Ret.) is the International President of the Armed Forces Communications and Electronics Association (AFCEA). Retiring from the Navy after a distinguished 34 year career, Admiral Boyes has experience in destroyers and submarines and in the technical and operational command, control and communications fields. His educational background includes graduate studies in communications and computer science, a Masters degree in International Law and a Ph.D. in International Affairs from the University of Maryland.

Colonel Alan D. Campen, USAF (Ret.), is currently serving as Director, C^2 Policy, Office of the Under Secretary of Defense, Policy. He was formerly Assistant Vice President, C^3I Programs, the BDM Corporation. Key assignments in a 25 year Air Force career included Commander, DCA-Europe; Chief of Staff, AFCS; Commander, UK Communications Region; and DCS C-3, Third Air Force. He served as President, London Chapter and European Region Vice President, AFCEA, and is currently on the *SIGNAL* Magazine Advisory Board. He was a radio operator in a Navy amphibious communications unit in the Pacific in World War II.

General Paul K. Carlton, USAF (Ret.), is currently a Consultant to the Martin Marietta Corporation, Denver Aerospace. Since his retirement, he has also served on the Board of Directors for Missouri Pacific Corporation and Union Pacific Corporation. Thirty-six years of active duty included wartime service in the B-29 program against Japan. Twenty-seven years of S.A.C. service in many great units ended as Commander, 15th Air Force. The highlight was four years as aide and pilot for General C.E. LeMay. General Carlton's last assignment was as Commander, Military Command, from 1972-77. General Carlton is a graduate of the National War College. He has logged over 13,000 flying hours on aircraft including the B-52, B-58, SR-71, C-5 and C-141.

Dr. John Clark has been Director, Space Applications and Technology, RCA Corporate Engineering, since 1976. For the previous 11 years he was Director, NASA Goddard Space Flight Center. He was Chairman of the FCC RARC-83 Advisory Committee and a member of the U.S. delegation to the ITU western hemisphere broadcast satellite planning conference. Dr. Clark was General Chairman of the 1984 AIAA Communications Satellite Systems Conference. He is a consulting editor, space technology, to McGraw-Hill. He holds a Ph.D. in Physics and a Professional E.E. degree. He was awarded the NAA 1974 Collier Trophy and NASA medals for Distinguished Service and Outstanding Leadership. He is an elected Fellow of AIAA and IEEE and a full member of the International Academy of Astronautics.

William Curry is Director, Satellite Operations, for GTE Spacenet Corporation which recently placed in revenue service the first two of five planned domestic communication satellites. He served 20 years as a Navy surface line officer and communications specialist where he played a key role in the early introduction of satellite communications to fleet operations. He holds degrees from the Naval Academy, the Naval Postgraduate School (Communications Engineering) and Southern Illinois (M.B.A.).

General Robert J. Dixon, USAF, served as Commander, USAF Tactical Air Command. Past positions include assignments as Division Commander, 45th Air Division, SAC; Assistant Deputy Chief of Staff, Personnel, for Military Personnel; and Commander, USAF Military Personnel Center. In 1969, General Dixon served in the Republic of Vietnam as Vice Commander, Seventh Air Force, and flew 36 combat missions. From 1970 to 1973, General Dixon was Deputy Chief of Staff, Personnel, HQ USAF. In 1973 he was assigned to the Air Force Tactical Air Command. He retired from the military in 1978, and then served as President and Chairman, Fairchild Republic Company from 1978 to 1982. He is now President, Confer Associates and a private aerospace consultant.

General Russell E. Dougherty, USAF (Ret.), has served as Commander in Chief, Strategic Air Command; Director, Strategic Target Planning; and Chief of Staff, Supreme Headquarters Allied Powers Europe. Currently he is Executive Director, Air Force Association; Publisher, Air Force Magazine; and a member of the Defense Science Board. In addition, General Dougherty is a Director of Internorth, Inc., and the Atlantic Council of the United States, as well as a Trustee of Aerospace Corporation.

Major General Robert L. Edge, USAF (Ret.), is an active Consultant with the U.S. defense industry. He retired from the position of Assistant Chief of Staff, Communications and Computer Resources, HQ USAF, in 1977. Prior to that, he had been Deputy Chief of Staff, HQ AFCS, and then Director of Command, Control and Communications, HQ USAF. In all three of these positions, he had important responsibilities for developing, defending and/or supporting Air Force requirements for what is now called C^3. He is also a Command Pilot with over 6,000 flying hours in various operational commands.

General Richard H. Ellis, USAF (Ret.), is currently the U.S. Commissioner on the US-USSR Standing Consultative Commission. He holds the rank of ambassador. In active duty, he flew over 200 combat missions in World War II. Later, his positions included Commander, 315th Air Division; Deputy Director, J-5 (Plans and Policy); Commander, 9th Air Force; Vice Commander in Chief, USAF, Europe; Commander, 6th Allied Tactical Air Force; Commander of Allied Air Forces, Southern Europe; Commander, 16th Air Force; Vice Chief of Staff, USAF; Commander, Allied Air Forces, Central Europe; Commander in Chief, USAF, Europe; and Commander in Chief, Strategic Air Command.

Lieutenant Colonel Gerald L. Fjetland, USAF (Ret.), is currently a Senior Systems Engineer with the Communication Systems Division of GTE Government Systems Corporation located in Garden Grove (Orange County), California. During his 26 years of active duty, through July 1983, Colonel Fjetland served in various communications and electronics positions ranging from Radar Maintenance Technician on B47 aircraft to R&D staff assignments in AFSC. In his last assignment, Colonel Fjetland served with the MILSTAR System Program Office at USAF Space Division. Colonel Fjetland is a graduate of the University of Arizona and the University of Southern California.

Lieutenant Colonel Alfred R. Garcia, Jr., USAF, is currently serving as Commander, 1936th Information Systems Squadron (AFCC) and Assistant Chief of Staff, Command, Control and Communications (J-6), United States Forces Azores, Lajes Field, Azores, Portugal. Previously he was Chief, Satellite Operations Division, Defense Communications Agency, Washington, D.C. He has also served in command and staff positions in Air Weather Service; the MAC Inspector General; Air Training Command; Air Force Logistics Command; and Air Force Technical Applications Center. He is a graduate of Squadron Officer School; Air Command and Staff College; Industrial College of the Armed Forces; and Air War College programs. He holds advanced degrees from the University of Southern California and Auburn University.

Lieutenant General Gordon T. Gould, Jr., USAF (Ret.) (d.-1979), graduated from the U.S. Military Academy in 1941. He served in China in World War II, where he was responsible for establishing a communications and control system for support of air operations. He also served with the Air Research and Development Command, the Strategic Air Command, and as the Director, Command Control and Communications for the USAF. General Gould concluded his military career by serving as Director, Defense Communications Agency (DCA).

Dr. Grant L. Hansen is Corporate Vice President of System Development Corporation (SDC), a Burroughs Company. Prior to joining SDC in 1978, he served with the U.S. Navy in World War II, worked for Douglas Aircraft and General Dynamics and, from 1969 to 1973, was Assistant Secretary of the Air Force for R&D. While at General Dynamics, he was responsible for the development of the Atlas/Centaur space launch vehicles. In 1975, he served as National President of the American Institute of Aeronautics and Astronautics.

General James V. Hartinger, USAF (Ret.), retired on August 1, 1984 as Commander, USAF Space Command and Commander in Chief, the North American Aerospace Defense Command. He was the first recipient of the USAF Master Space Badge, received the General Thomas D. White Space Trophy for his contributions to aerospace programs, and was presented the highest peacetime award, the Defense Distinguished Service Medal. As a career long fighter pilot, General Hartinger flew the F-84 in the Korean War, the F-4 in Vietnam, was the F-111 Test Director, and was Commander of the famous 23rd TAC Fighter Wing Flying Tigers.

E. Larry Heacock currently is Director of Satellite Operations, National Environmental Satellite, Data and Information Service, National Oceanic and Atomospheric Administration, Department of Commerce, operators of the GOES, NOAA and Landsat satellites. Prior to this position, he was Director of Systems Development in NESDIS (formerly NESS) for eight years and was with the European Space Agency as spacecraft manager for the development of the METEOSTAT, Europe's geostationary weather satellite.

Lieutenant General William J. Hilsman, USA (Ret.), is currently President and Chief Executive Officer, International Mobile Machines Corp. His military career included assignments as commander of operational communications units in Vietnam, Germany and the United States; assignments in the field of computer science; head of the U.S. Army Computer Systems Office; high technology project manager for the Army; Commander of the U.S. Army Communications, Research & Development Command at Fort Monmouth, New Jersey; Commander, U.S. Army Signal Center; and Manager, National Communications System.

Dr. A. Nejat Ince has served as Chief of the Communications Division at SHAPE Technical Centre (STC), the Hague, The Netherlands. He joined STC in 1961 and worked as a scientist and later headed their Radio Branch. Dr. Ince was involved in the development of almost every international communications system for NATO. He was the leader of a team that developed the concept for the NICS. Prior to joining STC, Dr. Ince worked at the engineering departments of The Turkish PTT and Broadcasting Administrations. He received his B.Sc. and Ph.D. degrees from the Universities of Brimingham and Cambridge, UK, in 1952 and 1955, respectively.

Nicholas P. Janus is currently involved in C^3 systems engineering for the Mitre Corporation. He continues to be actively involved in military command and control applications. Recent contributions include the upgrade of C^3 capabilities for major Army and Unified Command Headquarters, specifically in the area of both clear and secure telecommunications. Other contributions have addressed numerous issues concerned with the current proliferation of narrow-band secure voice terminals. Typical issues include assuring a high-level of multiple community, worldwide interoperability and compatibility. Examples of solutions include the development of both common terminal and RED subscriber enclave concepts, the latter possessing varied trunking termination capabilities.

David S. Johnson is President, Damar International, a firm involved in consulting in meteorology, remote sensing from space and technical management. When he retired from the U.S. Government in 1982, he was responsible for the civil operational environmental and land remote sensing satellite programs as Assistant Administrator for Satellites, National Oceanic and Atmospheric Administration, Department of Commerce. He was involved in the development of TIROS-1, the world's first weather satellite, launched in 1960, and led the subsequent development and management of the civil operational satellite system.

Dr. George A. Keyworth, II is Science Advisor to the President, and Director, Office of Science and Technology Policy in the Executive Office of the President, a position he has held since mid-1981. Dr. Keyworth earned his doctorate in nuclear physics at Duke University, where he did pioneer research in isospin conservation in nuclear reactions. From 1968 until 1981 he was on the staff of the Los Alamos Scientific Laboratory, where as Head of the Physics Division he strengthened the programs in basic experimental science and developed a comprehensive and imaginative program in weapons physics.

Arthur R. Kraemer is currently Vice President of Operations for XMR Inc., in Santa Clara, California. He has participated in the development of advanced electro-optics systems for 25 years. His previous position was Vice President and General Manager for GTE Lenkurt's Fiber Optics Communication Division. While at GTE Sylvania from 1968 through 1980, he was a major contributor to the development of spaceborne optical communication system concepts and hardware. Prior to that, at Lockheed Missiles and Space Company, he developed several spaceborne electro-optical systems. Mr. Kraemer has a B.E.E. from Cornell University, a M.S. from Stanford University, and completed graduate courses toward a Ph. D. at Stanford University.

Colonel John M. Kuhn, USAF, is currently the Chief, Engineering Support Division of the Joint Tactical Command Control and Communications Agency (JTC3A). Prior to assuming this position in May 1985, he was Chief, Operating Data Networks Engineering Division at the Defense Communications Engineering Center (DCEC) of the Defense Communications Agency (DCA). He served as the Assistant SATCOM Project Manager, Space Segment Acquisition at the NATO Integrated Communications System Management Agency (NICSMA) in Brussels. In his 21 years in communications with the Air Force, he has also served in the SATCOM office of AFCS; been an engineering instructor in Air Training Command; served with DCA in Vietnam; and managed the long haul communications for Tactical Air Command. Colonel Kuhn has a M.S. in Electrical Engineering from the University of New Hampshire and a B.S.E.E.

from Manhattan College where he received his commission through the ROTC Program.

Major General Doyle E. Larson, USAF, (Ret.) is currently President, Larson Lectronics, Inc. He served in the USAF for over 32 years, starting his career as a Russian linguist, attending flying school and spending three years in fighter interceptors. He has commanded three reconnaissance squadrons flying RC-135, RC-130, and EC-121 aircraft. General Larson served with the National Security Agency and HQ, USAF in the Pentagon and then was the Director of Intelligence, PACOM; DCS/Intelligence, SAC; Commander, Electronic Security Command; and Director, Joint Electronic Warfare Center before retiring in 1983. General Larson is currently President of the Association of Old Crows.

Robert S. Magnant has been involved for more than 21 years in the design and implementation of modern telecommunications systems and services and in the development of weapons systems within the defense electronics environment. His special interests include information policy, video telecommunications and the use of small earth stations for private networks. His work with select user communities has helped pioneer the application of domestic satellites within the military health care and training arenas. A graduate of Worcester Polytechnic Institute and the University of Colorado, he is currently with the U.S. Army Information Systems Command.

Major General John B. Marks, USAF, is currently Commander, Electronic Security Command and Director, Joint Electronic Warfare Center, both at Kelly Air Force Base, Texas. Earlier in his career, as a member of an RB-50 crew, General Marks flew 35 missions during the Korean War. Subsequent assignments included positions at Headquarters, Strategic Air Command; Electronics Engineer Staff Officer, Office of the Deputy Chief of Staff, Research and Development; Chief, Electronic Reconnaissance System Developments Branch; Chief, Electronic Warfare Division, J-3, Operations Directorate; Deputy Chief of Staff, Intelligence, Air Force Systems Command; Commander, Air Force Systems Command's Foreign Technology Division; and Director, Intelligence, J-2, Headquarters Pacific Command.

General Robert T. Marsh, USAF (Ret.), is currently an Aerospace Consultant with several U.S. aerospace companies and government agencies. He retired in August of 1984 from the position of Commander of the Air Force Systems Command. Prior to that, he commanded the Electronic Systems Command.

Lieutenant Colonel John J. McConnell, USA (Ret.), is currently with Rockwell International's Defense Electronic Operations in Washington, D.C. During his 20 years of active duty, Colonel McConnell served in a variety of airborne and tactical communications assignments. In his last assignment, Colonel McConnell served in the office of the Deputy Chief of Staff, Research, Development and Acquisition, Department of the Army, planning the development and justification of Army programs to the Office of Management and Budget and the Congress. He is a graduate of the U.S. Army Command and General Staff College and has a graduate degree from the University of Colorado.

Colonel Paul C. McLellan, USAF (Ret.), is currently employed by TRW in

Redondo Beach, California, as a Project Manager. Immediately prior to that, he served as System Program Director, Fleet Satellite Communications System, Air Force Space Division, Air Force Systems Command at Los Angeles Air Force Station. Past assignments include Maintenance and Weapons Control Officer, 868th Tactical Missile Squadron; Chief of Maintenance and Chief of Material, 505th Tactical Control Group; Staff Development Engineer, Defense Special Projects Group; Systems Officer, Headquarters Air Force Systems Command; Director of Engineering, Defense Satellite Communications Systems Program Office; and Manager, DOD Worldwide Communications Satellite System, Defense Communications Agency.

Dr. John L. McLucas is Executive Vice President and Chief Strategic Officer, Communications Satellite Corporation (COMSAT). His career in communications began in World War II when he spent three years in the Pacific as a radar officer in the U.S. Navy. From there, he worked for the Air Force Cambridge Research Center. In succeeding years, he became President, HRB-Singer; Assistant Secretary General for Scientific Affairs, NATO; President and Chief Executive Officer, MITRE Corporation; Under Secretary and then Secretary, U.S. Air Force; Administrator, Federal Aviation Administration; President, COMSAT General Corporation; and President, COMSAT World Systems Division. Dr. McLucas is also President and a fellow of the American Institute of Aeronautics and Astronautics and a past Chairman of the Board of the Armed Forces Communications and Electronics Association. He has also been appointed to the Aeronautics and Space Engineering Board of the National Research Council. Dr. McLucas earned his Doctorate in Physics from Pennsylvania State University in 1950.

Lee A. Meadows is Programs Manager of Offboard Countermeasure Systems for Raytheon Electromagnetic Systems Division, Goleta, California. During his 29 year association with Raytheon ESD he has managed the Division's pioneering application of state of the art technology to advanced systems for Air Force, Army and Navy operational ECM and ESM requirements. He earned a B.S. degree in Electrical Engineering from the University of California Berkeley and is a lifetime member of the Association of Old Crows.

Lieutenant Colonel Philip L. Noggle, USMC (Ret.), served in tactical air support operations and was special assistant to the Assistant Secretary of the Navy (Research and Engineering). His current activity at the E-Systems, Inc., Center for Advanced Planning and Analysis includes Tactical C^2 and C^2W areas of interest.

Lynn S. Nored is Chief Executive Officer and co-founder of Southwestern Management Associates, Inc. He received a B.S.E.E. degree from Texas Tech University and an M.S.E.E. degree from Southern Methodist University. Prior to his present position Mr. Nored was Group Head and Program Manager at Collins Radio Company and Director of Communication Systems for Electrospace Systems, Inc. He has worked on Intellsat Systems, LES-6, GAPFILLER, DSCS, FLTSATCOM, AFSATCOM and the Survivable Satellite System. Mr. Nored has since served as Manager of Engineering for Vara, Inc. — an Electro-

Optical Company — and as Vice-President of Transmission Engineering for NetAmerica.

Lieutenant General Lee M. Paschall, USAF (Ret.), served in the Infantry during World War II and subsequently with the U.S. Air Force until his retirement in 1978. His entire career was spent in a broad range of command, control and communications duties culminating in assignment as the Director, Defense Communications Agency (DCA) from 1974 to 1978. Prior to his position with DCA, General Paschall was Director, Command, Control and Communications, Headquarters U.S. Air Force. Retiring in 1978, he became a consultant to industry and government on telecommunications and information systems until assuming the position as President, American Satellite Company in 1981 and Chairman in 1984 until his return to retirement in 1985. He serves on the Board of Directors of several companies including the American Satellite Company Management Board. General Paschall is a member of the Advisory Board on Telecommunications for EIC/Intelligence, Inc., serves on the NASA Space Applications Advisory Committee, and the National Security Telecommunications Committee of the National Research Council. He also serves as a Director of Armed Forces Communications and Electronics Association.

Paul H. Phillips is currently Department Head for C^3I Test and Evaluation at Electrospace Systems, Inc. He has been actively involved in the technical and operational evaluation of airborne C^3 systems for more than ten years and has developed operational concepts, developed data acquisition systems, implemented operational and technical support capabilities, and performed trade-off studies for a number of strategic C^3 systems. He manages a courseware development group which utilizes computer-based instruction techniques to train communication operators in the use of strategic systems. In addition, he has managed all facets of test and evaluation, including establishment of test issues and criteria, test design, evaluation planning, data analysis, and reporting of test and evaluation results.

Wilbur L. Pritchard has been President of Satellite Systems Engineering, Inc., since 1974. He is also President, Direct Broadcast Satellite Corporation, conceived to construct and operate a direct-to-home TV service. Mr. Pritchard has worked with communications satellites since 1962 when he headed Aerospace Corporation's first COMSAT group. He became first Director of COMSAT labs; Vice President of Communications Satellite Corporation; and U.S. Delegate to the technical committee of interim INTELSAT. He is a Fellow of the IEEE and the AIAA and has received the U.S. Air Force Systems Command Award, the AIAA Aerospace Communications Award, and the Lloyd V. Berkner Award of the American Astronautical Society, Inc.

Dr. William N. Redisch is a Senior Systems Manager at the NASA/Goddard Space Flight Center. Most recently he has served as the Assistant Director of Networks for the Tracking and Data Relay Satellite System (TDRSS), being second in command for all aspects of NASA's operational multi-billion dollar Space Network. He has served as Deputy Project Manager for the NASA/NOAA Meterological Satellites Project, and was the initiator of the international Satellite-Aided Search and Rescue Missions. Previously, he was

Chief of the Communications and Navigation Division responsible for satellite communications efforts and development of RF remote sensors. He has headed Goddard's Projects Technical Support Office and in addition was in charge of the Delta PAC system. While Head of the Systems Engineering Office, Dr. Redisch performed as the Chief Engineer for the highly successful ATS communications satellite program.

Lieutenant General Kenneth W. Schultz, USAF (Ret.), is currently President, XonTech, Inc. Prior to his retirement in 1975, he served as Commander, Space and Missile Systems Organization, Los Angeles Air Force Station. Previous positions include Deputy, for Minuteman, SAMSO (AFSC); and Deputy, Balistic Missile Reentry Systems, SAMSO. General Schultz served in WW II in the European Theater of Operations as Commander, 95th Bombardment Squadron. He is a graduate of New York State University, and was awarded the Rensselaer Polytechnic Award for excellence in science and mathematics in New York.

General Alton D. Slay, USAF (Ret.), retired from the Air Force in February of 1981 after 38 years of service. His last assignment was Commander, Air Force Systems Command, Andrews AFB, Maryland. Prior to that, General Slay was Deputy Chief of Staff, Research and Development, Headquarters U.S. Air Force. Other assignments as a general officer included tours as Vice Commander of the Air Training Command, San Antonio, Texas; Commander of the Lowry Technical Training Command, Denver, Colorado; Deputy Chief of Staff, Operations, Seventh Air Force, in Southeast Asia; Director of Operations, Military Assistance Command, Vietnam; Deputy Chief of Staff, Operations, Air Force Systems Command; Commander of the Air Force Flight Test Center, Edwards Air Force Base, California; and assistant Deputy Chief of Staff, Plans and Operations, USAF in Europe. He is currently President, Slay Industries Enterprises, Inc., an industrial consultant firm, and serves as Chairman of the Board of EDTECH International Limited.

Marc Spellman is a systems engineer with Harris Corporation. He received a B.S.E.E. degree from Brooklyn Poly in 1965, an M.S.E.E. degree from Syracuse University in 1967, an M.S. degree in Management Science from Newark College of Engineering in 1971 and an M.B.A. degree in finance from NYU in 1975. Since joining Harris in 1975 he has worked on military satellite communications systems. He has published a dozen articles in the fields of spread spectrum communications, radio wave propagation, investment analysis, network synchronization and antennas and is also a member of Eta Kappa Nu, Tau Beta Pi and Beta Gamma Sigma.

Colonel Verl R. Stanley, USAF (Ret.), is currently working on C^2 and C^2W programs at the Center for Advanced Planning and Analysis, E-Systems, Inc. During his 25 years of active duty through September 1979, he served in various intelligence and R&D command and staff positions with DIA, SAC, USAFE, PACAF and AFSC. He served on battle staffs in USAFE and PACAF and on the Senior Battle Staff at SAC. Colonel Stanley is a graduate of George Washington University (M.B.A.), the Armed Forces Staff College and a distinguished graduate of the Air War College.

Major James V. Stiles, USAF, is currently the Air Force Program Manager for the Airborne Self Protection Jammer and the Comprehensive Power Management System in the Advanced Tactical Aircraft Protection Systems Joint Program Office at the Naval Air Systems Command in Washington, D.C. An active member of AFCEA for over twelve years, Major Stiles was awarded the AFCEA Meritorious Service Award at the June 1983 National Convention. Major Stiles is a graduate of the University of Arizona and holds an M.B.A. from Western New England College.

Major General Paul R. Stoney, USAF (Ret.), is currently in private business. During 32 years of military service General Stoney held a variety of staff and command assignments in communications and electronics, including tours on the Air Staff and OSD/DDR&E. His most recent assignments included Chief of Communications, Strategic Air Command, then Deputy Commander and later Commander, Air Force Communications Command.

Charles A. Taylor is a graduate of the Georgia School of Technology. He has served on the National Advisory Committee for Aeronautics on the development of research instrumentation for aeronautical research as applied to supersonic and hypersonic rocket propelled models and aircraft research. Following that, he worked for Burroughs Corporation Research Laboratories on the development of military electronic systems. His next position was for General Electric Co., Space Division, where he was involved in system design and conceptual studies of spacecraft for military application and planetary exploration. He concluded his career at NASA, where he worked on the development and operation of global tracking networks and communication systems for NASA Earth orbital and planetary exploration spacecraft.

Michael Thurbon is Director of Consultancy and Principal Consultant in Electronic Warfare at Systems Designers Scientific, a company in the Systems Designers International Group. He served as an Engineer Officer in the Royal Air Force for 20 years, retiring as a Wing Commander. He has spent most of his career specializing in Electronic Warfare and is the author of a number of technical and general papers on this and closely allied topics. During the last four years he has been responsible for the development of a wide range of consultancy services within Systems Designers Scientific.

Brigadier General John S. Tuck, USA, is currently serving on the Army Staff as the Deputy Assistant Chief of Staff for Information Management for Command System Integration. In his most recent assignments, he was Commander, White House Communications Agency and Commander, Pacific Area of the Defense Communications Agency. He is a graduate of Northeastern University and the Army War College.

Lieutenant Colonel Charles E. Whited, USAF, is currently Director of the Fusion Research Division, Office of Inertial Fusion, Defense Programs, Department of Energy. He was commissioned as a Second Lieutenant in January 1968 and has been involved in all aspects of research and development, with a particular orientation toward management of high-technology systems in the fields of lasers, communications, energy, nuclear weapons, and space. His assignments have included the Atomic Energy Commission's Nevada Test Site, the Air Force Avionics Laboratory, the Air Force Space Division (formerly Space and Missile Systems Organization), Rand Corporation, and the Department of Energy. He is a graduate of the Air Command and Staff College, the Air War College Program, and the Air Force Institute of Technology.

Captain (Major selectee) Edward Whittenberg, USAF, enlisted in the Air Force in 1971, upon graduation from Louisiana State University. His first assignment was as Ballistic Missile Analyst Technician on a Missile Combat Crew at Little Rock AFB, Arkansas. He attended Officer Training School, and upon his return to Little Rock AFB served as Deputy Missile Combat Crew Commander (DMCC); Instructor DMCCC; Missile Combat Crew Commander (MCCC); and Instructor MCCC. Subsequently, he served the Advanced Space Communications program office at Space Division, Los Angeles AFS, California. After attending the Air Force Institute of Technology (1981-82), he was assigned to the Air Force Technical Applications Center (AFTAC) at Patrick AFB, Florida. From January to June of 1985, he attended the Defense Systems Management College's Program Manager's Course, and upon graduation was assigned to his present position as a program manager in AFTAC's Acquisition Directorate.